STEPHEN WALKER has a BA in history from Oxford and an MA in the history of science from Harvard. His previous book, *Shockwave: Countdown to Hiroshima*, was a *New York Times* bestseller. He is also an award-winning documentary director. His films have won an Emmy, a BAFTA and the Rose d'Or, Europe's most prestigious documentary award. He lives in London.

'This remarkable account of the 1961 race into space is a thrilling piece of storytelling. Stephen Walker is a documentary film-maker as well as a historian, and it shows in the sophistication of the storytelling ... Walker handles details brilliantly – the book is shot through with Soviet colour ... Apollo 11 turned Gagarin into little more than the answer to a quiz question: who was the first man in space? This book ends that. It is high-definition history: tight, thrilling and beautifully researched' *Sunday Times*

'The thrilling ride to be the first man in space is vividly captured in this retelling of the career of Russia's favourite son ... [A] scintillating account of the race between the Russians and Americans to achieve manned space flight' *Financial Times*

'Stephen Walker has amassed a wealth of detail on this fascinating time, and on the engineers and experts who put the first human in orbit, but who were never allowed to speak about it. *Beyond* is a reminder of the fragility of history and a testament to the brilliance of a terrorised people' *New Statesman*

'Always thrilling ... brings a huge amount that is fresh and new to our understanding of the Space Race' *Daily Telegraph*

'Cinematic ... Walker develops a
theatre of space exploration'

'Many intriguing revelations [and an] extensive, blow-by-blow account of the race to put a human in space … *Beyond* tells the full story of the finest two hours of [Gagarin's] tragically short life'

Literary Review

'Walker mixes screenplay-like dynamism with forensic detail to produce an account that captures a pivotal moment in time'

BBC History Magazine

'A gripping story, rich in novelistic detail … Highly recommended'

BBC Sky at Night Magazine

'A fine account … told with verve … A cliffhanger'

Wall Street Journal

'Thrilling … A great introduction to the gripping tale of Gagarin's flight and its impact on space exploration history' *Library Journal*

'Gagarin's story gets a fresh look in Stephen Walker's riveting and thoroughly researched new book *Beyond*'

Air & Space/Smithsonian Magazine

'Vivid … Dramatic. This entertaining and carefully researched history achieves lift off' *Publishers Weekly*

'A beautifully structured and evocative piece of writing, told with empathy for the era and a corresponding eye for detail … An outstanding study of early space travel' *Soldier Magazine*

'Brings to life the space race and the extraordinary story of Yuri Gagarin. Using first-hand testimony, Stephen Walker evokes the personalities, the science and the atmosphere of the time in a history that reads like a thriller'

ANNE APPLEBAUM, author of *Twilight of Democracy*

'The very best account of Yuri Gagarin's pioneering space mission – vivid, thoughtful, and respectful of the characters in the story. A wonderfully rendered story of an epochal event'

ASIF SIDDIQI, author of *Challenge to Apollo*

'This book is simply incredible. Meticulously researched and beautifully written, *Beyond* is the gripping story of the unknown side of the space race'

SARAH CRUDDAS, author of *Look Up*

'Exciting, well-researched and fascinating. I loved every word'

LAURA SHEPARD-CHURCHLEY, daughter of Alan Shepard, America's first astronaut

'Dramatic and dynamic. Stephen Walker's passion for his subject along with his exceptional research … have brought my father's extraordinary journey vividly to life'

ELENA GAGARINA, daughter of the world's first man in space

ALSO BY STEPHEN WALKER

King of Cannes: Madness, Mayhem and the Movies
Shockwave: Countdown to Hiroshima

BEYOND

THE ASTONISHING STORY OF THE FIRST HUMAN TO LEAVE OUR PLANET AND JOURNEY INTO SPACE

STEPHEN WALKER

William Collins
An imprint of HarperCollins*Publishers*
1 London Bridge Street
London SE1 9GF

WilliamCollinsBooks.com

HarperCollins*Publishers*
1st Floor, Watermarque Building, Ringsend Road
Dublin 4, Ireland

First published in Great Britain in 2021 by William Collins
First published in the United States by Harper in 2021
This William Collins paperback edition published in 2022

1

A catalogue record for this book is
available from the British Library

ISBN 978-0-00-837254-5

Set in Minion Pro
Printed and bound in the UK using 100%
renewable electricity at CPI Group (UK) Ltd

MIX
Paper from
responsible sources
FSC™ C007454

This book is produced from independently certified FSC™ paper
to ensure responsible forest management.

For more information visit: www.harpercollins.co.uk/green

For my daughter Kitty,
the joy and light of my life

CONTENTS

AUTHOR'S NOTE

THE RACE BETWEEN the superpowers to put the first human in space was filled with drama, complete with nail-biting twists and turns and larger-than-life personalities on both sides of the Iron Curtain; so much so that the story that follows might sometimes appear to read like fiction. But it is not fiction. It is a work of history. In writing it I have tried, as far as possible, to sift and verify evidence from a multitude of written, visual and oral sources cited in the endnotes, in order to present the most accurate account I can. Even every line of dialogue has its source. Inevitably there will be errors for which I take full responsibility. But I hope that readers who tread the same path as I have will also find, as so often is the case, that sometimes nothing is more fantastic than reality – and certainly not this tale of humanity's first leap beyond the earth.

I have largely avoided giving both US and metric measures at the same time on the grounds that it can interrupt the flow of narrative. My guiding principle, I'm afraid not always consistently followed, is to use US measures when the action is located there and metric when in the USSR. To give the distance between, say, New York City and Cape Canaveral in kilometres strikes the same slightly discordant note for me as giving the distance between Moscow and Gzhatsk in miles.

On occasion I use the word 'Russian' when in fact I mean Soviet, a common practice in the west at the time. I have done so to help evoke a particular sentiment or mood. I have also largely avoided the use of the Russian patronymic, except rarely for added emphasis (for example, Yuri *Alekseyevich* Gagarin). Given that a substantial part of this story takes place in Russia I felt it was more practical for English-language readers if used sparingly.

PROLOGUE

FIFTEEN MINUTES BEFORE LAUNCH

08.52 APRIL 12, 1961

Mankind will not stay on Earth for ever, but in its quest for light and space it will first humbly penetrate beyond the atmosphere, and then conquer the whole solar system.

Konstantin Tsiolkovsky, Russian scientist and visionary, 1911

Control of space means control of the world.

Senator Lyndon B. Johnson, 1958

April 12, 1961
Scientific and Testing Range No. 5 (NIIP-5)
Tyuratam Cosmodrome, Soviet Republic of Kazakhstan

DAWN COMES QUICKLY in the Kazakh steppe but from one horizon to the other there is almost nobody to witness it. A combination of burning summers and long freezing winters has made this one of the most sparsely populated regions on the planet, a land of emptiness and silence. The wild horses, camels, poisonous spiders and ferocious scorpions have the place to themselves.

Or almost to themselves.

Bisecting this half-desert of wormwood and tumbleweed is the railway line from Moscow to Tashkent, a distance of nearly three thousand kilometres. Two days out from the Soviet capital the train stops briefly at a place called Tyuratam. Legend has it that this was where the body of one of Genghis Khan's descendants was discovered in the seventeenth century, floating in the mud-brown Syr Daria river after dying in battle. Before 1955 the only people living here were the stationmaster and his family, and the few Kazakh nomads who moved with the seasons across the steppe, for whom this was a sacred spot. But by 1961, just six years later, everything has changed. An area approximately four times the size of Greater London has been requisitioned by the Soviet ministry of defence for the USSR's most audacious, most expensive – and most secret – project. A new settlement called Leninsky has sprung up where there was nothing before, home to engineers, soldiers and construction workers and their families, its streets lined with new apartment blocks, administration offices, schools and a barracks. Thirty kilometres to the north are new complexes of assembly and test buildings, a rapidly expanding grid of roads and rail tracks, power pylons, tracking stations, control bunkers and now four giant missile pads. The nomads have long since been expelled. Out here in the middle of nowhere, as far as possible from spying eyes, the Russians are launching rockets.

One such rocket now stands gleaming and hissing as the sun edges above the horizon. Codenamed the R-7, it is the biggest rocket in the world and one of the USSR's most secret weapons. A CIA spy plane attempting to photograph its launch pad last May was shot down; the pilot, Francis Gary Powers, survived to be put on trial for espionage and sentenced to ten years in a Soviet jail. Towering more than 38 metres or 125 feet over the steppe, the R-7 is almost three times as powerful as America's largest missile, capable of carrying a thermonuclear warhead with the destructive power of two hundred Hiroshima bombs a quarter of the distance around the globe – all the way from here to New York.

But this particular R-7 is not going to New York today. Nor is it carrying a nuclear warhead. It has been adapted for a different purpose. At 09.07 Moscow time its five massive first and core-stage engines will ignite and, if nothing goes wrong, the rocket will launch into space. Instead of a warhead it will carry a spherical capsule inside which will sit – or rather lie – a man. The man will be short, no more than five feet five inches, and light too, since his capsule will be approximately the weight of the five-megaton nuclear bomb it is replacing. He will have trained for over a year to accomplish this mission. And he will have to be brave. No human has been in space before. Nobody knows exactly what will happen when he gets up there – if he gets up there. Will his eyeballs burst? Will his blood stop circulating? Will he survive the crushing acceleration forces of launch – enough to drain the blood from his head? Will the heatshield of his metal sphere withstand the 1,500-degree Celsius temperatures of re-entry? Will his rocket blow up, as some of its predecessors have done in earlier test flights? And if he does make it up there, divorced from life and existence as no human being has been before, will he go mad?

As Yuri Alekseyevich Gagarin – just turned twenty-seven, married with a daughter of almost two and a one-month-old baby, ex-foundry student, ex-fighter pilot and loyal Communist – steps out into the glaring sunshine in his orange spacesuit and towards the lift that will take him to the top of the rocket, nobody outside a tightly restricted circle knows what he is about to do. Even his wife, back at their home near Moscow, does not know that her husband has been chosen for this task or that today is the day.

And the rest of the world knows nothing.

Despite hints in the Soviet media, the formal existence of this programme is a state secret. The training of Gagarin and another nineteen cosmonauts is a secret. The man responsible for it is a secret, protected on his travels within the USSR by a KGB bodyguard in case CIA agents attempt to kidnap or assassinate him; the US agency has spent years trying, and failing, to find out his name. The several camera crews now filming the preparations for launch have all been sworn to secrecy: if one of their number breathes a word, he will face the consequences in this rigidly policed state. In the business of cold war the stakes are too high for the USSR to reveal its hand. For two years the Americans have been openly preparing to put their own man into space. The latest word is that they may make the attempt less than three weeks from now. Without publicising anything, the Soviets are going in the gap.

In America it is still the evening of April 11. Millions will be watching their youthful new president, John F. Kennedy, and his glamorous wife Jackie talk on TV about the challenges of bringing up their young children, Caroline and John, in the White House. In New York's West Village, an unknown nineteen-year-old Bob Dylan is about to make his professional debut at Gerde's Folk City club, opening for John Lee

Hooker. On the nightly news the big story is the first day of the trial of Adolf Eichmann in Jerusalem, the former SS officer charged with crimes against humanity for his part in the murder of six million Jews.

Meanwhile, thousands of miles to the east, Gagarin is strapped into his seat while the last bolt is fastened and his hatch is finally closed. Alone in his little sphere, he waits, and whistles a love song. The controllers and engineers also waiting in their underground bunker less than a hundred metres from his rocket can hear him over their headphones. In the next few minutes, he will either become the first human in history to escape the earth and look down on its majesty from space, or he will die in any number of gruesome ways. If he succeeds, by the time the sun sets he will be the most famous man on the planet, a victor in the bitter war of ideologies with the United States and its allies. If he fails, almost nobody will know he even existed.

ACT I

FOUR MONTHS EARLIER

DECEMBER 1960–JANUARY 1961

The day will come when we will embark on
interstellar flight
Who can prevent us from dreaming such dreams
When it was Lenin who first taught us how to
dream?

Nikolai Krivanchikov, Soviet poet, published within hours of the launch of Sputnik, the world's first satellite, October 1957

I'll be damned if I sleep by the light of a Red moon
... Soon they'll be dropping bombs on us from
space like kids dropping rocks on cars from
freeway overpasses.

Senator Lyndon B. Johnson, also after the Soviet launch of Sputnik

ONE

PALLO'S DOGS

December 24, 1960
60 kilometres west of Tura
Siberia, USSR

IN EVERY DIRECTION, for as far as the eye could see, there was nothing but taiga, the dense, dark Siberian forest of spruce and birch and fir trees unfolding to the distant horizon, and beyond. The helicopter flew low, the dull thud of its rotor blades the only sound to puncture the silence of this primeval landscape, otherwise bereft of humans and thickly wrapped in snow. And the snow was everywhere. It blinded the eyes and coated the tops of the trees and made it even harder, if that were possible, for Arvid Vladimirovich Pallo to find what he was looking for. Nor did it help that there was not much daylight left. In this region of Siberia, at this time of year, the days lasted barely four hours. They had been flying for nearly thirty minutes now, and in two hours the forest below would disappear into yet another long sub-Arctic night. But by then, Pallo and his team might be too late.

Somewhere out there, lying in the snow in one of the remotest places on the planet, was an empty aluminium sphere, slightly more than two metres in diameter, weighing two and a half tonnes. And somewhere close by it, Pallo hoped, was a sealed metal box containing two dogs. He also hoped the dogs were still alive, despite the fact that they would have had to survive the extreme cold of the Siberian winter for the past two days, in temperatures plunging to minus forty degrees Celsius, not to mention the severely traumatic events that had accidentally landed them in this abandoned corner of the world in the first place.

Pallo's adventures had begun two days previously on December 22, when an R-7 rocket had launched the sphere into space from the secret missile site in Soviet Kazakhstan. Inside the sphere was the metal box carrying the two dogs. Such was the secrecy surrounding this mission that even today their names remain uncertain. Depending on the source they are either Kometa and Shutka, or Zhulka and Zhemchuzhina, or, in Pallo's writings, Zhulka and Alfa. Since Pallo is our guide we will go with that. They were both stray mongrel bitches picked off the streets of Moscow. The purpose of the mission was to get the two animals into orbit and safely back to earth again. The sphere had a name – it was publicly a *korabl-sputnik*, a 'ship-satellite' or spaceship, although its designers called it Vostok 1, a name that means 'East' and was classified. The Vostok 1 was a prototype, in effect the first version of the type of spaceship that it was hoped would one day – and one day soon – carry a Soviet comrade into space, and before the Americans got there first. Zhulka and Alfa would help to pave the way.

But the precedents were hardly encouraging. By the time Pallo set off in his helicopter on December 24 to look for the two dogs in the freezing Siberian wilderness, five Vostoks had been launched since May. All but one of them had failed. Two dogs had been killed on the second Vostok flight when their rocket blew up 28.5 seconds after launch. Another two dogs had been destroyed on re-entry on the fourth flight. The single success had been in August, when two more dogs, named Belka and Strelka, along with forty mice, two rats, fruit flies and a rabbit, managed to complete an astonishing eighteen orbits and come back alive. This flight, the third Vostok to be launched, was a sensational achievement, the first time any living organisms had successfully reached orbit and returned safely to earth. The Americans had accomplished nothing that was comparable and naturally the Soviet media made the most of it. But away from the glare of the press the undisclosed truth was that the animals were lucky to be alive. During their fourth orbit, Belka could be seen via a TV camera vomiting and tearing desperately at her harness in obvious distress. Before re-entry the spacecraft's primary orientation system had failed and a back-up system had to be used instead. Without it, Belka and Strelka would have faced slow deaths, stranded helplessly in space.

And now, just as parts of the world were about to celebrate the first Christmas of the new decade, this latest, fifth Vostok flight carrying Zhulka and Alfa had gone wrong too.

Exactly 425 seconds after launch the third-stage engine of the R-7 rocket carrying the Vostok shut down too early. As a result, the spacecraft failed to reach orbit. Instead it jettisoned from the rocket automatically before streaking in a ballistic arc across several time zones of the Soviet Union. But even as the Vostok, with the two dogs strapped inside their little box, plummeted back into the earth's atmosphere at several thousand miles per hour, matters went even more wrong. To begin with, the on-board bomb did not go off.

This bomb was a feature of every Vostok dog flight, a tribute to the deep paranoia of a regime that did not want to reveal its technological secrets, especially not to the Americans. Known by its code letters A.P.O., or *Avariynyy Podryv Obyekta* – the translation is 'Emergency Object Destruction' – its function was to do exactly that, namely destroy the 'Object', or Vostok, in the event of its landing off-course in a foreign and possibly capitalist country, which was considered by the Soviet leadership to be an emergency even if the dogs inside might have had a different opinion. Indeed this is exactly what had happened just three weeks earlier on the fourth Vostok mission, when a problem with the braking engine meant that the capsule would end up landing outside the borders of the USSR. In that instance, a sensor on the bomb had detected that the capsule was returning to earth at the wrong point. The two dogs inside,

Pchelka and Mushka – Little Bee and Little Fly – were blown up, along with all traces of their spaceship. The official Soviet news agency TASS briefly announced that the craft had burned up as a result of a 'non-calculated trajectory' on re-entry. Nobody said anything about bombs.

Now, as Zhulka and Alfa initially hurtled to an unknown point on earth after the third-stage engine failure, the Emergency Object Destruction device did not activate, for reasons that are still not entirely clear. There was, however, a back-up sixty-hour timer, although here too its details remain obscure. What seems likely is that its countdown to detonation would begin as soon as the capsule had landed intact – if it landed intact.

As the curve of its ballistic arc steepened towards the ground in the final few minutes of flight, the Vostok was plunging towards the Podkamennaya Tunguska river in one of the most inaccessible parts of Siberia. In a certain respect there was a fitting irony here: the last time something had dropped from space in this area had been in 1908 when a meteor struck, flattening an estimated eighty million trees with the force of a nuclear bomb. Meanwhile the dogs inside their sealed box were experiencing a terrifying ride as their little sphere shook and shuddered in its headlong dive through the thickening atmosphere. But their nightmare was only just beginning.

At an altitude of seven kilometres, the Vostok's hatch was supposed to be blown off, followed two and a half seconds later by the dogs ejecting in their own container and parachuting separately to the ground. This was a similar system that would one day return a human cosmonaut safely back to earth; except that in this instance the system failed. Both the hatch and the dogs ejected at the same time, causing the dogs' container to slam violently against the opening, denting it and preventing them from getting out. They were now trapped inside the falling sphere. Even though the latter had braking parachutes to soften its impact with the ground, it was not designed for dogs to be inside when it did. At the very least, Zhulka and Alfa were in for a very hard landing, somewhere in Siberia and in the middle of winter. And with a live bomb on board.

Back at the Kazakhstan launch site and at a secret computation centre outside Moscow, nobody could work out what had happened at first. For several hours signals from the Vostok were dead. Radio data received from the rocket's third stage revealed that engine's failure, but of the capsule itself, or its two passengers, nothing was heard. Then, later that evening, long-range radar stations in Moscow, Krasnodar and Tashkent began picking up the capsule's faint beacon transmissions from somewhere in deepest Siberia, suggesting that it had somehow succeeded in landing, although in what condition was not known. Nor were the precise co-ordinates of the landing spot. Six search planes were immediately despatched to try to find it and recover the dogs, if they were still alive – an especially challenging operation given the remoteness of the area and extreme weather conditions. But this was not some mercy mission. Before a human could fly in space it was crucial to analyse these accidents and put them right. And that meant finding the Vostok in the short time left before it blew itself up.

Arvid Pallo and his rescue team had been waiting at the air base in Tura, an isolated settlement sixty kilometres east of the assumed landing area, when the news came in that one of the aeroplanes had found it. He commandeered a helicopter and headed towards the site. With him were a KGB officer and Anatoly Komarov, a senior engineer from the institute in Leningrad that had designed the on-board bomb. The KGB presence was no surprise. There were secrets here – not least the bomb itself, let alone the entire mission – that needed to be closely guarded.

Pallo was the perfect man for the job. Tall, lean and forty-eight years old, he had trained as an engineer before the war at an explosives factory – a valuable experience given what he might be about to encounter – before working on the first rocket-powered aircraft designed in the USSR. This was an advanced design with an engine that was both novel and highly unreliable. Pallo's own face was a testimony to just how unreliable, since on one occasion in 1942 the aircraft's rocket engine had exploded while it was still on the ground. Pallo had raced to rescue the pilot trapped in the cockpit, burning his face in the rocket fuel's nitric acid and leaving him with a scar for life. But the pilot had survived.

As Pallo's helicopter approached the capsule's location it became increasingly clear that it would be impossible to land nearby. The trees were too dense. Above them the search aircraft that had found the Vostok sphere still circled, keeping it in view. Pallo ordered the helicopter pilot to touch down on an open patch of ground approximately eight hundred metres away. Together with Komarov, he jumped out and found himself waist-deep in snow. It is not recorded whether the KGB captain joined them.

Carrying their tools and a radio, the two – or possibly three – men struck off into the forest. After less than sixty metres they lost their way. The snow obliterated all visible landmarks. The cold punched the breath out of their lungs. Above them the pilot of the search plane radioed to say that it would soon be dark and that he was returning to base. Pallo interrupted him. He ordered the plane to point them in the right direction by flying in a straight line towards the Vostok.

They trudged on through the snow, and then they found it. The battered sphere was sitting in a small clearing in the forest. Two parachutes hung limply from nearby trees. A thick bundle of burned wires dangled from the open hatch in its side. Its exterior had been scorched by the intense heat of its violent return to earth. Pallo had expected to find the dog container with its own parachute somewhere nearby. But there was no sign of it. Then he peered through the Vostok's hatch and saw that it was still inside, which meant that the dogs were still there. By now well over fifty hours of the bomb's sixty-hour timer had passed. If they did not disarm the bomb now, before dark, Zhulka and Alfa would shortly be blown up. Assuming, that is, they were still alive.

At this point, according to Pallo's own memoirs, the story takes on a truly surreal twist when he suddenly decided that he wanted to disarm the bomb himself, an especially odd decision considering his personal history with explosions. He told Komarov to hide behind a tree while he got on with it. Komarov refused, arguing that the bomb was his device and not Pallo's. Standing in the snow beside the charred capsule in the middle of the Siberian wilderness, they had an argument. In the end they agreed to draw lots using matches. Komarov won. Pallo stood behind the tree. We can only guess that the KGB officer joined him – if he was there.

Komarov approached the capsule and began to fiddle with various cables while Pallo watched from his tree. By the time the job was finished, it was almost night. With the capsule safe, Pallo leaned inside again. He tried to spot the dogs through their container's portholes, but the glass was thickly covered in frost. He knocked several times on its walls. There was no response. Then the helicopter pilot radioed to say they had to leave before it got too dark. There was no alternative but to abandon the capsule and return in the morning. It would mean yet another freezing Siberian night – the third – for Zhulka and Alfa. But everything by now suggested they must be dead.

The next morning Pallo flew back. Along with several members of his rescue team, he brought Armen Gyurdzhian, a veterinary surgeon. Once again the helicopter landed in the clearing and the team headed back through the snow to the spacecraft. As soon as they arrived, Pallo reached inside. This time he lifted the sealed container out. As he did so, he heard a weak bark from inside. The men raced to release the bolts and then pulled off the lid. There, lying still strapped to their couches, frightened and almost frozen after their three-day ordeal, were the two dogs, and incredibly they were both alive.

Gyurdzhian gently wrapped them both in his thick sheepskin coat and carried them to the waiting helicopter. Within half an hour they were at the airfield in Tura; by the following day they were back in Moscow, home in their kennels at the Institute of Aviation and Space Medicine where they had been trained. They were safe – although not a word of their ordeal nor even the existence of their mission was revealed to the press, and would not be for several decades. Meanwhile for Pallo, the job was only half done. In the race to put a Soviet citizen in space before the Americans, it was not enough just to get the dogs back to Moscow. He had to get the capsule back too.

IT WOULD TAKE Pallo almost three weeks, a heroic trek of over three thousand kilometres in the bitterest of Russian winters. By January 11, twenty days after it had first streaked halfway across the USSR, the Vostok was back in its closed compound at Kaliningrad near Moscow known as OKB-1, the place where Soviet missiles were designed and the Vostoks were built.

And not before time. Only six days previously on January 5, Konstantin Bushuyev, a deputy chief designer at OKB-1, had set out the latest timetable for the first Soviet human space flight.

By now too many things had gone wrong with too many Vostok dog missions. But that was not all that had gone wrong. Less than three months earlier, on October 24, a catastrophe had occurred at the rocket site in Kazakhstan when a fully fuelled R-16 missile exploded on the launch pad while it was being prepared for a test flight. The explosion happened at night. At least seventy-four people were killed in the resulting inferno, some of them leaping from the gantries to their deaths, some of them incinerated, some asphyxiated in thick clouds of toxic smoke. A number of very senior rocket personnel had also died, including the chief of the USSR's strategic missile forces, Marshal Mitrofan Nedelin, who had whipped on his teams through the night from a deckchair to meet the missile's tight launch deadline. What was left of his body was identifiable only by his Hero of the Soviet Union Gold Star medal, along with the half-melted keys to his office safe.

This was, and remains, the worst accident in the history of rocketry anywhere. It was also kept secret. Nedelin was reported in the Soviet media as having died in a plane crash. But even though the missile that blew up was not the same as the R-7 used to launch the Vostoks, the catastrophe had its own impact on the Soviet manned space programme. Several major design organisations contributed to both programmes. Nedelin himself had chaired key meetings on Vostok. The result was further delay, reflected in Bushuyev's new timetable.

There would now be two further test launches, each carrying a single dog, in February. For the first time both would utilise an upgraded version of the Vostok – known as the Vostok 3, this was the version that would one day carry a human. Sharing the cabin with those dogs would be a life-sized human mannequin wearing a spacesuit just as the real human cosmonaut would be. Both of these February flights would follow exactly the same profile as the one that would subsequently carry the cosmonaut: a single orbit of the earth, returning to a point inside the USSR. Given the string of earlier Vostok failures, completing more than a single orbit was considered too high a risk. But if these two test flights were successful, the mission carrying a human would go next. A secret tentative date was set for March.

Meanwhile, half a world away in the United States, NASA's own human space programme was also close to flying its first astronaut in space. Formed towards the end of 1958, it was called Project Mercury after the Roman messenger god whose helmet and sandals famously sported wings, an appropriate piece of branding if somewhat undermined by the fact that another of Mercury's roles was to lead the dead to the underworld. The Americans had had their own share of accidents, explosions and delays, although unlike the Soviets they did not keep them secret. Now NASA was hoping to launch its own animal proving flight carrying not a dog but a chimpanzee. A provisional date had been set for the end of January.

Over the New Year, six trained chimpanzees had already been flown from New Mexico to Cape Canaveral, the launch site in Florida. One of them would later be selected to fly. If successful, the likelihood was that a human would follow. In its publicity releases NASA had always proved reluctant to commit to specific dates, having too often failed to meet them. But the story splashed across the American press for any KGB agent or Soviet rocket engineer to read was that this first flight with an American astronaut could also happen as soon as March.

Everything was now at stake in this greatest of adventures. It was less than sixty years since the Wright brothers had clawed a few feet into the air in their wire and canvas biplane. Now a human would take the first step to the stars. He would break free of the planet to which all life had clung since life began. He would enter the most hostile and dangerous environment ever encountered. The risks were tremendous. It would be a leap into a long and terrifying list of unknowns. Whichever superpower got there first in this coldest of cold wars would score a massive technological, political and ideological victory over the other.

And as things stood in that first week of the year 1961, just as a new, youthful, dynamic president was about to enter the White House, both of those superpowers were hoping to do it in the month of March, and quite possibly at exactly the same time.

TWO

WHO LET A RUSSIAN IN HERE?

January 19, 1961
Washington, DC

THE NIGHT BEFORE President John Fitzgerald Kennedy was due to take the oath of office, the mother of all snowstorms hit the nation's capital. Within a very few hours, eight inches of snow had been dumped on the city, causing chaos on the roads and triggering the worst traffic jams in living memory. Thousands of cars bringing well-wishers hoping to get a glimpse of their new president and his wife Jackie at the inauguration ceremony were simply abandoned wherever they broke down or got stuck or their drivers had had enough. By early evening, fourteen hundred vehicles were stranded on Pennsylvania Avenue alone, the route of the next day's parade. National Airport was closed and the city's highway department was forced to bring out every one of the two hundred snow removal trucks in its fleet to cope with the conditions. And that was still not enough. As temperatures dropped even lower, the Army Corps of Engineers and teams of boy scouts were also drafted in to help sort out the mess in time for the following day's events. They worked at a frenzied pace throughout the night, even using flamethrowers to melt the snow. It was, said the New York *Daily News*, a 'nightmare'.

But with the forty-three-year-old president's inauguration scheduled for noon the next day the show had to go on.

While the city's organs battled through the night, the president-elect and Jackie plunged into a numbing series of pre-inaugural events that would have exhausted anyone, as many of Kennedy's aides would no doubt have agreed; except that this couple actually appeared to be having fun and, if anything, the snow just added to the fun. Earlier that day, 'Jack' Kennedy had met the outgoing president Dwight D. Eisenhower at the White House, where they had talked privately for an hour and Eisenhower had shown him how to summon a helicopter onto the lawn. The contrast between the two could not have been starker: the old wartime general, the grand veteran of D-Day born at the end of the Victorian age, and the much younger man beside him, handsome, lean and smiling his million-watt smile; an image of health and vigour that was familiar to all Americans, if one that also hid a rarely publicised history of illness, operations and severe back trouble.

As the first flakes fell, the young presidential couple had emerged from their Georgetown house to open a night of partying, Jack resplendent in white tie and tails, Jackie radiantly beautiful in a white silk floor-length ball gown, while a secret service bodyguard held an umbrella over her head; two figures caught in a hundred flashbulbs, and representing to millions of Americans, with a brilliance that was electrifying, the new decade and the future. By the time the pair got to Constitution Hall for the pre-inaugural concert half the National Symphony Orchestra was stuck somewhere in the snow, but the presidential couple did not seem to mind and chatted easily to the other guests until the musicians got there. After that was over, and now an hour late, they went on to Frank Sinatra's gala at the National Guard Armory, where not only Sinatra but Ella Fitzgerald, Nat King Cole, Gene Kelly and a host of other stars performed in what the press was calling the most glamorous party in inaugural history. The gala was supposed to recoup half of the $3 million deficit that had been run up by the Democratic Party in the presidential campaign. Some of the guests had paid $10,000 for a ticket but the weather stopped many of them from turning up. One White House correspondent, *Time* magazine's Hugh Sidey, described the performance as 'interminable', a sentiment with which Jackie Kennedy might have privately agreed, since she left for home to grab some sleep at 1.30 a.m.; but her husband stayed the full course, still smiling that dazzling smile, and then off he plunged into one of the worst blizzards in the city's history to yet another party, this one thrown by his father Joe Kennedy, at a fashionable new restaurant downtown. By 3.30 a.m. he was back home. By 4 a.m., while the flamethrowers were busy scorching the snow off Pennsylvania Avenue's trees and secret servicemen were battening down the manhole covers to guard against possible assassins, he was asleep. And he still had the oath of office, his address, the parade and another five inaugural balls the following night to get through.

LESS THAN TWO hundred miles south of the capital, on the same evening the Kennedys stepped out of their front door to brave those first flakes of snow, seven men sat waiting in a classroom at NASA's research centre in Langley, Virginia. The centre was part of an aviation and space testing site that stretched across several hundred acres of land on the south-eastern edge of the Virginia Peninsula near Newport News. Since its establishment in 1917 as the field station of NACA – the National Advisory Committee for Aeronautics – it had mushroomed into a labyrinthine complex of hangars, workshops, wind tunnels, engineering offices and laboratories, where some of the most advanced flying machines ever dreamed up were tested in every possible variety of conditions, sometimes deliberately to destruction. Langley was aviation's proving ground. Within this sprawl of buildings was to be found the cutting edge of American aeronautical innovation.

And not just aeronautical innovation. When the new US space agency NASA replaced NACA in 1958, it took over a number of Langley's locations and it was in one of them, Building 60, a nondescript two-storey redbrick structure flanked by a neatly trimmed lawn, that the seven men were now waiting early that evening just as the snow began to fall.

They sat at their metal desks like a class of schoolboys, but they were all in their thirties, most of them wearing Ban-Lon shirts and all of them in excellent physical shape. To the American public their faces were already famous, since they had regularly been plastered across the pages of one of the country's most widely read weeklies, *Life* magazine, with which they also had an exclusive, and highly lucrative, contract. Readers by now knew – or thought they knew, since most of the warts were airbrushed out – everything about them: their hobbies, their families, their life stories, their previous careers as military test pilots, their fears and dreams, the cars they drove and the clothes their perfectly groomed wives liked to wear. From the moment the seven men had stepped onto the world's stage at a packed press conference at NASA's Washington headquarters on April 9, 1959, they had become celebrities, and for one simple reason: these were the men chosen to be America's first astronauts. Dubbed the 'Mercury Seven' after the manned space programme Project Mercury, every one of them was a volunteer, picked after a rigorous and ruthless medical and psychological testing programme from hundreds of qualifying military test pilots in early 1959. To the press, if not in reality, they were the country's seven bravest and best pilots. Each of them was ready and willing not only to fly but if necessary to die for his nation up there in the uncharted territory of space. They were America's gladiators in the cause of freedom. Little wonder that almost every American returned the compliment and was instantly star-struck.

That press conference was now twenty-one months ago, and despite an intense and demanding training regime over each of those months, none of the seven men had yet flown anywhere near space. The provisional dates kept being postponed. The latest talk of March hinged on the chimpanzee proving flight scheduled for the end of January in less than two weeks. Only if that were successful would one of the seven men sitting in their classroom hope to go next.

All of them were vitally aware that the Soviets were also planning to put a man in space – and soon – even if there were no *official* Soviet acknowledgement of that plan, or of any Soviet astronauts in training like themselves, or of anything concrete on the subject at all. But everybody knew it just the same. And in case people needed reminding, the Soviet press was always happy to drop heavy hints from time to time, especially after stellar successes like Belka and Strelka's orbital dog flight the previous August. 'Cosmonaut Get Ready to Travel,' boasted one Russian headline when that had happened, and the highly popular illustrated magazine *Ogonyok*, a sort of Soviet version of *Life*, had been spilling over with excitement: 'Space, Expect a Visit from Soviet Man'. But unlike *Life*, the *Ogonyok* reporters did not go on to describe who that Soviet Man would be or what kind of car he drove, or even if he existed. All of that was just a blank.

And yet it was inevitable that a Soviet manned visit to space would happen soon. It was five months since Belka and Strelka's flight. Meanwhile the Americans were still waiting for their chimpanzee to fly. And Belka and Strelka had flown eighteen orbits of the earth. It was therefore surely inevitable that the Soviet manned visit to space would be orbital too; whereas any American visit, at least initially, would have to be brief and suborbital. To hurl a man into orbit required big, powerful rockets, which the Soviets already had and the Americans were only still developing. Quite how and why this alarming imbalance had come to pass was a question that long exercised many American newspaper editorials and congressional hearings, but what it meant in practice was that the first American astronauts, like the first American chimpanzee, would only be capable of flying a simpler, ballistic flight, punching up above the atmosphere into space for just a few minutes before gravity yanked the capsule back down to earth. The trajectory would be more like the arc of an artillery shell, except with a chimpanzee or human sitting inside the shell. But even if this up-and-down, suborbital affair was technologically less impressive than circling the entire globe at nearly 18,000 mph, it nevertheless counted for everything.

To the American public and the media, space in the end was still space, whether a man was up there for a few minutes or a few hours – and the only thing that mattered was who got there *first*, preferably without being killed in the process.

To the seven would-be American astronauts there was an additional and perhaps more immediate question, namely which one of *them* would get there first? Like most of the country, they were frustrated by all the delays and postponements in this programme, but they also understood that there was no textbook for a mission of this scale and ambition; that book was being written and re-written as things went along. As test pilots, they recognised that many of these delays were necessary, not least for the very practical consideration that it was their own hides that would be exposed if things fell apart up there. Nobody actually wanted to die; but that did not stop each of them from hoping that his hide would be first.

Up to now, none of them had been told who had been selected for that honour but the decided press favourite was John Glenn, a former marine pilot who had flown fifty-nine missions in the Second World War, had shot down three Communist aircraft in the Korean War, won five Distinguished Flying Crosses for bravery, captured the world record for the fastest jet flight across the continental United States in 1957 and appeared several times as a contestant on the hugely popular CBS TV show *Name That Tune*, sporting his uniform, a big smile and a chestful of medals. People loved the fact that Glenn's nickname was 'Ol' Magnet Ass' because of the number of times his fighter had been shot up by enemy fire. All put together it was quite a résumé; and with his wide, freckled, grinning face, his committed Presbyterian church-going, his regular Sunday-school teaching and his strongly expressed family values, Glenn looked and sounded every inch not only like an apple-pie American astronaut, but like the *first* apple-pie American astronaut. Clearly he thought so too. 'Anyone who doesn't want to be first,' he once declared, 'doesn't belong in this programme.'

Running a likely second behind him was Alan Shepard, a thirty-seven-year-old former navy pilot who had never fought in combat

– perhaps a sensitive issue compared to Glenn – but who had test-flown some of the toughest and most challenging jets at the navy's Test Pilot School at Patuxent River in Maryland, fast, unfriendly planes with names like Banshee, Demon or Panther, in which, as the saying went, you had to be afraid to panic. Shepard's many specialities included landing on an aircraft carrier in heavily pitching seas at night, just about the most dangerous kind of flying there is, and he had taught himself to land a jet from 40,000 feet without an engine. He was a brilliant pilot who for fun had once looped the loop underneath and around the span of the Chesapeake Bay Bridge, a stunt that had almost cost him his career. On one occasion he was flying high-speed rolls in one of those vicious Banshees when both wing fuel tanks abruptly sheared off, and still Shepard managed to bring the plane safely home to Patuxent. His piloting skills were legendary, as was his unflappability when his planes suddenly turned rough. 'He could fly anything,' said a colleague.

Shepard made no secret of his ambitions as an astronaut: 'I want to be first because I want to be first,' he told a reporter, and he was certainly smart, maybe the smartest of all the seven men, with indisputable leadership skills along with a prodigious talent for technical analysis. Apart from flying the nation's hottest jets at terrific speeds and sometimes upside down, Shepard's passions were water-skiing, golf and very fast cars; and some said women too. He had bold blue eyes and a wicked, toothy grin that many women found irresistible; a face that was not handsome, according to a *Life* magazine profile, but still boyish despite the widow's peak, and packed with character. Yet people could sometimes be wary of him. He was fun and funny but he could also turn, one minute all smiles and the next clipped and aloof – a 'cool customer', according to *Life*. One of his nicknames was the 'Icy Commander'. 'I have never been my own favourite subject,' he said in that same profile, shutting down any probing questions about his personality. A few, like the later Apollo astronaut Gene Cernan, were able to 'crash through that barrier' and 'let me realise what a tremendous man he really was'. But not so John Glenn. Shepard and Glenn were not close friends.

These two were the front-runners but there was still no clue which of them – if either – would be the first to fly. Only a fool would have discounted the others. There was Leroy 'Gordo' Cooper, thin, handsome, a downhome farmboy from Oklahoma and the youngest of the seven at thirty-two. A Buck Rogers fan, he once confessed that one of his reasons for wanting to become an astronaut was because of his interest in UFOs. There was Donald 'Deke' Slayton, also a farmboy from Wisconsin, a quiet, leathery-faced man and another outstanding test pilot and keen hunter; and Slayton's best friend and hunting partner Virgil 'Gus' Grissom, the shortest of the group, veteran of a hundred combat missions in Korea with his own Distinguished Flying Cross, and famously not much given to talking. There was Walter 'Wally' Schirra, the joker of the pack, who loved pulling pranks, a fast car fanatic like Alan Shepard, and one of the Mercury Seven's few remaining hardcore smokers. And there was Malcolm Scott Carpenter, an ex-Navy pilot whose inclusion in this elite group of test pilots was a mystery to all the others since most of his career had been spent flying slow and boring propeller patrol planes. That was bad enough; even worse was Carpenter's hobby, which was playing the *guitar*, sitting on cushions at home with his beloved wife Rene like some beatnik rather than a true-blooded hotshot pilot.

Competitiveness boiled off these seven men like steam. It was why they were there in that room in the first place; it was what had taken them to flight school, to the navy or the marines or the air force, to fly – in all but Carpenter's case – some of the hottest jets on the planet and in some cases to risk their lives flying in combat; it was what two years earlier had driven them to apply for this new and untested career to become astronauts – literally *star-voyagers* – flying in strange, thrilling new machines that looked nothing like any aeroplane any of them had ever flown, and flying them in places none of those aeroplanes could ever go. And for all the same reasons, it was what made each of them want, perhaps more than anything else in their lives, to beat both the Russians and each other and go down for ever in history as *The First*.

Before Christmas, the astronauts had been asked to nominate one of their number for that position in a secret ballot, but the result had not been revealed. From mid-January, with the chimpanzee flight now imminent, the question was beginning to take on urgency. Then, on the day before President Kennedy's inauguration, the astronauts had been told to clear their diaries and be in their classroom by 5 p.m. Their chief, Robert Gilruth, wanted to see them.

Since November 1958, Gilruth had been head of the Space Task Group based at Langley, the organisation set up by NASA to run Project Mercury. He was a superb appointment: as one of the most outstanding aeronautical engineers in the country he had worked on everything from British Spitfires during the war to advanced pilotless rocket-planes in the 1950s. He was a true innovator, still reasonably young in his mid-forties, fired by his new responsibility of putting an astronaut into space; and behind a misleadingly quiet exterior he could be decisive and tough. For fifteen minutes the astronauts waited for Gilruth in their classroom while the snow fell more thickly outside the windows. Normally they would have spent the time chatting or teasing. Now they hardly spoke. The tension, as everybody afterwards remembered, was electric. Grissom, the 'little bear of a man' as *Life* described him, who almost never opened his mouth, made a joke. 'If we wait any longer,' he said, 'I may have to make a speech.' But before he did, Gilruth entered the room.

Shutting the door quickly behind him, he wasted no time. Telling the astronauts that this was 'the most difficult decision I've ever had to make', he said that Alan Shepard would make the first flight. Gus Grissom would go next and John Glenn third. Glenn would also be the back-up astronaut to both Shepard and Grissom. The other four – Deke Slayton, Gordon Cooper, Scott Carpenter and Wally Schirra – would go on later flights. Gilruth did not offer any reasons for his choices and never would, at least not to Shepard himself; although he later acknowledged that Shepard's sharp intelligence was a key factor. Before he left, Gilruth said that his decision was not to be made public yet. The press intrusion on Shepard would be too extreme. For now, the astronauts were to keep the news to themselves. Gilruth thanked them, wished them luck – and left the room.

There was a stunned silence. Shepard later recalled how he had stared at the floor and tried to hide his elation while Gilruth had been speaking. He only looked up once Gilruth had disappeared. Six faces were looking back at him. Then Glenn stepped forward and shook his hand. One by one the others did the same. Within a few moments they were all gone, leaving Shepard by himself in the room.

He did not stay long. As the full realisation of what had just happened hit him, he knew that he was going to have to tell his family, whatever Gilruth might say. Out he went into the night and the falling snow and drove to his home in Virginia Beach. The moment he opened the front door his wife Louise threw her arms around him. She saw the grin creasing his face and she knew. 'You got it!' she said. Whatever fears she might have had she kept firmly to herself. She had been a test pilot's wife far too long to parade those fears to her own husband. 'Lady,' said Shepard, 'you can't tell anyone, but you have your arms around the man who'll be first in space.' Louise pulled back, glancing mischievously around the room. 'Who let a Russian in here?' she said. Her words were meant as a joke; but they were closer to the mark than either she or her husband could have imagined.

THREE

THE HOUSE IN THE FOREST

January 18, 1961
Military Unit 26266, 41 kilometres north-east of Moscow

DEEP IN A birch forest in the Shchyolkovsky District north-east of Moscow, far from the main highway to the city and hidden from prying eyes, stood a small, old-fashioned two-storey building half-buried in the snow. Here the snow lay even more thickly than it did eight thousand kilometres away in Washington. It piled on the building's steeply pitched roof, on the overhanging porch of its front door, on the tangled branches of the birch trees, accentuating the heavy silence of the forest.

Apart from its curious location, there was nothing especially remarkable about the building itself. Perhaps its very anonymity helped to disguise its true purpose. For this nondescript edifice sitting in the middle of the forest in fact housed one of the most secret institutions in the USSR. Its code name was Military Unit 26266, but it was also known by its initials TsPK or *Tsentr Podgotovki Kosmonavtov*: the Cosmonaut Training Centre. And it was here on this particular Wednesday, two days before President Kennedy's inauguration in Washington and the day before Alan Shepard's selection as America's first astronaut, that six men were competing to be the first cosmonaut of the Soviet Union, if not the world.

Like the Mercury Seven astronauts in Langley, Virginia, the six men were also sitting in a classroom. But the similarities ended there. They were all obviously younger than the Americans, in their twenties and not thirties. They were all wearing military uniforms, not casual Ban-Lon shirts. And with the exception of Gus Grissom, they were all shorter, short enough to fit inside the Vostok spherical capsule replacing the thermonuclear warhead on top of the R-7 missile that they all hoped one day soon to fly in space.

And unlike the Mercury Seven too, these six men were not waiting for somebody to arrive. Instead, they were sitting an examination. Later that afternoon they would each be interrogated by a committee responsible for their training. The same committee had already assessed their performance the previous day in a crude simulator of the Vostok capsule. This simulator, the only one in the USSR, was not located in the forest but in a palatial pre-revolutionary building in Zhukovsky forty-five kilometres south-east of Moscow named LII: otherwise known as the Gromov Flight Research Institute. For up to fifty minutes, under the watchful eyes of their examiners, each of the six men had practised procedures in the mock-up of the Vostok's cockpit on the second floor of what had once been a tuberculosis sanatorium under the jurisdiction of the People's Commissariat for Labour.

The building in the birch forest where the men were now sitting their examinations was the first structure in what would become, over time, a vast, heavily guarded complex closed to the outside world and dedicated solely to the training of the USSR's cosmonauts. In 1968 it would change its name to Zvyozdny Gorodok – Star City – but that was still far in the future. Until then the area itself was known as Green Village, a nod perhaps to the acres of trees that shrouded it. When the new Cosmonaut Training Centre was established on January 11, 1960, by order of the commander of the Soviet Air Force, Chief Marshal Konstantin Vershinin, Green Village was chosen as the best site for its construction. It had many advantages: it was shielded by its forest but not far from Moscow. It was also only a few kilometres from Chkalovsky air base, the largest military airfield in the Soviet Union. And it was close to OKB-1, the secret design and production plant at Kaliningrad where Arvid Pallo's battered Vostok had been returned only the week before the men sat these examinations – and where the next-generation human-carrying Vostok capsules were also being built.

The six men had been training for ten months since March 1960, although the classroom building had only been in use since June. They had also used other anonymous centres in Moscow and continued to do so. In years to come, many of them would move into specially built residential blocks in Star City, but in January 1961 they were living near the Chkalovsky air base in basic two-room apartments with their wives and children where they had them, or in even more basic bachelors' quarters where they did not. These were a far cry from Alan Shepard's comfortable house in Virginia Beach's leafy suburbs or indeed anything the American astronauts and their families would have willingly put up with, but by Soviet standards they were privileged accommodations. Meanwhile nobody in Chkalovsky outside their closed circle knew why the six men were there or what they were training for. Nor did their parents, nor their friends, nor their former colleagues in the air force. Even their own wives were not encouraged to ask too many questions. Unlike the Mercury Seven, who were famous throughout America if not the world, the six men existed only in the shadows.

And there was another key difference from the Mercury Seven. These six were not the only cosmonauts being trained. There were fourteen more. In a selection process that was even more ruthless than that undergone by the American astronauts, twenty men had been chosen from an initial pool of almost 3,500 military pilots. The Soviet space programme had ambitions to conquer the cosmos, or at least that part of it encircling the earth, and they needed the manpower to do it. All twenty men had begun their training in the spring of 1960, just two months after Soviet Premier Nikita

Khrushchev had urged his chief space engineers that they 'should quickly aim for space. There's broad and all-out levels of work in the USA and they'll be able to outstrip us.' By then the Mercury Seven had already been training for almost a year. The Soviets needed to catch up fast and these twenty cosmonauts were the answer. But other than those broad hints in the tightly controlled press about 'Soviet Men' soon visiting space, the facts would be either hidden or deliberately misrepresented. The rockets, the capsules, the designers, the engineers, the training centres and launch locations, and of course the cosmonauts themselves – all of it would remain behind closed doors.

By the autumn of 1960 the Soviet manned space programme had become a top national target, not least because back then NASA appeared to be aiming to send an American into space as early as December that year. To speed things up and prioritise training on the single simulator, a shortlist of six front-runners was filtered from the original twenty. They would fly first before the others. In essence, they were the premier league team that would take on the Mercury Seven. To those permitted to know, this group of six was given a name. Perhaps in deliberate imitation of the Mercury Seven, they were called the Vanguard Six.

Depending on the source, the criteria for selecting the Vanguard Six appear to be quite different. Dr Adilya Kotovskaya, a specialist in pathophysiology who supervised the cosmonauts' centrifuge tests, claimed that the men were chosen simply in order of their ability to endure terrific accelerations, similar to those they might one day experience in a rocket launch, without passing out. One of the six, Andriyan Nikolayev, a quiet, heavyset, dour bachelor, proved the most resilient at this. He sailed through Dr Kotovskaya's crushing centrifuge rides.

Another cosmonaut, Aleksey Leonov, who four years later in 1965 would perform the world's first spacewalk, recalls a different rationale behind the selection: it was based entirely on weight and height. As both were critical for the Vostok, only the shortest and lightest men were chosen. Leonov himself was slightly too tall to make the grade. Everybody in this first group had to be under five foot seven. The smallest, at just five foot five, was an open-faced, married twenty-six-year-old former foundry student with blue eyes and a winning smile reminiscent of John Glenn. His name was Yuri Gagarin. He too made it to the Vanguard Six.

A third was Grigory Nelyubov. A dark-haired, handsome pilot the same age as Gagarin, Nelyubov was one of the very few among the twenty cosmonauts to have flown the USSR's first supersonic jet fighter, the MiG-19. He possessed a remarkably sharp mind – possibly the sharpest in the entire cosmonaut group – as well as lightning reactions. In some respects he resembled Alan Shepard, including perhaps a tendency to arrogance and lack of self-criticism that was often noted by his instructors. Some thought him narcissistic. Many actively disliked him. A few believed he would be chosen to fly first. But Nelyubov had not initially been picked for the Vanguard Six. He had replaced another cosmonaut, Anatoly Kartashov, who had been badly injured on Dr Kotovskaya's centrifuge in July 1960. For reasons that were never explained something had gone wrong and Kartashov was subjected to excessive accelerations as he spun too rapidly in his gondola. A fellow cosmonaut, Dmitry Zaikin, remembered the result. 'The blood vessels in his back,' Zaikin said, 'just blew off.' Kartashov was suspended without appeal. Nelyubov was put in his place.

These decisions could be brutal – and final. In the same month as Kartashov's accident, another original member of the Vanguard Six, Valentin Varlamov, was enjoying a rare day of relaxation in Medvezh'i Ozera, a stretch of shallow lakes near the training centre. With him was Valery Bykovsky, a cosmonaut whose competitiveness was notorious. Varlamov challenged Bykovsky to dive into one of the lakes. Bykovsky took up the challenge at once, grazing his head on the bottom. He cautioned Varlamov. But when Varlamov dived in he hit the lake bed hard. He was raced to hospital where he was diagnosed with a displaced cervical vertebra in his neck. He was placed in traction – and suspended from the programme without appeal. Bykovsky stepped into his shoes.

Two other men completed the Vanguard Six. Pavel Popovich was a Ukrainian ex-fighter pilot and by almost all accounts one of the most popular members of the cosmonaut corps. Everybody loved Pavel. His humour was infectious, and he had a terrific sense of fun. 'He made you want to live,' said his daughter Natalya, who was four when Popovich sat his cosmonaut exams. If he had an opposite number in the Mercury Seven it would probably have been Walter Schirra, the man who liked playing pranks. Popovich loved singing too, mostly Ukrainian folk songs, and at the drop of a hat he would happily croon in his beautiful baritone the lush, sentimental lyrics he had learned as a child in his tiny village of Uzin near Kiev, including his favourite 'The Night is so Moonlit, so Starry, so Bright':

The night is so moonlit, so starry, so bright
There's so much light you could gather the pine needles
Come, my love, weary with toil
If just for one minute to the grove ...

'We all wanted to be first,' said Popovich years later, just as Alan Shepard and John Glenn had said. But it would take more than love songs for Popovich to become the USSR's first cosmonaut. Two issues held him back. To begin with, he was Ukrainian, not Russian. Even as the Soviet Union's propagandists paid lip service to the socialist ideals of ethnic equality, Popovich's origin was a handicap. The first to go up would have to be a Russian. Only nobody would officially say so.

His other issue was his wife. Marina Popovich was a brilliant pilot in her own right who had attempted to join the Soviet air force when she was just sixteen to fly in one of three female pilot regiments that then existed, a state of affairs almost unimaginable in the United States at the time. By 1961 she had already been an outstanding flying instructor for three years. Later she would become one of the USSR's top test pilots. She was glamorous, beautiful, clever, headstrong, sometimes hot-tempered and also a mother; and all together this was a problem. Lieutenant General Nikolai Kamanin, the chief of cosmonaut training, wrote that Popovich 'gives the impression of being tough but is too soft with his wife. Family troubles hold him back.' And, Kamanin added somewhat darkly, 'We will take measures to help him.'

Gherman Titov was the final member of the six. A man with movie star looks, Titov was athletic, excellent at games and gymnastics, an amateur violinist and a passionate pilot who had graduated with top grades from aviation academy in 1957 and who went on to fly fighter jets in Leningrad. It was there that he met his wife Tamara, who was a waitress at the base cafeteria. They got married within four months. 'He was a fighter pilot,' she says. 'He did everything fast.' Born in a log cabin in Siberia, Titov had been brought up by a strict, demanding father who was also the local schoolteacher. His father is said to have named his son Gherman after the lead character in Pushkin's story 'The Queen of Spades', and the father's love of Pushkin was comprehensively inherited by the son. Titov could recite whole passages of Pushkin's works from memory in the same way that Popovich could sing Ukrainian love songs. But his habit of doing it in the middle of training sessions was also guaranteed to exasperate his hard-bitten proletarian instructors. His erudition was both impressive and a liability. He was almost *too* educated, too bourgeois, too clever, and prone to question rules. Colonel Yevgeny Karpov, who had been appointed to run the training centre, was often driven to distraction by Titov's resistance to taking orders, especially from doctors like himself. 'He would hear the doctors out, then often do things his own way or ignore their instructions altogether,' Karpov said. 'When they insisted … he became irritated.' But Karpov also admired Titov's directness and his refusal to invent excuses for himself whenever he got into trouble. It was a constant refrain: hot-headed, sceptical, a man who sometimes broke the rules if he thought they were petty or stupid; but also brilliant, attractive and consistently able. He was another prime candidate for the first flight.

The Titovs lived next door to Yuri Gagarin and his wife Valentina in their Chkalovsky apartment block. The two cosmonauts often spent time with each other, climbing over the partition between their narrow balconies on the fifth floor as the quickest, if not always safest, way to move between apartments. Their friendship deepened later in their training when, in the autumn of 1960, the Titovs' baby son, Igor, died at the age of seven months from heart complications. Gagarin and Valentina, who had a one-year-old daughter of their own, did everything they could to help the couple cope with a loss so traumatic that Tamara Titova was unable to speak about it to the author almost sixty years later. In an interview in the 1990s, Titov paid tribute to Gagarin's warmth and support at that time: 'Without being mawkish or sentimental he simply behaved like a truly close and real friend. I was grateful to him and, although I did not know him very well at the time, began to like him a great deal.' As they sat their examinations in January 1961, the bond between the two cosmonauts was very strong, even while they were both competing for the same ultimate prize.

'All six cosmonauts are great guys,' wrote Lieutenant General Nikolai Kamanin, their chief of training, on the day the Vanguard Six took their final exams. But nobody else saw these words, nor his words about Popovich's difficult marriage; nor would they for several decades. Kamanin was a legendary aviator and Hero of the Soviet Union who was now fifty-one, a stocky man with thinning hair and penetrating eyes. In photographs of the period he is rarely smiling. Some of the cosmonauts regarded him as a martinet and Stalinist of the old school. But this Stalinist also had a secret. He kept a hidden diary. He did this in defiance of every rule and at great personal risk. To write a diary, especially at his level of responsibility in the Soviet space programme, was a serious criminal act. And yet he continued to write it from 1960 until the late 1970s. Its entries throw a rare spotlight onto a world that is often impenetrable and closeted in myths, falsifications and conspiracy theories, even to this day. Kamanin is like a hidden camera at the very centre of events, a fascinating, if flawed, fly on the wall, his observations occasionally distorted by personal prejudice and bitterness but also touched by a genuine personal regard, and sometimes deep affection, for the cosmonauts under his command. We will meet him often in the pages that follow.

THE FINAL SCORES of the Vanguard Six were awarded the same day that they finished their written papers. All of them were graded 'excellent'. Kamanin was one of the examiners. 'Who among these six will go down in history as the first person to fly in space?' he wrote in his secret diary that night. 'Who will be the first of them to pay with his life in making this daring attempt?' While making the questionable assumption that the USSR would win the race against the Americans, both sentences betray the implicit paradox at the heart of this adventure: to fly in space was quite possibly to die in space. 'There is still,' continued Kamanin in the same entry, after reviewing all the failures of the various Vostok dog flights, 'no guarantee of a safe landing.'

Back in December, in one of many striking parallels with the Mercury Seven, all twenty cosmonauts were asked to vote on which of their peers should fly first. A majority had voted for Yuri Gagarin – one source claims twelve of them did so, another as many as seventeen. But even after these examinations were over and the results known, the decision was still not made. Given the tortuous and often murky hierarchies of Soviet political life, such a decision would anyway have been unthinkable. In Langley, Virginia, Bob Gilruth, head of the Space Task Group responsible for Project Mercury, could summon his seven astronauts into a classroom and tell them, simply and on his own authority, who would be first. Things were done differently in the USSR. Unlike the Mercury Seven, the Vanguard Six would have to wait – and they would have to wait until almost the very last moment.

But even as they waited, unsure which one of them it would be, the examining committee 'tentatively', and without telling the cosmonauts, recommended a ranking. In third place was Grigory Nelyubov. In second place Gherman Titov. And in first place the examiners chose Gherman's close friend, next-door neighbour and confidant, the winner of the peer vote with the winning smile: Yuri Alekseyevich Gagarin.

NINE DAYS EARLIER on January 9, the head of Military Unit 26266, Colonel Karpov, signed off a comprehensive clinical and training report on Gagarin before his exams. After noting his excellent blood pressure (110/65), the fact that he did not smoke, his infrequent use of alcohol ('tolerance is good') and, perhaps peculiarly, the quality of his skin ('white and delicate'), Karpov plunged into the nitty-gritty of Gagarin's character. A key trait was his 'high intellectual development', and it was observed that he was 'balanced in difficult conditions' – a euphemism perhaps for the wide variety of ways a rocket flight into space could unhinge less balanced minds. Among Gagarin's leading traits Karpov listed optimism, accuracy, courage, self-discipline, decisiveness and fearlessness. No doubt Bob Gilruth at NASA's Space Task Group was looking for equivalent qualities from Alan Shepard. But at this point the similarity abruptly ends. Because another of Gagarin's qualities that Karpov added to the list was his 'collectivism' – not a word one would associate with any of the Mercury Seven – and the report concluded with the ringing endorsement that Gagarin was also 'ideologically consistent [and] devoted to the aims of the Communist Party and the Socialist Motherland and ... is capable of guarding military secrets'. In other words he was fit, friendly, able to keep his mouth shut and unlikely to panic if things got really ugly. And if things went well, he would make a terrific poster boy for the USSR. In short, Gagarin was pretty well perfect for the role of first cosmonaut. But who, exactly, was he?

He was born in 1934 in the village of Klushino, a collection of cottages straddling a single road and lying in flat, rural farmland in the Smolensk region 190 kilometres west of Moscow. It is a landscape of wide skies and distant horizons, short hot summers and long deep-chilled winters. His father, Aleksey, was a carpenter and handyman of considerable skill who built the family house out of wood with his own hands. 'He could do everything,' Gagarin remembered.

'Make furniture, sew *valenki* [traditional felt boots], repair shoes.' He could also play the accordion, a talent which in combination with his lucrative carpentry skills made him very marriageable, and he married Gagarin's better-educated and highly literate mother, Anna, in 1923. If Yuri inherited his love of craftsmanship and precision from his father then from Anna he inherited both his prodigious capacity for learning and the meticulous, starched smartness that would one day serve his ambitions so well.

By the time Gagarin was born they already had two children, a son, Valentin, and a daughter, Zoya. A fourth child, Boris, would arrive two years later in 1936. But when Gagarin was just seven everything in his world changed. On June 22, 1941, Hitler invaded the Soviet Union. Three million German troops poured across the Soviet borders along a 2,900-kilometre front, punching eastwards at terrifying speed with the largest invasion force in history. Stunned by the scale and ferocity of the assault, Soviet positions were swiftly overwhelmed by the advancing German army. Gagarin's village of Klushino lay directly in their path. On October 12 the Germans reached the hamlet. Their first acts were to burn down the school and slaughter the cattle for food. Over the following days they burned down another twenty-seven houses in the village. They also seized the Gagarin family house. Aleksey, Anna and their four children were forcibly ejected.

Aleksey's ability to build anything now became useful in a way that he and his family could never have imagined. With the permission of the Germans he constructed a dugout behind the house in which the whole family would live for the next two years, including two further freezing Russian winters. The dugout exists still: a tiny, low-ceilinged, claustrophobic space with bunkbeds, a table and a dirt floor, lined with the same spruce Aleksey had used to build his own house. Food was always scarce. Gagarin and Boris could not go to school. But Anna did her best to school them herself. The older Valentin remembered how rapidly Gagarin transformed from a fun-loving little boy into a quiet, introverted child. 'He smiled less frequently in those years, even though he was by nature a very happy child. I remember he seldom cried out at pain or about all the terrible things around us … Many of the traits of character that suited him in later years as a pilot and cosmonaut all developed around that time, during the war.'

Later Soviet-era biographies of Gagarin – and even some modern Russian ones – often overlook the darker sides of his character and past. The impulse to sanctify one of Russia's iconic figures is too strong. But there is one horrific wartime experience that none of them omit and which stamped its mark on Gagarin for life. His younger brother, Boris, was just five when a particularly sadistic German soldier they called 'Albert' accused the boy of acts of sabotage. He grabbed hold of Boris, tied a scarf round his neck, strung him up from a tree and walked away. The seven-year-old Gagarin saw it happen. He rushed to his brother and desperately tried to loosen the knot; but it was too tight. He yelled for his mother, and Anna came running from the dugout to find her son hanging limply from the tree. Frantically she cut him down. Boris was unconscious but alive. For the next month he could not walk. He howled in his sleep. And for years afterwards the memory haunted him. While Gagarin went on to achieve worldwide fame, Boris found it hard to make his way in life. In a tragic postscript he died in 1977 at the age of forty-one. Sick and in terrible pain from stomach cancer, he hanged himself at his home.

Klushino was liberated by Soviet forces in the spring of 1943, although the Germans seized Valentin and Zoya, Gagarin's elder siblings, as slave labour. Sick and emaciated, they did not return home until after the war, by which time their parents thought they were dead. By then Gagarin was eleven. 'He grew up very quickly,' says his daughter, Elena. 'Whenever there was a movie on TV about World War Two, he always left the room. He didn't want to see it.' But the war also gave Gagarin his passion for flying when, just before the Nazi occupation, he watched a Soviet fighter crash-land near Klushino. A second Soviet plane landed nearby to pick up the downed man. Its pilot let Gagarin climb inside the cockpit and pointed out the controls. The boy was entranced. One of the men gave Gagarin a few bars of chocolate – which, his Soviet hagiographers assure us, he shared with his friends. But it was more than chocolate that he took away when the pilots flew off. Now he dreamed of becoming a pilot.

In the immediate aftermath of the war Aleksey Gagarin's carpentry skills were highly sought after. He moved his family to the nearby town of Gzhatsk where, with so much destroyed, there was plenty of work. He moved his entire house too, dismantling it in Klushino and transporting it log by log on horse-drawn carts the twenty-five kilometres to Gzhatsk, where he then set about reassembling it. After schooling, the young Gagarin went on to study foundry work at a technical college in Saratov, a city on the Volga river eight hundred kilometres south of Moscow. The toughness and self-reliance that he had learned during the war were now allied to an inexhaustible energy and drive. He studied everything the college had to offer, borrowing over five hundred books from its library. He played the trumpet, took up a lifelong hobby of photography and acted in a number of plays – including the hit musical *Great Construction Sites of Communism*. On top of all that he found time to participate in almost every sport the college offered, including skiing, athletics and basketball, whose team he captained even though he was by far the shortest member. He dated girls and read papers to the college's physics club; notably one on the Russian scientist and rocket visionary Konstantin Tsiolkovsky, whose writings anticipated a future of space travel and left Gagarin, according to his ghosted autobiography *Road to the Stars*, with 'a new illness, one which has no name in medical science – an irresistible urge towards space'.

He also learned to fly. As a child after the war he had joined the Young Pioneers, the first stage on the road to becoming a Communist Party member, and his mother Anna often recalled how he wore his immaculate uniform and red scarf with pride. At Saratov his political credentials opened an opportunity for him to join a local flying club run by the Soviet military. 'He's crazy about flying,' said his instructor. 'Plus, he's a hard worker and fun-loving too.' He was also praised for his tidiness, another useful asset. And here too we find his almost uncanny knack of standing out from the crowd, when on his first solo flight he was featured on the cover of *Young Stalinist* magazine, sitting in his cockpit under the headline 'To Fly Higher, Faster, Further!' – a gift for future Soviet biographers, except that by then Stalin had been officially denounced, and so it was never mentioned.

A career in the air force beckoned. Graduating from college with predictably top marks in thirty-one out of thirty-two subjects, Gagarin was offered the chance to train as a smelting specialist in Tomsk. Unsurprisingly he chose to train as a fast jet pilot instead. He was flying MiG-15s in the Arctic Circle when the initial secret selections for the USSR's manned space programme were made over the summer and autumn of 1959. Gagarin was a natural choice. His infectious vitality, his gift for making a pleasing impression, his leadership skills and intelligence, his immaculateness and martial punctuality, his peasant background and even his face, all helped to open this new door to the cosmos as they had opened every other door for Gagarin since the war. In another person the entire ensemble might have been repellent, but Gagarin had a genuine talent for making friends. Perhaps it was that sense of fun his flying instructors noticed or that striking smile; or perhaps it was something deeper that his friend and rival Gherman Titov envied in him when he said that Gagarin 'could talk to anybody … in every situation he would find the key to people'. The unspoken question now was whether he would also find the key to immortality.

FOUR

INAUGURATION

January 20, 1961
The Capitol, Washington, DC

THE DAY AFTER his secret selection Alan Shepard and his wife Louise braved the snow and the traffic jams to watch President Kennedy take the oath of office. When the new president mounted the podium at the recently extended East Front of the Capitol at 12.50 p.m., the night before's blizzard was long gone and the sun was blazing out of a clear blue sky. Snow blanketed the Capitol's grounds but everything, as *Time* magazine's Hugh Sidey later wrote, 'felt fresh and promising'. Along with the Shepards an estimated one million spectators watched Chief Justice Earl Warren administer the oath. Another eighty million Americans were watching on TV; this was the first time the inauguration was televised live and, for those few who had the appropriate sets, in colour. Despite the cold, Kennedy was hatless and wore no coat, unlike the seventy-year-old Eisenhower who was thickly wrapped in a white scarf. There had been a slight kerfuffle a few minutes earlier when smoke had started issuing from the podium while the Archbishop of Boston, Cardinal Richard Cushing, was delivering the inaugural invocation. The cardinal later claimed that he had deliberately slowed down his prayers – they lasted a full twelve minutes – so that he could use his body to shield the president from the blast in case the smoke turned out to be a bomb. In the light of Kennedy's assassination in November 1963 there is a frisson of fear in the moment; but in this case an electrician dived under the podium, yanked out a couple of wires and the smoke stopped. Then the president spoke:

> Let every nation know, whether it wishes us well or ill, that we shall pay any price, bear any burden, meet any hardship, support any friend, oppose any foe, to assure the survival and success of liberty.

His address lasted just two minutes longer than the cardinal's, but even today, sixty years later, its impact still reverberates with its ringing peroration to his fellow Americans to 'ask not what your country can do for you – ask what you can do for your country'. But Kennedy's words also signalled a trumpet-call to defend freedom 'in its hour of maximum danger'. Without specifying the USSR by name, he offered the hand of negotiation in the quest for world peace; but it was a hand implicitly backed by the threat of force should America's 'foe' choose not to comply. Sidey could not help glancing at the Soviet ambassador, Mikhail Menshikov, at his place in the stands near the president. He sat impassive, his hat low over his brow, his gloved hands clasped in front of him.

The world in which Kennedy became president on that cold, bright January day in 1961 was arguably more dangerous than at any time in history. In his campaign Kennedy had spoken energetically about reducing the so-called 'missile gap' between the US and the USSR, a gap that did not in reality exist, even if the myth that it did helped him to win office. But the very existence of nuclear stockpiles and missiles, along with the fear and anxiety that they engendered, created an almost permanent state of national anxiety, underpinned by repeated 'duck and cover' drills or cartoons in which a friendly turtle called Bert showed little American boys and girls how to avoid being blown to bits by bombs several hundred times bigger than the one dropped on Hiroshima, if they only just hid under a table.

Kennedy's opposite number, Premier Nikita Khrushchev, First Secretary of the Communist Party of the USSR, was a force to be reckoned with, as he had proved on a number of occasions, not least in sending Soviet tanks to crush Hungary's bid to free itself of the Soviet yoke in 1956, then finishing the job by ensuring the execution of that country's leader, Imre Nagy. Squat, thickset, loud and given to irate, but no doubt carefully aimed, outbursts, often combined with some vigorous fist-pumping, Khrushchev would do things like tell western ambassadors at a diplomatic reception that he was going to 'bury' them, or gleefully brag that his country was churning out nuclear missiles 'like sausages', even if in truth they were not; or – on one memorable occasion in October 1960 – take off his shoe at the United Nations and bang it repeatedly on the table to interrupt a delegate who happened to be saying something unkind about the Soviet Union. To this day there is some dispute as to whether he did actually bang his shoe (although his interpreter who was present claimed that he did so with such force that his watch stopped), but the bigger point is that none of the facts really mattered. What mattered was the legend, especially if it made the opposition think twice.

Khrushchev's latest harangue had come on January 6, the day after Bushuyev had set out the secret new timetable for a Soviet manned space flight and just two weeks before Kennedy's inauguration. With its talk of 'bringing the aggressive-minded imperialists to their senses', his speech was clearly intended to fly a warning flag at the incumbent president. 'There is,' declared Khrushchev, 'no longer any force in the world capable of barring the road to Socialism.' With the reality of a hot war meaning, more or less, an end to the world, the alternative strategy for either side in the cold version was to stamp its influence and ideology on those parts of the globe simmering with instabilities. And by January 1961 there were plenty of places to choose from: the Congo, which was in a ferment of civil war and where both the Soviets and Americans were playing a hand; Laos, the landlocked South-east Asian country next door to Vietnam, where the CIA was busy propping up a puppet government while the Soviets were busy delivering arms to the guerrillas fighting it; Cuba, whose revolutionary socialist leader, Fidel Castro, was actively being courted by the Soviets; and Berlin, the city divided by ideologies and soon by a wall, a place that Khrushchev, in another of his pithy discharges, described as the 'testicles of the West. When I want the West to scream,' he added, 'I squeeze on Berlin.'

There was, however, one other crucial arena of competition between the two superpowers, one that was keenly felt by many Americans because it appeared to capture the spirit of the new decade and its dawning technological possibilities so exactly. And that arena was space.

In fifties and early sixties America space was everywhere. It was in big-budget Hollywood epics and rock-bottom-budget B-movies, in countless TV shows and sci-fi books, in magazines like *Collier's*, *Astounding* and *Galaxy*, in a thousand children's toys and their parents' rocket-finned automobiles. It was in *Science Fiction Theater*, *Buck Rogers*, *Flash Gordon*, *The Twilight Zone*, *Invasion of the Saucer Men*, *The Brain Eaters* and *I Married a Monster from Outer Space*, and in Arthur C. Clarke, Robert Heinlein, Ray Bradbury and Isaac Asimov. And if plenty of it involved apocalyptic fears of the earth or decent American citizens being attacked, invaded, kidnapped, brainwashed or eaten by a wide variety of thinly disguised Communists, that was because the fears these tales reflected were all too real. Space was exciting, hyper-modern, *cool* – but it was also a new battleground.

Lyndon Johnson, Kennedy's new vice president who had just sworn his own oath of office a few minutes before his chief, grasped all this perfectly when he once said that 'control of space means control of the world'. But for Johnson, who had previously chaired the Senate Committee on Aeronautical and Space Sciences, control of space was not only a military matter, a business of spy satellites and exotic cosmic weaponry; it was also an ideological matter, a moral battle between two entirely opposed systems, sets of values and ways of life. Right now, as of January 1961, with the race to put the first human being in space apparently accelerating to the finish line – and possibly only two months ahead – it was for many a major, if not *the* major, battleground between the superpowers. Khrushchev certainly thought so. 'He realised,' said Boris Chertok, a senior Soviet rocket engineer at the time, 'that space victories could be of greater importance in politics than the threat of a club with a nuclear bomb on its end.' The space race was here, now, in the ether. To borrow Khrushchev's own phrase, whoever won it would prove to the world that history was on their side.

And yet, in his first address to eighty million of his fellow Americans, as well as to the entire world, the new president barely mentioned the subject at all.

For a man perhaps chiefly remembered today both for being assassinated and for his bold and impassioned commitment to putting an American on the moon, it is remarkable, and distinctly odd, to read Kennedy's inaugural speech and find just one, brief, easily missed and wholly anodyne reference to space. Two-thirds of the way in, after a plea to 'both sides' to agree nuclear arms inspections and controls, he invites the same both sides to join together to 'explore the stars', along with also jointly conquering the deserts, eradicating diseases, tapping the ocean depths and encouraging the arts and commerce. By any standards it is quite a list, not to mention one mired in a thick ooze of platitudes. In particular, the three words that comprised the entire space element of his address were hardly likely to set pulses racing, if they were noticed at all. This is quite a change from the Kennedy who, just four months later in May, would ask God's blessing to go to the moon before the end of the decade 'on the most hazardous and dangerous and greatest adventure on which man has ever embarked', and to agree to do it for some $20 billion of taxpayers' money (or approximately $174 billion in 2021), and possibly much more.

On his campaign trail Kennedy had often scored points against his opponent Richard Nixon concerning America's lag in the space race, just as he had over the so-called, if non-existent, nuclear missile gap, even suggesting in one televised debate that he did not want 'to look up and see the Soviet flag on the moon'. Such posturing played well with parts of the American public. And perhaps Kennedy really believed his own rhetoric. But there is evidence to suggest that he did not, and not simply from his inaugural speech.

Ten days before the inauguration, on January 10, Kennedy had sat down with his vice president-elect Lyndon Johnson, along with Jerome Wiesner, a forty-five-year-old electrical engineering professor at the Massachusetts Institute of Technology who had once worked on radar systems and electronic components for nuclear weapons. Kennedy had got to know Wiesner during the presidential campaign when he had needed advice on nuclear arms control and test bans. Wiesner had clearly impressed him; so much so that he had then asked the professor to set up a task force to investigate the US position on 'outer space'. The meeting that Kennedy attended in Johnson's office was to hear what Wiesner now had to say.

What he had to say was not good. While admitting that his report had been hastily composed, Wiesner did not let that stop him from 'feeling compelled' to attack almost every aspect of America's space endeavours, and most particularly its manned programme, Project Mercury. Detailing what he called 'serious problems within NASA', he described Project Mercury as a 'crash program' and frankly 'marginal', arguing that it could not 'be justified solely on scientific or technical grounds', and roundly questioning the ranking of manned space flight as the highest national priority:

> A thorough and impartial appraisal of the MERCURY program should be urgently made. If our present man-in-space program appears unsound, we must be prepared to modify it drastically or even to cancel it.

And there was more. In addition to convincing the new president to commission this 'impartial' appraisal, Wiesner forcefully urged him to distance himself from *any* endorsement of manned space flight in case an American astronaut became stranded up there and died. That, wrote Wiesner, 'would create a situation of serious national embarrassment' in which Kennedy would inevitably 'take the blame'. Wiesner's report landed like a bomb in Lyndon Johnson's office. Kennedy's response was swift. The very next day he named Wiesner as his new science and technology advisor.

With the advantage of hindsight, and not least the fact of Kennedy's subsequent championing of the lunar project, the reasons behind his thinking at this juncture might seem perplexing. But this is to mistake the mood of the moment. Even as the teams of men and women across the country involved in Project Mercury were working night and day to make manned space flight a reality, even as its first astronaut-in-waiting Alan Shepard was ready to risk his life for his ambitions, his place in history and his nation, much of that nation had lost confidence in its own ambitions for space. And for very good reasons. Because at the time the country's thirty-fifth president was sworn into office on that memorable January afternoon, America's space programme had become, in many American eyes, nothing less than one long, humiliating, national embarrassment.

FIVE

SHAME AND DANGER

A LITTLE MORE THAN three years earlier, on October 4, 1957, Lieutenant Commander Alan Shepard – then a Navy Aircraft Readiness Officer attached to the staff of the Atlantic Fleet – took his ten-year-old daughter Laura to the end of their drive on Brandon Road in the Bay Colony suburb of Virginia Beach and pointed to a brilliant star arcing slowly across the sky. This was nothing unusual for Laura. Her father often took her outside to point out the stars and planets and constellations. Back then there were no streetlights and sometimes, on really clear nights, it seemed to Laura like whole galaxies sparkled from one horizon to the other. Shepard would become animated as he explained to his daughter, and sometimes to her younger sister Julie and their orphaned first cousin Alice who lived with them, how one day humans would travel up there in that big black void. There would come a time, he said, when people would visit the moon and the planets and perhaps even the stars, and they would see wonders beyond anyone's wildest dreams. And that time might come sooner than they thought.

But with this particular star it was different. For one thing it was not even night-time; it was afternoon and yet the star was brilliantly visible. For another, instead of her father being excited as he usually was, Laura remembers he was 'gruff'. And as the gleaming star swung across the sky he became gruffer still. She wondered whether she had done something to annoy him. Only later did she realise the truth. Her father was not mad at her. He was mad at the Russians.

He was not alone. A substantial percentage of his 172 million fellow Americans were feeling exactly the same way. They were almost all of them mad at the Russians, and not just mad but also frightened, panicked, shocked, bewildered, envious, appalled and frankly amazed. For those same Russians, a nation popularly made up of hopelessly enslaved Communists who were supposed to be so backward that they could not even build a working refrigerator, a nation that had also lost some twenty-seven million of its citizens by the end of the Second World War only twelve years before, with many of its cities and much of its economy left devastated, had just succeeded in putting the world's first artificial satellite in space – and, more to the point, they had done it before the Americans.

The satellite had a name, one that the Soviet press advertised to the world after its successful launch on that epic October day. It was a *Sputnik*, a 'travelling companion' to the earth, a highly polished, pressurised aluminium sphere weighing 184 pounds and speeding in elliptical orbits around the globe every ninety minutes at an unimaginable 17,800 mph. Some of those orbits took it high over huge swathes of the United States, including Alan Shepard's front drive, serenely sailing across American skies, undisturbed, unreachable – and uninterceptable; a brutally convincing and maddeningly *visible* demonstration of Soviet technological supremacy at the height of the Cold War. No wonder Shepard was gruff. What was worse, as soon as this Sputnik reached orbit four whip antennas had sprung open and two radio transmitters on board had begun broadcasting a non-stop, and even more maddening, beeping signal to anyone down on earth who happened to have the right listening equipment. It seemed that the Soviet designers must have helpfully selected frequencies that were easy for any amateur ham to monitor, because within hours almost every TV and radio station in the US and across most of the planet was re-broadcasting that endless *beep-beep-beep* noise as the little metal ball repeatedly wound its way around the world, like a Soviet call of victory. And it went on transmitting it for three long weeks, until its batteries ran out.

Sputnik's success unleashed an unprecedented orgy of self-laceration and hysteria in the American press, not only because it was so manifestly a technological triumph but because, quite literally overnight, it appeared to leave the United States' citizens hideously vulnerable to the possibility of other, more unambiguously hostile, Soviet machines that might very soon be coming their way, flying at terrific speeds and unassailable altitudes way above their heads. Lyndon Johnson, then the Democratic majority leader of the Senate, painted a shocking spectacle of Russian bombs being dropped 'like kids dropping rocks' from space without anyone being able to do a thing about it, and a future president, Gerald Ford, then a Republican congressman, ran with the same theme when he raved about Soviet thermonuclear warheads 'flashing down at hypersonic speed' at any spot anywhere on earth. Another senator, Henry Jackson, declared that Sputnik's achievement was nothing less than 'a life and death matter for our country and the free world', and – working himself up into a real lather of disgust – demanded that the United States government instantly proclaim a 'National Week of Shame and Danger'. As the little beeping Soviet sphere crossed the continent of America four times in its first *day*, the panic only multiplied. Perhaps the celebrated science fiction writer Arthur C. Clarke, who would later go on to write the novel and co-write the movie *2001*, best summed up the national mood – or more accurately, the national trauma: the day Sputnik began orbiting the planet, he declared, was the day the United States became a second-rate power.

What made things even worse was that this whole idea of getting satellites in space was supposed to be an American one, since it was no less than President Eisenhower himself who, back in 1955, had announced that the US would launch a satellite as the American contribution to something called the International Geophysical Year. This was a project scheduled for 1957/8 involving sixty-seven countries – including, portentously as it turned out, the USSR – that would each contribute to a tapestry of worthy scientific projects, including studies in oceanography, meteorology, cosmic rays, the aurora borealis and the earth's gravitational field. Eisenhower had given the task of launching this American satellite to the navy, despite misgivings from the other two services which argued, unsuccessfully, that they had much better and more reliable rockets than the navy's 'Vanguard' to do the job. The navy was still working on that job when Sputnik was launched, and it did not help that Eisenhower not only hopelessly underestimated the public's anger at the whole affair but also hopelessly underplayed its significance. Airily claiming that 'Sputnik does not raise my apprehensions, not one iota', and wondering what all the fuss was about over 'one small ball in the air', he went on to enjoy a game of golf when the story first broke.

On November 3, 1957, just four weeks after Sputnik had launched into space leaving many Americans, apart apparently from their own president, feeling that this had possibly been the worst humiliation for the United States since the Japanese attack on Pearl Harbor, the Soviets did it all over again with Sputnik 2. Only now with a dog inside. Compared to Sputnik 1, Sputnik 2 was huge, weighing more than six times as much as its predecessor, containing a padded compartment for a female mongrel with a dash of husky and a smattering of terrier in her genes called Laika. The American press promptly gave her another name: *Muttnik*.

This was the first time any living being had ever left the earth to orbit in space, with the spacecraft at the apogee of its ellipse pitching Laika to the astonishing height of over a thousand miles. It was a truly remarkable success, even if the technology did not yet exist to bring Laika back home, meaning that she would go on to win everlasting fame not only as the first dog to go into space but the first to die there too. The Soviet press claimed that she survived in orbit for a week before her oxygen ran out (a claim that the Russians continued to maintain until 2002), whereas the truth was that she died, perhaps mercifully, probably within the first few hours from overheating, when the temperature inside her sealed cabin soared to almost 43°C. Her death earned those unidentified Soviet scientists who put her there the condemnation of dog lovers all over the world, but others saw matters from a more geopolitical perspective, noting with alarm that the launch of Laika was clearly timed to coincide with the fortieth anniversary of the Russian Revolution. With Laika's flight it seemed that Khrushchev was making, arguably with some justification, another of his pithy points about Soviet superiority over the decadent west, a notion buttressed later in his own memoirs when he wrote that these Sputniks 'made our potential enemies cringe with fright, but made many other people glow with joy' – by which, presumably, he also meant himself.

Fifteen days after Laika had first arrived in orbit, when Sputnik 2 with its dead dog inside had already orbited the globe some two hundred and fifty times, *Life* magazine ran an op-ed piece entitled 'Arguing the Case for Being Panicky'. 'In short,' wrote its author George Price, 'unless we depart utterly from our present behaviour, it is reasonable to expect that by no later than 1975 the United States will be a member of the Union of Soviet Socialist Republics.' That might have been pushing it a touch, but the sentiment offers a vivid glimpse into the way many were thinking at the time, and helps explain why, on December 6, 1957, the long-overdue launch of the navy's Vanguard rocket carrying America's first satellite – named TV-3 – was watched with more than several pinches of anxiety.

Jay Barbree was a young reporter who covered that launch. Back in the summer of 1956 and at the height of the civil rights troubles he had left Albany, Georgia, after secretly filming a Ku Klux Klan rally for a local TV station and getting caught. The Klan had beat him so badly that he ended up in hospital, at which point he decided that it might be a good idea to 'relocate' – his own word – before matters got even worse. The place he relocated to was Cape Canaveral on the east coast of Florida. Here was the beating heart of America's attempt to get back into the space race with the Soviets, a vast triangular-shaped missile range that had been established in 1950 on sixteen thousand acres of alligator- and rattlesnake-infested scrub and salt-marshes between the Banana river and the Atlantic Ocean.

Barbree had once flown F-86 fighter jets for the air force and had a powerful conviction, especially after Sputnik, that space in the late fifties was the most exciting place in America for a reporter to be. It was a conviction that would lead him over the next several decades to cover 166 rocket launches, more than anyone ever has. But Vanguard, with its tiny TV-3 satellite perched on top, was his first.

Tiny being the operative word. Compared to Sputnik 2's huge, 1,120-pound hulk, TV-3 weighed just over three pounds. It was barely bigger than a grapefruit. Still, even a grapefruit was better than nothing, and while Barbree and his fellow newshounds watched from distant sand dunes or boats – the navy had refused to let any of them inside the Cape Canaveral compound – the three-stage Vanguard rose its first few inches on the way to orbit with a tremendous roar; then, as if already giving up the game, suddenly fell back onto the pad after just two seconds and promptly exploded.

Worse still, the little grapefruit-sized satellite at the top was somehow thrown clear of the resulting fireball, bouncing – as Barbree remembers it – 'into hiding in the Cape's wilderness, its small transmitter broadcasting its lonely distress'. The sound was 'mournful', a long, unending string of plaintive beeps, prompting the famous game show panellist Dorothy Kilgallen to speak for every American when she asked, on CBS's *What's My Line?*, 'Why doesn't someone go out there, find it, and kill it?' Meanwhile the press had another field day. The whole thing was a *Flopnik*. Or a *Dudnik*, an *Oopsnik* and a *Stayputnik*.

It was a dark hour for the United States. With the navy smothered in shame and firmly out of the running, hopes quickly turned towards a man who had graced the cover of that same *Life* edition of November 18, 1957, that had suggested now might be a good time to panic. His name was Dr Wernher von Braun, and he was pictured in full colour in his elegantly cut, but reassuringly sober, grey suit beside a breathtakingly futuristic model of the sort of rocket that might one day go to the moon; a portrait of manly strength, good looks and luminous intelligence, suggesting to every reader out there, as the awe-struck reporter wrote, that here was the 'seer of space' and a 'practical prophet'; and that here, at last, was the man who could get America out of its mess.

The facts that von Braun had also once been a member of the Nazi Party, as well as a *Sturmbannführer*, or major, in the SS, had designed the V-2 missile that had destroyed substantial chunks of London and Antwerp, killing many of their inhabitants, and had once worked Hitler up into such a froth of excitement about his V-2 that the dictator made von Braun a professor on the spot, were not heavily, or at all, dwelt upon by the magazine's reporter. What was important was not his past but the present: and von Braun's acknowledged brilliance as both a rocket designer and a space visionary, if forged in the morally muddy business of a war that was over, was simply beside the point. The point was that von Braun, who now worked for the US Army Ballistic Missile Agency in Huntsville, Alabama, already had a rocket called a Jupiter-C that, with a few modifications, was essentially all ready to go, one that would have a much better chance of lobbing a US satellite into space before the Soviets pulled another of their rabbits out of the hat.

And anyway, by now he was no longer a German but a fully fledged American. In a special ceremony in 1955, on a day he described as the happiest of his life, von Braun had become a naturalised US citizen, along with his wife and over a hundred of his fellow German rocket designers who had also worked under him on the Nazi V-2 missiles, and who were now busy designing more advanced missiles for the US. These men had surrendered to American forces in the very last month of the war, after first burying fourteen tons of V-2 design blueprints in a mineshaft in the Harz mountains as useful collateral; thus providing them, under a top-secret US War Department operation called Paperclip, with a free passage across the Atlantic where they could go on building rockets for their new masters. The more dubious aspects of their war records were simply erased from their files by US officials. The fact that their V-2 missiles had been built under horrific conditions by concentration camp inmates, thousands of whom had died of sickness or hunger or had been executed in the process, sometimes being hanged by their SS guards from the overhead cranes used to lift the V-2s into position – all of this was swept to one side.

For von Braun, none of these matters appeared to present any difficulties. 'War is war,' he once said, 'and because my country found itself at war, I had the conviction that I did not have the right to bring further moral viewpoints to bear.' And if challenged he could always point to the fact that he once spent almost two weeks in a Gestapo jail for being overheard declaring that the war was lost. A journalist for the London *Daily Express* who interviewed him in 1945 shortly after he had surrendered to the Americans wrote: 'He feels no guilt at all … His one passion in life was the success of his rockets. It was immaterial to him whether they were fired on the Moon or on little homes in London.' Or, as the singer Tom Lehrer would later put it:

Once the rockets are up, who cares where they come down?
'That's not my department,' says Wernher von Braun.

That singular purpose, untroubled by conscience and combined with both a striking charm and a pronounced aptitude for personal promotion, was exactly what his adopted country now needed. Von Braun's modified Jupiter-C rocket – renamed a Juno – was rapidly enlisted to send a new American satellite into space, an object called Explorer 1. This one weighed in at approximately thirty pounds, which was barely a fortieth of the weight of Sputnik 2 even without its dog. But at least von Braun's Juno succeeded in not blowing itself up when it launched into Florida's skies on January 31, 1958, and Explorer 1 actually made it into orbit. The press trumpeted the occasion as a huge national achievement but a better depiction was probably one of huge national relief. Von Braun was hailed as his country's saviour. And this time he made it to the cover of *Time*.

That was in its issue of February 17 and the artist's portrait of von Braun, again wearing his reassuringly sober but elegant suit, has him perceptibly smiling as one of his rockets blasts off in a fury of flame behind him. Perhaps in that smile there lies a hint of a smirk too, a confident certainty that with himself at the helm America's space programme was finally in the best of hands. He was now rich and famous, and he bought himself a white Mercedes. And he had become a committed Christian. His particular brand of faith was perhaps less of the soul-searching, conscience-examining kind and more of the socially conservative, militantly anti-Communist variety. 'The secret of rocketry,' he once announced, 'should only get into the hands of people who read the Bible.' Many of his fellow Americans would have agreed.

Unfortunately the godless Soviets already had the secret. And in May 1958, just three months after Explorer 1 had made it into orbit, they pulled out yet another of their rabbits and flung it across the world, this one their biggest yet. Sputnik 3 was a monster weighing a flabbergasting three thousand pounds, almost a *hundred times* the weight of little Explorer 1, and it carried virtually a laboratory-load of scientific instruments on board. Despite a problem with its data recorder, the satellite went on to make forty thousand optical observations in orbit, thus conclusively ensuring its fame as the world's first scientific space observatory and comprehensively trouncing America's standing in what the media was now openly dubbing the space race.

There was just no catching up with the Russians. Their mighty intercontinental ballistic missiles, built to hurl hydrogen bombs halfway across the world, were now doing an excellent job of hurling heavy payloads up into space. The details of these missiles were, of course, kept strictly secret, even though US intelligence agencies were frantically trying to penetrate those secrets, as we shall discover. But to ordinary Americans the Soviet advantage was simply incomprehensible. How could the richest and most advanced nation on the planet possibly be falling behind? How could it be always coming in *second*? In truth, it was the very backwardness of Soviet technology that gave them their massive advantage. Both the electronics and mechanical equipment needed for their hydrogen bombs were crude and heavy, while the more advanced American hydrogen bombs were light. And the heavier Soviet bombs required bigger, more powerful rockets to carry them. From the beginning, the US space programme faced a paradoxical handicap: the nation's technological supremacy in bomb-building *facilitated* their lack of supremacy in space. They were winning the arms race for the same reason that they were now losing the space race.

In an attempt to compete, von Braun and his team of German and American engineers at the Army Ballistic Missile Agency in Huntsville were already busy designing a very big rocket of their own – the Saturn – but in 1958/9 the concept was still barely off the drawing board, not to mention short of government funding. The Americans therefore had no choice but to get on as best as they could with their Redstones, their Junos, Jupiters and Atlases, grand names for rockets that were, in reality, puny compared to the Soviet competition.

And so the humiliation went on. America's place in the space race always seemed to be an embarrassing and distant also-ran. Between 1958 and 1960 there were eight separate American efforts to get a space probe called Pioneer to the moon; all but one of them failed. Meanwhile the Soviets kept chucking up lunar probes like confetti. The fact that some of these also failed was simply never publicised. The key was to *look* invincible and in this the Soviets were pre-eminently successful. So when Luna 1, launched in January 1959, missed the moon by mistake and ended up orbiting the sun instead, Khrushchev brazenly claimed that this had been its purpose all along, crowing that 'even the enemies of Socialism have been forced, in the face of incontrovertible facts, to admit that this is one of the greatest achievements of the cosmic era.'

But often Khrushchev did not even have to lie, or at least not that much. Luna 2, launched in September that same year, really *did* reach the moon, the first artificial object ever to make contact with another celestial body. Shattering on impact, its sphere split open into a shower of glittering pentagons, each marked with the emblem of the Soviet Union and the letters *CCCP* – for USSR. It was a tremendous publicity stunt and it prompted another when Khrushchev asked the designers for a model replica to give as a gift to the White House. 'Just make sure,' he told them, 'they put it into an attractive box.' They did, and it was duly presented to Eisenhower by a smiling Khrushchev himself on his visit to the United States two weeks later, with a gaggle of photographers to capture the moment for the whole world to see. In the photographs Eisenhower is also smiling – through gritted teeth.

Just a month after Luna 2's shower of Soviet pentagons, its successor Luna 3 was beaming back photographs of the far – or, more popularly, dark – side of the moon, a region no eyes had ever seen. By now the US was so far behind in the space race it was barely off the blocks. And in case there were any Americans left who still thought otherwise, Khrushchev was happy to educate them. 'We have beaten you to the moon,' he said, visiting a hot dog factory in Des Moines on that same American tour. 'But you have beaten us in sausage-making.'

BY THIS TIME, a new space agency was orchestrating the American space effort. Stung into action by an ever-lengthening catalogue of bad news, Eisenhower had signed the National Aeronautics and Space Act in July 1958, forming NASA. The new National Aeronautics and Space Administration was explicitly a civilian body, much to the anger and dismay of some sections of the military which held that space should be a military concern. NASA brought together a number of key organisations under one umbrella to co-ordinate the nation's space exploration programme. But some cynics might have argued that it had, essentially, one goal: to beat the Soviets.

Not that NASA spelled out its goal as such. Indeed its first administrator, Dr T. Keith Glennan, went out of his way to spell out the exact opposite, claiming – with a perceptible sniff of distaste – that 'we are not going to attempt to compete with the Russians on a shot-for-shot basis in attempts to achieve space spectaculars', which, under the circumstances, was just as well considering that up to now the Americans were almost always spectacularly losing. And yet, even in this crisis, NASA did have one ace left up its sleeve; and one which, with the assistance of another of the versatile Wernher von Braun's rockets, might finally break the curse and pull off the greatest of all spectaculars. And that was to put the first human being in space.

There were already blueprints for doing this, exciting, futuristic winged machines that would transport humans into orbit and even beyond to the moon and planets. The problem was that they only existed on paper. But nobody could deny they had some terrific names, many of which chimed with the argot of the day such as Boeing's X-20 Dyna-Soar, a hypersonic glider that sounded like something out of *The Flintstones* – first broadcast in 1960 – and which would *soar* into the heavens and then glide *aerodynamically* back to earth, landing like an aeroplane on a runway. Based on the existing North American X-15, a highly experimental rocket-plane that could fly at several times the speed of sound to heights of fifty miles or more, the Dyna-Soar was, before NASA's formation, the US Air Force's manned space vehicle of choice, in some ways anticipating the Space Shuttle by two decades. But that was just the problem: it was too advanced and would take too long to develop. What was needed was something that could be done *fast* – something that could beat the Soviets, and soon. A much better, if less exotic, solution was simply to insert a man in the nose cone of a ballistic missile and shoot him upwards into suborbital space, detach the cone with the man inside and then let the whole contraption fall back to earth again. It was crude, ugly, unaerodynamic, dirty and inelegant; but it was also, at least compared to something like the Dyna-Soar, simple.

Along with Dyna-Soar, the US Air Force backed this simpler up-and-down idea too in a programme they called Man In Space Soonest, which must have sounded like a winner of a title until somebody pointed out too late that its acronym was MISS. But when NASA took over the space programme in the autumn of 1958, a more sophisticated version of this up-and-down man-in-suborbital-space concept was adopted. It would form the basis of Project Mercury's first-phase programme, with manned orbital flights to follow as a second phase later as the technology was developed. All that was needed at this point was to design a human-carrying capsule and find a human who would agree to sit inside it and blast off, successfully one hoped, into space. And come back alive.

SIX

THE AMERICAN TEAM

WHEN DR T. Keith Glennan introduced America's first astronauts-to-be to the world in Washington, DC on April 9, 1959, Project Mercury, NASA's manned space programme, was just six months old. The introduction took place at two o'clock in the afternoon in a press conference at NASA's headquarters in Dolley Madison House. Every inch of the ballroom was crammed with reporters, photographers and cameramen who had all come to see America's new heroes in the race to space. And there they were, seven men in sharp suits and ties – and in two cases bow-ties – lean, groomed, all in their thirties, all white and all grinning diffidently while a hundred or more flashbulbs popped and crackled in their faces. They sat at a table on a stage with NASA's zippy new logo behind them, a model of a rocket and a space capsule in front of them and, in case nobody got the real point of all this, the Stars and Stripes to the left of them: the Mercury Seven.

The question that electrified the atmosphere in that ballroom, and indeed large swathes of the American population, was simply this: who exactly were these seven men, and out of hundreds of contenders how did they get to be picked?

The original astronaut specification had called for the sort of people used to seriously hazardous activities, a list that included submariners, deep-sea divers, mountain climbers and Antarctic explorers. In the end President Eisenhower ordered that only military test pilots should be considered, not because of their piloting skills – in the early days they were not expected to *fly* the prospective spacecraft, just *endure* it – but because Eisenhower believed they would best be able to cope with whatever unknown and terrifying hazards a space voyage might throw at them. They also had to be in excellent physical shape, under forty, jet pilots with at least 1,500 hours' flying experience and graduates of a test pilot school, all of which meant that they also had to be men, since no woman in the United States in 1959, however superb and experienced a pilot, was in any position to meet all those criteria at the same time. A potent stench of male chauvinism permeates the thinking behind the selection. 'There is,' said Brigadier General Donald Flickinger, one of the flight surgeons involved, 'the aggressive response to stress, as we find in the tiger, and the docile response, as exhibited by the rabbit.' And he added, 'We are looking for tigers.' Charles Donlan, assistant director of the Space Task Group which ran Project Mercury, was more blunt. What they were actually looking for, he said, were 'real men'.

Alan Shepard clearly fit their definition. The selection committee pulled the records of every graduate of every test pilot school in the country going back over the past ten years, whittling down the list initially to 508 candidates, and then further down to 69. Shepard was one of them. In January 1959 he was invited to a briefing at the Pentagon and asked if he wanted to volunteer as an astronaut. He said yes. By February, he was in the final 32 selected for two brutal weeks of medical and psychological tests. Since no human had ever been in space before, nobody knew what to expect. Everything that was possible to test was therefore tested. 'We had to ensure,' said Dr Robert Voas, a psychologist on the programme, 'that after we had put something like a million dollars onto each of these guys, we had a good long-term investment.'

At the Lovelace Clinic in Albuquerque, New Mexico, a hospital that also conducted medical research on the CIA's high-altitude spy pilots, Shepard and his fellow candidates were subjected to a battery of tests for fourteen hours every day for a week. They arrived in batches over February. After checking in anonymously at a local motel, the men were welcomed on their first evening by the clinic's director, William Randolph Lovelace, and his wife Mary, at a cocktail party. There were no more cocktail parties after that. Identified from then on only by a number and never by their names, the candidates embarked on a relentless, painful and often degrading schedule of examinations. Their livers were injected with dye, their intestines filled with radioactive barium, their anuses explored with a metallic instrument called a 'steel eel' which snapped open once inside their rectums. They had seventeen separate eye tests and up to six enemas a day. Electrodes were inserted into the muscles of their hands to induce a succession of electric shocks, their prostates were squeezed, their stomachs were pumped, they were required to masturbate in cubicles to have their sperm counts checked, and they were photographed naked from every conceivable angle, including one with them squatting over the camera. They were prodded and poked and probed in every orifice and elsewhere. 'I didn't know,' said John Glenn, 'the human body had so many openings to explore.' In one peculiarly unpleasant test they were blindfolded while a nurse pumped water into their ears with a hose until their eyes felt like they were floating out of their sockets. 'It was,' said Deke Slayton, one of the final seven, 'your worst nightmare.' Even the clinic's director and owner, Dr Lovelace, agreed. 'I just hope,' he later confessed, 'they never give *me* a medical examination.'

One of the chosen astronauts sailed through these medical tests above all the rest. Scott Carpenter, the guitar-playing navy patrol pilot, proved to be an almost perfect specimen of the human male, which was also why he ended up making the final grade despite his relative piloting inexperience. Carpenter even *enjoyed* his time at Lovelace. 'I found it fascinating,' he later said, 'it gave us a new appreciation for what a marvellous machine the human body is.' But his own marvellous machine of a body had everybody else 'psyched out', according to Bob Solliday, a candidate who failed: 'He had the lowest body fat, the best treadmill test, cycled for ever, held his breath longer, never lost his cool. We were afraid to arm-wrestle the guy.' Wally Schirra, the Mercury Seven's humorist, hated the whole experience. 'We were well patients being looked at by sick doctors,' he wrote later, and the night before the doctors were scheduled to examine his intestines he made sure to eat the fieriest Mexican dinner it was possible to find in the whole of Albuquerque.

But Lovelace was not the end of it. In the selection committee's determination to find tigers and not rabbits, thirty-one of the men – one had already been rejected – were next sent to the Wright Air Development Center in Dayton, Ohio, to undergo what one of the candidates, John Tierney, categorised as 'a whole new host of torture machines.' Here their bodies were assaulted with every possible stress and trauma the examiners could dream up. Merely the names of some of these tests were enough to induce fear in the fiercest of tigers: the Cold Pressor test, the Tilt Table test, the Equilibrium Chair test, the MC-1 Partial Pressure Suit test, the Flack and Valsalva Overshoot test. The list was endless, or at least it felt that way over the week it lasted. There were twenty-five psychological examinations and psychiatric interviews – Shepard, the 'cool customer' who was not given to self-analysis, predictably loathed these the most, while Schirra, the class clown, fell asleep in one of his – there were Rorschach tests and a 566-item personality questionnaire (Scott Carpenter enjoyed *that* one too); there were treadmill tests and 'high-energy' sound tests and centrifuge tests where the men were spun faster and faster, racking up the crushing G-forces on their bodies to see just how much they could take before they lost consciousness. In one test Shepard and his fellow candidates were forced to sit with their bare feet in a bucket of ice for seven minutes; in another they were placed in a 'bake-chamber' and baked at 55°C for two whole hours; in a third they were asked, in some nightmarish version of a children's party, to blow up balloons over and over again until they almost passed out. 'These guys,' said John Tierney of his torturers, 'loved to press the limits. They were scientists, or maybe sadists, in a theme park of their dreams with no rules.' Tierney was not selected.

Shepard was. By the time he and the six other Mercury astronauts were paraded at that Washington press conference – resembling, as one NASA employee remembered, 'a beauty pageant' – public interest in the manned space programme was running at fever pitch. To some of the astronauts the sheer high-voltage intensity of that public interest was almost too much and it began when the entire press corps rose as one, applauding the seven men grinning uncertainly up there on their platform. When a reporter asked them which, of all the tests they had undergone, was the worst, the tongue-tied Grissom answered, 'This is the worst, here.'

They were supposed to be pilots, not celebrities. But nobody seemed to care a jot about their piloting careers. The very first question was about their wives, except that the word used was not wives but 'good ladies': what did their *good ladies* think about their husbands' career move, a question which was really another way of asking – and the suggestion hovered thrillingly in the room – what did their good ladies think about a career move that might, and probably would, *kill* them? Shepard was typically unforthcoming: 'I have no problems at home,' he said, adding, 'My family is in complete agreement', an answer that drew a big laugh from the largely all-male press corps. But for Gordon Cooper, the handsome farmboy from Oklahoma and quite possibly the best pilot of the seven, any questions about his wife were always going to be a challenge because he and Trudy had separated the previous year when she had caught him cheating on her, and not for the first time. Only when it looked like he might be selected as an astronaut had he driven to San Diego where she was living with their two children and begged her to come back to assist his chances. She did in the end, and amazingly she even agreed to keep their separation a secret – but Cooper was always going to struggle with any questions about his wife. 'Mine,' he said briefly, 'is enthusiastic.'

But the real eye-opener of that press conference was John Glenn, the man who had already charmed TV audiences with his performances on *Name That Tune*, and who now went on charming and dazzling every reporter in just the same way. 'I got on the project because it will probably be the closest to heaven I'll ever get,' he joked, and the room dissolved. Glenn was an instant star. He was perfect media fodder for the role of astronaut: an all-American, all-Presbyterian test pilot and multiply medalled war hero with easily the sunniest smile in the room, cheerfully discoursing for minutes on end about his church boards and his faith and God and heaven and country and how his wife Annie backed him 100 per cent in his ambition to get to space. The reporters lapped it up while the other six astronauts gaped. With Glenn everything, as they were soon to learn, was 100 per cent, including his ability to keep on talking. 'Who is this boy scout?' wondered Cooper, while Grissom, who had probably not set foot inside a church in years, found himself limply admitting to the swarm of reporters that 'I am not real active in church, as Mr Glenn is.' Glenn's charm was simply terrific. Just by itself it looked enough to take on the Russians and beat them all the way back to Moscow. When a reporter asked how confident they were that they would come back from outer space alive, all the astronauts put up a hand. Glenn put up both hands.

THEY WOULD GET an opportunity to test that resolve, at least in theory, only a month later in the early hours of May 19, after the seven astronauts had flown down to Cape Canaveral to attend their very first launch, in this case of an unmanned Atlas rocket similar to a type that some of them would one day fly on the later orbital missions in space. Sixty-four seconds after launch it exploded right over their heads. 'It looked like an atomic bomb went off,' said Glenn. 'We were looking at this thing and looking at each other and deciding we wanted to go back and talk to the engineers a little more before we go further.' NASA may have got its new astronauts, but America's jinx was not over. Their rockets kept on blowing up.

And they blew up in so many different ways. 'I saw a lot of rockets launched,' wrote Chris Kraft, a brilliant aerodynamicist who was often down at the Cape at the time developing the then totally novel idea of mission control, 'and I'd say that somewhere between thirty and forty percent of them failed. A lot of them came off the pad and went in the opposite direction. Some of them got halfway off the pad and blew up. Some of them got up to 10,000 feet and turned the other way and blew up.' 'Boy,' said Glynn Lunney, who was on Kraft's team, 'those things were really scary! It was frightening to watch.' The pad blockhouses had steel-reinforced walls and windows a foot thick but that did not stop Ike Rigell, a launch engineer who had seen action in the Pacific during the Second World War, from worrying that the ceiling might cave in. Huddling behind his panel could sometimes feel like being under attack. Ed Murrow, the CBS reporter who had famously covered the London Blitz during the war, was in the blockhouse in 1959 when the Juno rocket he was filming for a documentary about the American space programme suddenly chose to do a U-turn seconds after launch and nose-dived straight back towards the pad just a hundred metres away. Rigell was there too. 'We fried a lot of rattlesnakes that day,' he said.

Jay Barbree, the journalist who had watched America's first satellite bounce into the Floridian undergrowth instead of into orbit when the Vanguard rocket exploded, now found himself fascinated by the astronauts every time a rocket blew up. 'We got a big kick out of watching them,' he wrote afterwards. 'It was great looking into their eyes.' But what was worse, if anything could be worse than witnessing the many different ways one of those rockets could kill you, was the fact that all the while the Soviets went on amazing and dumbfounding the world with a whole new series of space spectaculars, the most spectacular of them all being that orbital flight of Belka and Strelka in August 1960. Here were two cute little dogs who had effortlessly flown eighteen times around the globe without apparently a scratch while American rockets were still doing U-turns in the sky. The standing ovation at their Moscow press conference almost rivalled that of the Mercury Seven's. Moreover, NASA had not yet managed to get *anything* alive into actual earth orbit yet. The humiliation was complete and inevitably Khrushchev made a meal of it. In a pattern of behaviour that was becoming familiar, he would later give one of Strelka's puppies, Pushinka – otherwise known as

Fluffy – to President Kennedy's wife Jackie as a gift. She accepted, after security officers had first thoroughly searched Fluffy for electronic surveillance bugs.

The fact that Belka and Strelka's flight had very nearly gone disastrously wrong, not to mention all the other things that had gone wrong with every one of the other Vostok dog-carrying flights – as Arvid Pallo and his Siberian search team were painfully aware – remained of course strictly secret. There were no Soviet Ed Murrows filming Soviet rocket disasters for Soviet TV. To the architects of the Soviet space programme, this could only be a boon. If a future American astronaut were to be blown to bits eighty million viewers would get to see the whole thing live – a consideration that understandably created a culture of caution. If the same happened to a future Soviet cosmonaut nobody would get to see anything.

A remarkable document from 1960 provides a fascinating insight into just how strictly controlled information was in the USSR. Its sixty pages minutely list every type of information 'forbidden to be published in regional, city, and large-circulation newspapers and radio programmes', including 'any information about military accidents or catastrophes'. And since 'satellites and rockets' came under the auspices of the ministry of defence, they were considered to be military vehicles, whether they carried dogs or nuclear warheads. The result was that the Soviet space programme continued to look invincible while the US just kept lagging behind. But even without the distorting filter of Soviet censorship, the truth was that the US *was* lagging behind. Boris Smirnov, one of the very few Russian photographers at the time allowed access to Soviet space technology, remembers seeing pictures of American rockets back then in international magazines. Compared to the Soviet versions, he said, 'they were garbage buckets'.

SO IT WAS perhaps hardly surprising that Jerome Wiesner gave his damning report on the US space programme to President Kennedy, or that the president made him his science advisor the very next day. Perhaps the final straw had already come a couple of months earlier on November 21, 1960, just thirteen days after Kennedy was elected, when the nation's media, with its hustle of reporters, photographers and cameramen, gathered at Cape Canaveral on a sunny morning to attend a very special launch. Also watching from the blockhouse were Wernher von Braun and some of the astronauts, while Alan Shepard watched from the VIP area and locals and tourists packed the nearby shorelines at Titusville and Cocoa Beach with their binoculars, transistor radios and cameras. Off the coast in the Atlantic other eyes were watching too – from fishing trawlers anchored in international waters and regularly used as cover by Soviet agents to monitor American rocket launches.

By then the Mercury astronauts had been in training for nineteen months. Over that time all of them had been subjected to a battery of exotic training devices designed to condition their bodies – and their minds – to the extreme stresses of launch and re-entry as well as to the strange, and possibly deadly, sensations of weightlessness. There were further multiple centrifuge rides at the Naval Air Development Center in Pennsylvania, where each astronaut was pushed to withstand higher and higher gut-wrenching G-forces strapped inside the rapidly spinning gondola, along with multiple sessions in much weirder spinning contraptions with names like the Multiple Axis Space Test Inertia Facility or MASTIF – a truly horrible carnival ride designed to simulate an out-of-control spacecraft that was far faster, wilder and crazier than any carnival ride ever invented, in which the astronaut was placed in a cage and rotated very violently along all three axes and all at the same time. Riding the MASTIF felt like your body was twisting around your eyeballs. A mop and a bucket were conveniently placed on the floor nearby and almost everybody who went on it ended up being sick in that bucket. Some of the astronauts got sick just *watching* others do it. It was, said John Glenn, 'the original vomit machine'.

There were rides on another vomit machine too – the 'Vomit Comet' – a big KC-135 jet plane the size of an airliner that flew aerobatic parabolas repeatedly in the sky, climbing at scarily steep angles before plunging sickeningly down to simulate thirty seconds of weightlessness. There were exhaustive sessions in two state-of-the-art Mercury capsule simulators where the astronauts got to practise every aspect of a space flight, including every kind of emergency, from launch to splashdown, in an exact replica of the real capsule, with all the correct sounds piped into their helmets at all the correct times and even with a little movie of what the earth might look like from out of the portholes. There were survival training exercises in the Nevada desert, in case the capsule landed off course by mistake, and water-survival exercises in the Gulf of Mexico, and throughout it all mountains of bookwork and lectures. And in the brief intervals when they were not training or studying there were visits to the many facilities across the country involved with the business of putting an American in space – to Chrysler and Convair that were building the rockets, to McDonnell that was building the capsule, a non-stop circus of glad-handing, flesh-pressing, public speaking and meals of rubber chicken which every one of the astronauts loathed except Glenn who seemed to love every minute of it and who was always adept at saying all the right things to all the right people.

As test pilots, each of the seven men were also actively put to work advising on some specific engineering aspect of the programme – for Glenn it was the design of the instrument panel, for Shepard it was liaison with tracking and recovery procedures, for Schirra it was the spacesuit; and of course they all regularly came down to Cape Canaveral to watch rocket launches. And it was on this late November morning that they had come to watch this very special launch above all.

The clue to the flight's significance was in its designation, Mercury-Redstone 1 or 'MR-1'. This would be the first unmanned test of the rocket together with its Mercury capsule that, it was hoped, would soon carry the first American astronaut into space. The rocket itself was a Redstone, one of von Braun's designs, intended to carry a nuclear warhead over a short range of about two hundred miles; except that in this case the warhead would be replaced for the first time by the Mercury capsule. MR-1 would blaze the trail for the chimpanzee and manned suborbital flights to come, flying the up-and-down, ballistic profile that would take the capsule briefly – but decisively – into space. The first Redstone had flown back in 1953, and by 1958 the missiles were fully operational with the US Army as tactical battlefield weapons. For these Mercury space flights, von Braun and his engineers had added some extra punch in the shape of a bigger fuel tank, along with eight hundred other modifications designed to transform a tactical nuclear weapon into a man-carrying space rocket. But the really key point about Redstones was that they were dependable, so much so that they enjoyed the nickname 'Old Reliable'. And Old Reliables were exactly what was needed when it came to the hazardous business of lobbing live flesh-and-blood objects like chimpanzees, and ultimately human beings, into space.

Jay Barbree was also attending this launch and standing not far from Alan Shepard – no doubt to keep a good watch on his eyes – when, at 09.00 and with a tremendous roar, the Redstone's engine burst into life. Belching flame and smoke, the 83-foot-tall rocket lifted off the pad on its journey to space. It got no further than four inches – 3.8 inches to be exact – before its engine abruptly stopped and the whole rocket fell back to the pad, denting one of its fins on the way. At this point the capsule's escape 'tower', a bright red pylon clamped to the top of it and designed to pull the capsule clear of the rocket in an emergency, shot up on its own into the sky spewing flames before diving straight towards the VIP area like an incoming shell. Loudspeakers blared warnings at everyone to take cover. Reporters and dignitaries jumped beneath benches and threw themselves under cars. Barbree wrapped himself round a pole. The pylon just missed them, slamming instead into the sand approximately 350 yards from the launch pad on which the Redstone with its Mercury capsule was now sitting again quietly, definitively not behaving like an Old Reliable. Alan Shepard grinned with gallows humour at Barbree. 'Are we having fun yet?' he said.

But the show was not over. As Wernher von Braun, the Pacific war veteran Ike Rigell and all the other launch personnel in the blockhouse were trying to make sense of the 'crazy indications' on their consoles, the capsule's drogue parachute inexplicably chose that moment to pop open, followed in short order by the main parachute and, a few seconds after that, the reserve parachute. The Redstone, topped by its capsule and the three hanging parachutes began to tilt slightly on the launch pad, partly because the parachutes were beginning to fill in the sea breeze. If the wind picked up much more they threatened to bring the whole assembly of rocket, capsule and eighteen tons of fuel and cryogenic liquid oxygen crashing to the ground and very probably blowing up, a catastrophe that millions of TV viewers would get to watch on the nightly news.

At this point Albert Zeiler, one of the German engineers who had worked with von Braun on the Nazi V-2s, suggested getting a hunting rifle and shooting live bullets at the rocket's liquid oxygen tanks to make holes in it, thus allowing the contents to boil off into the outside air and prevent a possible explosion. The idea was rejected. 'Even to a rookie like me,' said Gene Kranz, then a member of the mission control team and who would one day help bring the stricken Apollo 13 spacecraft home from the moon, 'shooting a hole in the tanks did not seem to be a sound plan.'

Fortunately the wind did not pick up. In the end the Redstone's batteries were allowed to die overnight while a relief valve in the liquid oxygen tank let the gaseous oxygen vent slowly and safely into the outside air. Engineers subsequently discovered that the Mercury capsule had in fact behaved exactly as it was supposed to once the rocket's engine had cut off, following a precise and pre-programmed automatic sequence of events right down to the deployment of its parachutes. The problem was that the engine cut-off signal had come at the wrong moment – not when the rocket was well on its way to space but when it was less than four inches off the ground. The cause was finally traced to two electrical cables plugged into the rocket's base that were supposed to unplug as it lifted off the pad but had done so in reverse order, thanks to a tiny modification made weeks earlier to a single electrical connector less than the size of a beer can.

A month later, on December 19, the flight was re-run – NASA called it MR-1A, presumably with the hope that there would not also be a B or a C. This time it worked; but the greater damage had already been done. And the timetable for America's first manned space flight was postponed yet again. Back in June 1960, NASA had been talking about a possible flight in October. In October they were talking about December. And now here they were in December, and the date being bandied about was March 1961. A new moniker was doing the rounds about NASA: it stood for Not Absolutely Sure about Anything. But what became known afterwards as the 'Four-Inch Flight' seemed to symbolise everything that was wrong with the US space programme, and had been wrong since Sputnik first blazed a trail around the globe. What those Soviet agents watching from their fishing trawlers must have made of it remains, perhaps mercifully, a mystery to this day.

And conceivably too the Four-Inch Flight cost at least some of the confidence of the president. His three brief words about space in that inaugural address on January 20, 1961, were only the tip of an iceberg of doubt and despair where America's space programme was concerned. Standing in the crowds on Capitol Hill watching the inaugural proceedings, Alan Shepard's own shot at immortality, handed to him by his chief Bob Gilruth only the day before, was arguably now imperilled by a double threat: from his own president and the president's new science advisor as it was from some nameless Soviet cosmonaut half a world away. If, that is, there even *was* a Soviet cosmonaut. Because neither Shepard nor his fellow astronauts, frankly, had any clue whether there really was.

But of course the opposition did exist, all twenty of them; and now six of those twenty, the Vanguard Six, their final examinations completed, their tentative order of precedence decided, were waiting to fly. The Soviets had built their own team as a direct response to the American team. The two existed in strange, parallel universes, one public, one secret, each a distorted mirror of the other. Having traced the birth and infancy of the American side, we must briefly turn the clock back to the summer of 1959 and a remote air base above the Arctic Circle, and tell the story of how the Soviet side too was born.

SEVEN

THE SOVIET TEAM

Late Summer 1959
769th Fighter Aviation Regiment, 122nd Fighter Aviation Division
Luostari, near Murmansk, USSR

DESPITE ITS LOCATION in the far north of the USSR, some two hundred kilometres above the Arctic Circle, the Soviet air base of the 122nd Fighter Aviation Division near the small town of Luostari was a busy place. Most days and sometimes nights, throughout the long winters and short summers, the little MiG-15 jet fighters with their five-point red stars screamed off the runway to fly patrols and protect Soviet submarines in the nearby Barents Sea. Just ten miles to the west, one minute's flying time over a wild landscape of a thousand lakes and empty tundra, ran the border with Norway. Luostari lay in one of the most inhospitable parts of the USSR, but it was also a strategic nerve-centre. This was the closest Soviet air base to any NATO country.

Conditions were primitive and harsh, but for a young, ambitious fighter pilot the flying was exhilarating, challenging and even addictively dangerous; the extreme weather, the winds, the snow and the ice were enough in themselves to test the skills of the men in their cramped MiG cockpits, but there was also the visceral thrill of being so close to a cold war enemy, a hair's breadth away across the Iron Curtain. If that war were ever to turn hot, Luostari would be at the very front of front-lines.

It was for all of these reasons that the then twenty-three-year-old Yuri Gagarin volunteered for this bleak posting, after finishing his military flying training at Orenburg in the south of the country. But for his young wife, Valentina, whom he had met and married in Orenburg where she lived with her parents, it was especially tough. She had to complete her medical studies at home before joining him in Luostari nine months after Gagarin had moved there. The change came as a shock. Not only was she suddenly cut off from her family home and everything she had grown up with, but she was also, unlike her outgoing, sociable husband, naturally far more reserved and shy. Now she had to share her kitchen with another family in a dank, tiny apartment in one of the most isolated regions of Russia and learn to cope with the realities of being married to a husband who was away on the base for much of the time and might at any moment be killed in the air; both lessons that would later come in useful. The winters too were appalling, especially from mid-December when for six interminable weeks the sun never rose at all and the days were like nights, with only artificial lighting to pierce the darkness. And soon she had something else to cope with: their first child, Elena, had been born in April 1959, just as spring was waking the surrounding tundra all too briefly into life.

It was a few months after Elena was born and barely still summer, towards the end of August or early in September, when the base began to buzz with rumours. A special commission had just arrived to interview a number of pilots for a mysterious and secret assignment. Along with eleven of his fellow pilots, Gagarin was invited to meet the two middle-aged members of the commission. They all waited in the squadron's headquarters before being called into an office, one by one. We do not have Gagarin's testimony as to what happened next – the *Pravda* journalists who subsequently wrote his 'autobiography' never disclosed this part of the story – but we do have the recollection of one of the other pilots, Georgy Shonin:

> They asked me to sit down and began asking questions. We talked about the usual, perhaps I should say, boring things: how I was enjoying the air force, did I like flying, had I become adapted to the Far North, what did I do in my free time, what did I read, and so on.

The interview was soon over, leaving Shonin none the wiser: 'I left the office and met my comrades' questioning looks, but as I didn't have any sensible answer, I could only shrug my shoulders.' Two days later Shonin, along with Gagarin and fewer of the pilots this time, was called back. This time the tone was different. The two men began asking very detailed and penetrating questions about Shonin's flying experience, beginning with his earliest training. Then one of the men asked how he would feel about flying a 'different' kind of aeroplane. Shonin thought he meant helicopters, a far less appealing prospect than the fast jets he was currently flying, but the man cut him short. 'No, no,' he said. 'You don't understand. What we're talking about are long-distance flights, flights on rockets, flights around the world.' Shonin was stunned: 'Even though there were quite a few satellites in space by then, manned flights were still an idea from the realm of the fantastic. Even among us pilots, no one spoke seriously about such a thing.'

Gagarin too was called back for a second interview, and we can assume that he was asked the same question. But we do not need to assume what his answer was. Like his friend Shonin, he said yes.

THE TWO MYSTERIOUS men were doctors, and teams of them had begun travelling to major air force bases across western parts of the USSR from the middle of August 1959, four months after the Mercury Seven were introduced at their Washington press conference. The doctors always travelled in pairs and they asked the same questions of selected groups of pilots on each base. In every case they already knew plenty about the men they were interviewing because staff on the bases had been asked to provide the necessary files, without being told why. Just as the men they were interviewing were not always told why: 'The conversation,' recalls one of those doctors, Nikolai Gurovsky, 'had nothing whatsoever to do with space. Some officers had no idea what we were getting at and why we had come.' Others guessed; some, like Georgy Shonin, were told. A few 'asked permission to consult with their family. We had to absolutely forbid this: it was a new, top-secret project, and the candidate had to make the decision himself, without outside assistance.'

There were 352 such candidates and they had been whittled down from an initial pool of 3,461 for the role of cosmonaut – a word that had only come into being in 1955 and, unlike the American astronaut or 'star voyager', suggested something more than a merely physical journey; to sail into the *cosmos* would be a philosophical leap for humankind too, a trip into the beyond. But there were a number of major challenges to overcome before such a journey could be accomplished, and one of them was finding the best kind of person – or rather, as in the United States, the best kind of *man* – to do it.

The central figure in this search was Colonel Dr Vladimir Yazdovsky, a medical professor at Moscow's Institute of Aviation and Space Medicine. A tough, bald army doctor in his mid-forties, Yazdovsky's official title was director of medico-biological research, which effectively meant he ran the space dog programme. He also conducted early research on human volunteers. At the institute he was both respected and greatly feared by his staff. Valentina Bykovskaya was working as a nurse at the institute at the time and she loathed him: 'He was unbelievably harsh and we tried to sneak past him in the corridor or run past as quickly as possible. He was not always fair either. A severe and arrogant man.' In his memoirs Yazdovsky emerges as a man ruthlessly committed to his sometimes deadly canine experiments and almost wholly lacking in human compassion; and even when he does appear to express compassion the result is strangely dislocated. His description of Laika, a 'charming dog' is one example; there is something distinctly unsettling in his account of how he brought her home and 'showed the children' a few days before her one-way orbital flight in November 1957. 'They played with her and stroked her. I wanted to do something nice for the dog because she didn't have long to live.'

Having selected dogs to fly, and sometimes die, in space, Yazdovsky was now asked to turn his attention to humans. With the Mercury astronauts so prominent in the news after their April 1959 press conference, the Soviet Academy of Sciences had already begun receiving piles of letters from citizens willing to risk their lives on a future Soviet space mission. In his careful and precise way, Yazdovsky cites some of them in his memoirs. They included two second-year students from the mining faculty of the Central Asian Polytechnic, a forty-nine-year-old 'worker' from Riga, a middle-aged graduate from the Leningrad Communist Institute of Journalism who 'offered his life for science', and two convicted criminals. Clearly there would be no difficulty finding enough people willing to kill themselves for their country or just to get out of jail, but finding the right candidates was going to require a better strategy.

Initially the thinking centred on a wide range of possible contenders, just as it had for the American selectors: fighter pilots, submariners, racing car drivers or, as Yazdovsky recounts, 'members of other physically challenging professions'. In the end, the choice narrowed down to fighter pilots only. The ostensible reason was that these men would best cope with the acceleration forces likely to be experienced in a rocket launch. They also had the advantage of ejection-seat training, which would be an essential part of how they were going to get back to earth. But there was another consideration too. The Americans were using pilots. The Soviets would now do the same. In the end, the contest would be between two sets of pilots from either side of the Iron Curtain. They would not be shooting bullets at each other in the skies, but they would compete instead for primacy in the grander arena of space.

By June 1959, just two months after the Mercury astronauts had begun their training, the criteria for a Soviet team were secretly set in place. But there were notable differences from the American criteria. The candidates were required to be both younger and significantly smaller – a maximum height of five foot seven rather than five foot eleven. And they were not required to be test pilots. Unlike the Mercury astronauts with their minimum of 1,500 flying hours, the Soviet cosmonauts needed only to be serving fighter pilots. Experience was essentially irrelevant. The selectors were interested in young, obedient, superbly fit men who would cope with sitting in a fully automated capsule without touching the controls and without panicking.

In its earliest stages NASA had had very similar ideas about its own would-be astronauts, despite picking men with far more flying experience. But the astronauts themselves soon changed the game, insisting on a genuine measure of piloting *control* over their vessels. Their impressive résumés of test-flying skills helped them to win their case; more than that, they had fame, they had stardom, they had *Life* magazine, and that meant they had the muscle to get their way, or at least some of their way, with any sceptical aerospace engineer anxious about letting a human loose in one of his precious Mercury capsules. But the new Soviet cosmonauts would have no fame, at least not until after they flew. And unlike the American astronauts they would not be civilians working for a civilian entity like NASA; they would be soldiers serving in a military programme. Their job was to do what they were told – or face the consequences. In this light, and in pointed contrast to the US programme, it was hardly surprising that Dr Yazdovsky had been asked to run the selection process to find obedient Soviet humans. He already had plenty of experience selecting obedient Soviet dogs.

Yazdovsky's team set to work in that summer of 1959. The initial pool of 3,461 candidates whose records were inspected was approximately seven times the size of the initial American pool. The chief designer of the Soviet space programme, to whom Yazdovsky reported, had his reasons for that too, as Yazdovsky described in his memoirs:

> He was asked how many cosmonauts should be chosen. 'A lot,' he replied with a smile. 'The Americans have chosen seven men – and we need many more.' His answer created some bewilderment, but nobody made any comment. Everybody understood that it was not one or two flights being planned, but a much greater number.

Once the original files of thousands had been reduced to the more manageable second round of 352, Yazdovsky sent his pairs of doctors out into the field: 'In a short time, they were supposed to find several dozens of absolutely healthy … disciplined, professionally promising, young, short, thin, fighter pilots.' And also politically reliable ones. Supervising Yazdovsky's medical commission was a further commission known by its initials GMVK, one of those many Soviet mouthfuls that, when translated, stood for the State Inter-Agency Departmental Commission. A central purpose of the GMVK was to filter candidates 'on the basis of their political reliability'. 'It was done by the KGB,' said Pavel Popovich, one of the candidates. 'We were screened and checked. And they checked our whole family history and our relatives.' Every one of the 352 men the doctors interviewed at their air bases would need to pass this political test in order to continue to the next stage, including, of course, Gagarin. As someone who was obsessive about cleaning his red Pioneer scarf when a child and who in college had captained the Young Communist basketball team for the vast Saratov region, this was one test Gagarin was unlikely to fail.

From the 352 candidates interviewed at their air bases, the list was further narrowed down. Over the autumn and winter of 1959/60, 134 men were sent in batches of twenty to the Central Scientific-Research Aviation Hospital in Moscow to undergo a rigorous medical and psychological testing regime under the code name 'Theme Number 6'. Gagarin went there in early October, his friend Georgy Shonin at the end of November. There were no welcoming cocktail parties when they arrived. The men were stripped of their uniforms, issued with hospital pyjamas and placed in wards. They were warned not to tell anybody else in the hospital what they were doing or why they were there; not that all of them quite understood yet why they were there themselves. 'We were instructed not to talk,' said Aleksey Leonov, who arrived in the same batch as Gagarin, even though 'we had our suspicions that the programme would involve space flight'. And, he added, 'we were excited'.

Political credentials aside, many of the tests suffered by the Soviet candidates were remarkably similar to the American versions from several months earlier; not accidentally perhaps, since some of the tortured details of the astronauts' experiences had already been published in the American press. But the Soviet testing regime was, if possible, even more unpleasant and more prolonged. Some candidates would remain in the hospital for almost a month. All of them were put through the same gruesome ear test as the Mercury candidates in which water was injected directly into their ear canals. They were also forced to endure the heat chamber where the temperature was racked up to an even higher 70°C while they baked inside for as long as they could bear it. But instead of doing this once like the Mercury candidates they had to do it three times, with Yazdovsky meticulously noting down the effects in each case: 'After forty minutes the face and the visible mucous membranes acquired a purple-red colour along with a clear bluish hue.' On coming out of the chamber the men were then immediately ordered to perform a series of squats.

As with the Mercury candidates, the Soviet pilots were sealed into a high-altitude pressure chamber and progressively starved of oxygen while Yazdovsky and his team observed them through portholes. 'Many people fainted,' said one of the future cosmonauts, Dmitry Zaikin. 'They just could not stand this test. They simply fell on the floor.' There were also sickening vibration tests designed to assess how far a man could withstand the severe jolting and jarring sensations of launch. 'You sit on a seat and it starts to vibrate,' said Boris Volynov, who was also selected. 'The main thing is the amplitude of the vibrations. When the amplitude gets high, all your teeth start shaking and rattling.' One notable difference between the American and the Soviet testing regimes was the antiquity of some of the Soviet equipment, which lent an unintended but additional layer of terror to the tests. 'The centrifuges,' remembered Volynov, 'were not, shall we say, up to date. In fact they were quite old. Sometimes they had to tie the seat down with extra chains so that the pilot didn't fly off into the wall.'

Scores of men started dropping out. 'There were lots of doctors,' wrote Gagarin in later years, 'and each one was as stern as a state prosecutor. There was no appeal against their sentences.' The testing was so thorough that some men were grounded from flying altogether, their air force piloting careers suddenly and bewilderingly in tatters. A psychologist on the programme, Rostislav Bogdashevsky, noted coolly that the goal was to 'get a comparative estimate – that is, to try to see which of these selected people was best able to sustain all these abuses we invented for them'. The whittling down was pitiless. In the Mercury tests at Lovelace and Wright-Patterson all the candidates who actually attended also stayed the whole course and learned the results afterwards. In Moscow they were summarily ejected along the way. 'Our numbers went down and down,' said Boris Volynov. 'Acquaintances I'd seen just disappeared.' Among them were twenty candidates who left of their own accord. They had simply had enough.

Unlike Gagarin, who never breathed a word of what he was doing to his wife Valentina – and who, in turn, never asked him – his friend and future rival Gherman Titov broke all the rules and told his wife, Tamara, exactly what he was up to in Moscow. This was the kind of character trait his future instructors would hold against him, along with his resistance to taking orders from doctors and perhaps his penchant for quoting Pushkin at inappropriate moments. But Tamara was also heavily pregnant with the child who would die within a few months of his birth, and Titov wrote secret letters to her from Moscow every day to cheer her up. Most of these letters have never been published. Calling her his 'Tamarochka', he filled them with sweet and loving exhortations to watch her health – 'you have to walk three hours a day minimum and it doesn't count if you walk to the shops' – along with accounts of the various medical tortures he and others were daily enduring:

> January 13, 1960
>
> Sweet Tamarochka,
>
> I'm happy to say that the psychologist didn't send me to the madhouse. This means I keep going on the project. Today they irradiated me with ultraviolet rays. By the way, don't show these letters, especially the ones concerning my mission, or talk about them to any outsider … Boys have been dropping off like flies … My little girl, don't be upset and don't miss me. Right now, you need more life, more exercise and fresh air. There's two of you now. Sending kisses …

'I would stand by my post box,' said Tamara in an interview decades later, 'and my heart would skip a beat.'

To his own amazement Titov stayed the course. By the last week of January, every one of his squadron mates had been eliminated. 'Of the old guard,' he wrote to Tamara nearly a month after he had started, 'I'm the last one standing.' This was especially surprising for Titov because as a teenager he had badly smashed his ankle in a bicycle accident and incredibly the doctors had still not discovered it. 'We're examined so thoroughly,' he wrote, 'that even if we were ill as a child they'll find out.' But they never did find out about that ankle. A few days after Titov had been hung upside down in another fiendish experiment by the nurses he called 'the Gestapo' and 'had worried my poor heart would jump out of my chest', he was told he had passed.

But what exactly for? By now most of the candidates who had clung on this far knew the truth, even if it was still not officially acknowledged. That acknowledgement came only once they had passed. Aleksey Leonov had become friends with Gagarin from the moment they had first met on the ward. Together with eight other successful candidates they were all summoned into a senior air marshal's office.

> He addressed us in a fatherly way. He told us we had a choice to make. Either we could continue our careers as fighter pilots in the air force, or we could accept a new challenge: space. We

had to think it over. We left the room briefly and talked among ourselves in the corridor outside. Five minutes later we filed back into his office. We wanted to master new horizons, we said. We chose space.

THE FIRST TWELVE of the twenty cosmonauts finally selected were formally inducted into Military Unit 26266, the Cosmonaut Training Centre, in the first week of March 1960, eleven months after the Mercury Seven had begun their own training. The eight remaining cosmonauts would trickle into the programme over the next three months. There were no press conferences, no public appearances, no dashing profiles in *Pravda* or *Ogonyok* magazine, the Soviet version of *Life*. There was nothing. 'We knew,' said one of the chosen, Dmitry Zaikin, 'that the Americans were preparing for a manned flight. This explains the great secrecy: *not* to let the Americans know we were preparing too.'

The very different selection criteria from those applied to the Mercury astronauts were equally reflected in the mix of the final twenty. It was not just that the former were on average approximately a decade older. The differences in flying experiences between the two groups was perhaps even more striking. Alan Shepard had 3,600 hours in the air and had not only tested some of the most advanced and dangerous aeroplanes ever built but instructed other test pilots to do it too. John Glenn had only recently flown a combat jet across the continental United States in less than three and a half hours, smashing the record. Even the least experienced Mercury astronaut, Scott Carpenter, had 2,800 flying hours.

Gagarin had just 230 hours when he was selected. Most of that time had been spent in either simple propeller-driven training planes or the subsonic MiG-15, a fighter that by 1960 was well on its way to obsolescence. His level of experience was fairly representative of his peers. Pavel Popovich, the Ukrainian folk-singer enthusiast, and the talented, if narcissistic, Grigory Nelyubov were the only two cosmonauts who had piloted the more modern MiG-19, a Soviet fighter capable of breaking the sound barrier. And of course not one of the Soviets had been test pilots. But none of this counted. What counted was not how well they could fly but that they were younger, fitter and stronger than the American opposition. 'We were united by one thing alone,' said Popovich. 'That at the start of 1960 we were the twenty healthiest men on the planet.'

FROM THE BEGINNING conditions were spartan. It would be several weeks before tiny shared apartments could be found for the new cosmonauts in Moscow and months before they were moved into marginally bigger apartments in Chkalovsky, where Gagarin and Titov would later climb between balconies. At first they had to live in an army sports club attached to Moscow's Frunze Central Aerodrome. Aleksey Leonov and his wife slept on bunks in a corner of the volleyball court: 'We had to drape newspapers over the net in order to get some privacy, because another pilot and his wife were sleeping at the other end of the court.' Popovich and his wife Marina also bedded down there. 'We didn't bring anything with us,' he remembered. 'There was a toilet outside. We sat on newspapers and ate off them.' Some of the men did not bring their wives and families – Boris Volynov packed his off to Siberia – but those who did were still not supposed to tell them why exactly they were now being made to live in a gym and eat off the floor. Gagarin, as always, said nothing to Valentina, for whom probably anything was better than living north of the Arctic Circle. Titov, as always, told Tamara. She supported him as she did in everything, even if the risks terrified her. 'Since I'd already married a pilot,' she said, 'that's how I was going to go on – for life.'

The new cosmonauts-to-be had barely settled in before their training began. At nine o'clock on the morning of March 14, 1960, they gathered in a classroom at the Central Aerodrome's former meteorological office for their first lesson, a lecture from the forbidding Dr Yazdovsky on the physiological effects of space flight. The choice of subject matter exposes the priorities of this programme: if these young front-line fighter pilots expected a class on the techniques of flying spaceships they were in for a disappointment. Yazdovsky's discipline was space medicine. The uniformed men sitting at their desks in front of him were human guinea pigs in a grand experiment. And they had all just committed themselves to being strapped inside a tiny metal biosphere one day and flying in a weightless, airless, pressureless, irradiated, variously freezing or broiling vacuum around the globe at ten times the speed of a rifle bullet.

They had also just committed themselves to being part of a secret elite. And after several more months six of them, the Vanguard Six, would go on to become an elite within this elite. But in those first weeks there were still deeper secrets that were not yet revealed to them. Three days after Yazdovsky's lecture they had a visitor. 'He came inside,' recalled Pavel Popovich. 'He had a big head. A high forehead. And he was very energetic, with an alert and mischievous glint in his eyes. He greeted us warmly with the words, "Greetings, my little eagles".' The man did not identify himself. Leonov remembered him too, standing before them and calling out their names, one by one, from a list:

> He was a man of average height with brown eyes, dressed in a hat and a marengo coat. He looked at us and smiled. It was immediately clear that he was a man of business. He had a long sheet of paper with him with a list on it. 'A – Anikeyev. B – Bykovsky …' And in this way he got to Gagarin. Yuri stood up. And suddenly the man's expression changed. He was looking at Gagarin attentively, he saw something that interested him. He said, tell me about yourself. And then he checked himself and said, next please. And so he went through all of us.

The man's name was Sergei Pavlovich Korolev; and without their knowing it the young cosmonauts had just witnessed the Soviet space programme's greatest secret of all.

EIGHT

THE KING

January 27, 1961
Scientific and Testing Range No. 5 (NIIP-5)
Tyuratam Cosmodrome, Soviet Republic of Kazakhstan

NINE DAYS AFTER the Vanguard Six cosmonauts had completed their final examinations, the same Sergei Korolev sat at his desk in the secret missile complex in Kazakhstan to write a letter to his second wife Nina, two and a half thousand kilometres away in Moscow:

> My loveliest, dearest girl,
>
> I am using this opportunity to send you my tender heartfelt greetings! Please get better soon, my dearest kitten, and take care of yourself. Be at peace, my beloved. I am with you with all my heart and soul, till my last moment on Earth.
>
> I have no news from here over the last couple of days. Except plenty of work, worries and difficulties. We are getting ready and we believe totally in our cause.
>
> I embrace you very very tightly and I send you my kisses.
>
> Yours always,
>
> Sergei

Korolev's office at the rocket complex was a room in a single-storey cottage, and the cottage also served as his accommodation whenever he came there. With its steeply pitched roof and rustic wooden porch it was an incongruous sight given its location just two kilometres from one of four missile launch pads. Like the man himself, Korolev's office was almost ascetic in its simplicity. An ordinary wooden table doing duty for a desk, a telephone, a standard lamp; the walls painted a pale institutional green, unadorned with paintings or decoration; thin cheap striped green curtains hanging at the window. Nothing in there hinted at luxury and nothing at power.

And yet this austere, anonymous room, like the equally undistinctive bedroom next door with its narrow couch doubling up as a bed, perfectly expressed the central paradox of the man who wrote those loving, anxious lines to his wife. For Sergei Pavlovich Korolev, just turned fifty-four, thickset, still dark-haired and with eyes as brown, alert and sometimes mischievously glinting as Popovich had described him in that first meeting with the cosmonauts ten months earlier, was the most powerful man in the Soviet space programme. In some respects he *was* that programme; and much more besides. Nobody who worked for him used his full name: to them he was the Chief, or the King, or known only by his initials, S.P. To the outside world he was entirely unknown, his identity rigorously protected as a state secret until after his death in 1966, despite years of effort by the CIA to uncover it.

Korolev's official title was Chief Designer of OKB-1, the name for the special bureau tasked with building the USSR's long-range missiles. In this respect he resembles his great rival Wernher von Braun, who was also designing missiles for the US Army. But where von Braun was concerned exclusively with rocketry, whether of the warhead or human-carrying kind, Korolev's remit ranged across a truly astonishing number of projects. His energy and talents were awe-inspiring – literally so, since many of the men and women who worked for him often regarded him as a kind of god, and decades later still do.

Korolev was the reason why there were even cosmonauts at all to remember that mischievous glint in his eyes, since he was the driving force behind the USSR's manned space programme, and without him its very existence would have been in doubt. He was the man behind every one of those space spectaculars that had so fascinated the world and rocked American confidence, behind all those wondrous Sputniks and orbiting dogs, behind the shower of glittering Soviet pentagons on the moon and the sensational first photographs of its dark side. He was the magician who dazzled and disturbed all those millions of Americans like Alan Shepard and Senator Lyndon B. Johnson who simply could not understand how a nation so shattered by war and Communism could ever have achieved what the United States, rich, powerful and democratic, could not. He was the man behind the Vostoks that would very soon attempt to carry a human into space, the power behind the enterprise to find and train those humans, and the guiding hand behind the biggest rocket yet built that would launch them. Even as he wrote his letter to Nina, he was preparing to launch another unmanned probe, this time not to the moon but tens of millions of kilometres across the solar system to Venus. The rocket was expected to depart in a week from the area called Site 1, whose launch pad was almost in sight of Korolev's cottage window.

Adilya Kotovskaya, the doctor who supervised the cosmonauts' centrifuge training sessions, grasped Korolev's core when, in a rare interview almost half a century after his death, she said: 'Take a good look at his eyes. You can see for this man his lifelong dream was human space flight.' Over Korolev's desk at the home he shared with Nina near Moscow hung a portrait of Konstantin Tsiolkovsky, the self-taught turn-of-the-century space and rocketry visionary whose works would also fascinate Yuri Gagarin as a student in Saratov. Under the portrait were Tsiolkovsky's words: 'Mankind will not stay on Earth for ever but in its quest for light and space it will first penetrate humbly beyond the atmosphere and then conquer the whole solar system.' This single sentence summed up Korolev's whole being; it was the central source of inspiration for his life. His only daughter Natalya never forgot how he told her, shortly before his death, that his greatest dream was one day to fly in space himself. It was a dream he would never realise, but others would realise it for him. Perhaps his mother Maria Balanina sowed the first seeds with the stories she told him as a small boy growing up in Nyzhin in the Ukraine:

> He liked my stories … We 'flew' together on a fairy-tale magic carpet and saw Konyok Gorbunok (the humpbacked pony), the grey wolf, and many other miracles beneath us. This was enthralling and wonderful. My son would press close to me, would look wide-eyed into the sky, while the silvery moon peeked out from amid small clouds …

That same delight in flight and freedom rings through the biography of Wernher von Braun too, five years younger and growing up in Berlin and later at boarding school near Weimar in Germany. Korolev's assistant, Antonina Otrieshka, remembered her chief looking at a newspaper photograph of von Braun one day and saying, 'We should be friends.' Von Braun, for his part, would never know the name of his great rival until after Korolev's death. The first course of their lives would follow similar paths, both experimenting with amateur rocketry in the early 1930s, both fuelling their dreams of space flight by building missiles of war for their masters, in von Braun's case for Hitler, in Korolev's case for Stalin. But everything abruptly changed for Korolev, and literally overnight, when, on June 27, 1938, he was arrested by Stalin's NKVD secret police and accused of anti-Soviet sabotage at the rocket research institute where he worked. Two policemen searched his flat for almost seven hours while he sat holding hands with his first wife, Kseniya. Then he was driven away, leaving Kseniya alone. His daughter Natalya was just three. The penalty, if he was convicted, was death.

This was at the height of Stalin's Great Terror, a campaign of brutal and sustained repression, of mass imprisonment and executions in the second half of the 1930s. Attempts to determine the enormous numbers of victims with any precision are impossible. Official figures from subsequently published NKVD documents suggest that in

1938, the year of Korolev's arrest, nearly 1.9 million prisoners were held in the Main Directorate for Corrective Labour Camps or *Gulag*, the notorious network of forced labour camps across the USSR. That figure may well be misleading and is probably an underestimate. Nor does it include other victims of Stalin's Terror who never made it to the camps. Many were shot in the back of the head in the NKVD's cells after interrogation followed by a forced confession. Across a very wide spectrum of targets these included Communist Party officials, military officers, artists, musicians, writers, academics, scientists and rocket engineers. From the distorted perspective of a paranoid police state, the Reaction Propulsion Institute, known as RNII, where Korolev had been working on early rocket guidance systems since 1933, was a seething nest of sabotage. Three of its members had already been arrested. Two were shot. The third, Valentin Glushko, a brilliant rocket engine designer, was sentenced to eight years' imprisonment. It was on the testimony of all three that Korolev was taken.

It took a month to extract his confession. Repeatedly denounced by his interrogators as an enemy of the people, he was regularly beaten and on one occasion smashed over the head with a jug when he asked for water. They also broke his jaw. In the end he had little choice but to sign a statement admitting his 'guilt'. Before a closed military tribunal he was sentenced to ten years' penal confinement and his property was confiscated. But at least he was alive. His mother pleaded for clemency first in a letter and subsequently in a telegram sent directly to Stalin: 'I beg you to save my only son, a young, talented rocket expert and pilot, and urgently entreat you to investigate his case.' Unsurprisingly, her pleas were not answered. Korolev was moved from prison to prison until June 1939 when he was transferred to the Gulag. In the Kolyma area of eastern Siberia, one of the most desolate regions of Russia, he was put to work digging for gold in the Maldyak minefields. It was the worst of places. Huge numbers would die there, from hunger, beatings, executions, exhaustion, tuberculosis and cold. The harshest tasks were given to political prisoners rather than convicted criminals. Korolev was made to hack blocks of gold-bearing ore with a pickaxe in poorly maintained mineshafts a hundred feet underground. For the rest of his life he would never wear or own anything made of gold.

He spent five months in Kolyma. Emaciated and exhausted, he lost all of his teeth to scurvy and he suffered from heart problems that continued to dog him in the decades afterwards. He very rarely spoke about his experiences in the camp but his daughter Natalya remembered how he always kept his aluminium mug, his name scratched on the handle with a nail. He continued to protest his innocence, even writing his own, unanswered, letters to Stalin. His exit from Kolyma was quite probably secured by the famous Russian aircraft designer, Andrei Tupolev, who had once taught Korolev as a student in Moscow. That intervention almost certainly saved Korolev's life. He was eventually sent to a very different and uniquely Soviet penal institution known colloquially as a *sharashka*, a less brutal prison where, in Korolev's case, engineers like himself could be put to work in the service of the state. The NKVD police later moved him to a second *sharashka*, designing rocket engines. In July 1944, thirty-five of its prisoner-engineers, including Korolev, were released. One of those also released was the man who had originally denounced him, Valentin Glushko. He now became Korolev's chief in a new bureau tasked with continuing the development of missiles for the military. Their tortured, and sometimes openly hostile, relationship would have its own impact on the future Soviet space programme. But at last Korolev was free.

Korolev's prison experiences very nearly destroyed him. But they also stamped certain essential qualities on his personality that would help drive his astonishing future successes. Survival brought its own lessons, even if the costs were horrific. In prison Korolev learned to compromise where necessary, to exploit others if required, and sometimes to lie, cheat and deceive to get his own way. He learned to be pragmatic and *political*. Andrei Sakharov, who helped to design the nuclear warheads carried on Korolev's rockets, once described him as 'brilliant', but also 'cunning, ruthless and cynical' in the pursuit of his ambitions. Without those characteristics his ambitions would undoubtedly have failed. But perhaps above all Korolev learned that the system that had once taken away his freedoms, smashed his jaw and left him almost broken in body and mind, could do it again. Another contemporary, Josef Gitelson, who worked with Korolev, described the thinking that enabled him to go on serving that system, first under Stalin and subsequently under the less repressive Khrushchev, in language that could speak for almost every Soviet citizen at the time, if not for every subject of a totalitarian state:

> We developed a kind of voluntary schizophrenia in those days. You separate your soul into two independent parts. You can adapt yourself to the practice of living in one part of your soul while you keep the truth in the other part. Both halves of the soul can be in absolute contradiction to each other, each remaining absolutely logical within itself ... The truth part you shared with your family and a very few close friends.

Perhaps in this respect the parallel with von Braun, who made his own moral compromises, was close; but in von Braun's case it led him beyond mere doublethink to the teetering edge of actual crimes against humanity, which was never the case with Korolev.

All these hard-won lessons helped to propel Korolev's career to stratospheric heights, initially building replicas of von Braun's V-2 missiles in immediate post-war Germany, working with German engineers who, either willingly or not, had ended up on the Soviet side. He then moved rapidly up the ladder to emerge by the late 1950s as the chief of chief designers in the USSR's strategic missile development programme, now ahead of his former denouncer Glushko. Over those years he was responsible for a series of ever more ambitious ballistic missiles: the R-2 in 1950 with twice the range of the V-2, the R-5 in 1953 which could reach targets as far away as 750 miles, and finally the enormous R-7 in 1957, the world's first truly intercontinental ballistic missile and the rocket that would also launch his trailblazing Sputniks and Vostoks into space. His personality stamped its mark on everyone who worked for him as well as on his political and military masters; an unforgettably forceful and charismatic presence that some contemporaries described as 'epic'. Part of it was physical: his bull-neck, his big head, his prominent forehead and of course those fascinating eyes. 'He had lively black eyes,' wrote Boris Chertok, a rocket engineer who first encountered him in Germany. 'He looked directly at you as though X-raying you.' And his energy was almost inhuman. He worked all day and half the night and demanded the same from everyone around him. That energy spilled over into every aspect of his life, even to his love of fast cars – another trait he shared with von Braun – driving a big black German Opel at breakneck speed. He appeared to live as if he were making up for all the time he had lost. Oleg Ivanovsky, who became one of his most senior engineers, first saw him in 1947:

> We were standing near the gates ... and a powerful German trophy car came in and passed us at high speed. Inside was a guy in a brown jacket who looked very intense. Who is this crazy guy? 'That is the King,' we were told. 'He can't drive slower.'

Along with all this energy came a truly terrifying temper, especially when subordinates were careless or failed to take responsibility for their mistakes. 'For Korolev,' remembered one shaken witness, 'rage was an art form.' But his tempers never lasted very long and he would rarely follow up on any of his dire threats to kick miscreants off the programme to make, as he said, 'toilet sinks or frying pans'. If his rages were legendary so too was his kindness. On work trips to Leningrad he would often give his teams time off or take them on a personal tour of the Hermitage museum; they discovered that he was, perhaps unexpectedly, a great lover of paintings. He organised aeroplanes or his own personal transport for engineers whose parents were seriously ill or dying. He appeared to inspire both fear and adoration, often at the same time. Frequently he would drop into the staff canteen, his collar unbuttoned, and encourage open and lively discussions among his team members. Some people noticed how quickly he ate and how, after finishing his food, he would clean his plate with a piece of bread, scooping up every crumb just as he had once done in the Gulag.

While Korolev's energies never appeared to flag throughout these years, the terrible harm done to his body in the Gulag did not disappear. His heart problems worsened even though he kept this private. But the toll of his labours was more than physical. His marriage to Kseniya had fallen apart in the late 1940s when he fell in love with a much younger woman, who would become his second wife in 1949. Nina Kotenkova had studied English and Korolev had employed her as his translator, keeping him up to date with the latest American scientific journals, a job she would continue after their marriage. Where von Braun could only guess at what his anonymous rival in Russia was up to, Korolev could ask his wife to read to him about von Braun in the American press. It was only to Nina that Korolev could pour out all the self-doubt of his soul, as he did in the many beautiful letters he sent her, especially in the anxious period before rocket launches. But for his first wife, Kseniya, who had stood firmly by him during his years in prison and brought up their only child Natalya alone, it was a terrible blow. She never forgave him and forbade her daughter to see her father whenever he was with Nina. Today, over half a century after Korolev's death, it is moving to find that Natalya has transformed almost every room of her apartment in Moscow into a museum dedicated to her father's life and work.

EVEN AS KOROLEV continued to build his ever bigger missiles for the military, those dreams of space travel that Dr Kotovskaya was later to glimpse in his eyes never left him, although he was also careful not to reveal them too much. But after Stalin's death in 1953, Korolev found himself perfectly positioned to exploit for his own purposes the USSR's new leader Nikita Khrushchev's weakness for bragging on almost every possible occasion about the Soviet Union's superiority over the United States. This was the key that would enable Korolev to realise his dreams. His handling of the new Soviet leader proved to be nothing less than a master-class in manipulation.

Like most braggers, Khrushchev's bragging was founded in pathological insecurities, in his case about the USSR's – and Russia's – standing in the world, and most especially the fear of being seen to be technologically backward compared to America. His own son, Sergei, recalls how excited he was when the giant new Soviet Tu-114 airliner which flew him to the US on his 1959 tour was too big to land at Washington's National Airport. 'There could be only one conclusion: we were ahead of them. Let them envy us,' wrote Sergei. And when they did eventually land at a bigger airport, Khrushchev 'was pleased as a child when they couldn't find steps high enough to reach the door of the plane'.

Yet at the same time Khrushchev was green with envy about America, this 'fantastic country,' as Sergei wrote, 'which had gripped our curiosity all through the preceding years'. A visit to IBM's headquarters in California left the Soviet premier stunned – not by the computers, about which he understood almost nothing, but by the self-service cafeteria. 'He was amazed by the shiny plastic surface of the tables,' said Sergei. 'Father pushed his tray … with enthusiasm', and was deeply embarrassed when Vyacheslav Yelutin, the minister for higher education, dropped his own tray with a crash 'and stood staring at it' like a clumsy Russian peasant. Khrushchev was thenceforth determined that the Soviets must immediately introduce cafeterias based on IBM's model. He was also extremely disappointed that he was not able to visit Disneyland. He had been hugely looking forward to going, but a hostile crowd had gathered there and somebody had thrown a tomato at a police officer, so to Khrushchev's great dismay the trip was cancelled at the last moment. He never did get to meet Mickey.

Such, then, was the cocktail of pride and envy that stirred the premier's soul, and which Korolev also deliberately stirred in 1956 when he invited Khrushchev and a group of leading political figures to his headquarters at OKB-1 in Kaliningrad near Moscow. In his last years the increasingly paranoid Stalin had kept the work of his missile designers a secret even from many of his most senior colleagues. But Stalin had been dead for three years. The ostensible purpose of Korolev's invitation was to show off his enormous new R-7, a missile none of these politicians had yet seen and which would not be ready to fly for another year. Khrushchev's son, Sergei, was invited too:

> We were entering the sanctum sanctorum. This new hall-hangar was even more impressive … It stretched upward like a many-metered glass tower-aquarium. The windows were covered with a thick layer of white paint – protection from curious eyes.
>
> The size of the object inside amazed us. A single missile occupied the brightly lit well of the hangar. It reminded me of the Kremlin's Spasskaya Tower, with its combination of surging movement toward the sky and a heavy, earthbound, foundation. There it stood, at its full height, the 270-ton [245-tonne] 'Seven' or R-7. Crowding together at the entrance, we all stared silently at this miracle of technology. Korolev enjoyed the effect and didn't hurry to begin his presentation.

When he did begin his presentation it left his audience even more stunned. They were staring at a missile unlike anything anywhere else in the world. At over thirty-one metres tall even this first prototype was gigantic, with its four huge 'strap-on' booster rockets, each shaped like colossal artillery shells nineteen metres high and clustered symmetrically around a central fifth engine. The strap-on concept was a brilliant innovation, one that gave the missile its distinctive, notably 'un-American', profile that still lives on in Russia's Soyuz rockets to this day, sixty-five years later. All five engines would fire together at launch, punching the 'Seven' at terrific speeds into the sky until the fuel in the boosters was spent and they were jettisoned, enabling the lighter missile to streak on across a fifth – and in a later model a quarter – of the planet. At its tip would be a thermonuclear warhead destructive enough to obliterate entire cities.

This was a *monster* Korolev was showing off to his political masters. The power it appeared to place in Khrushchev's hands was formidable. Even better, the Americans had nothing like it. Moreover it was, relatively speaking, cheap. Khrushchev was already in the process of reducing the USSR's cripplingly expensive army by millions of men. An intercontinental ballistic missile like the R-7 was the answer to his prayers, offering all the muscle of a standing army and far more – at less of a price. 'How dare any country … attack us when we are literally able to wipe these countries off the face of the earth,' Khrushchev would later say. And now, or at least very soon, Korolev had handed him the means of doing just that. 'I don't want to exaggerate,' wrote the Soviet leader in his memoirs, 'but I'd say we gawked at what he showed us as if we were a bunch of sheep seeing a new gate for the first time … We were like peasants in a market-place. We walked around and around the rocket, touching and tapping it to see if it was sturdy enough – we did everything but lick it to see how it tasted.'

Korolev watched it all – and grasped his moment. He next led Khrushchev to a corner of the assembly hall and showed him a gleaming model of a prototype satellite he and his team had been quietly working on. Telling Khrushchev that the United States was at this very moment developing a satellite of their own, he set out the tempting possibility of 'thumbing his nose at the Americans' by launching the first one into space *before* them. All that was required was to 'remove the thermonuclear warhead [from the R-7] and put a satellite in its place. And that's all.' Electrified by the missile Korolev had just thrown at his feet, Khrushchev agreed to let him go ahead, on condition that this did not interrupt his work on the missiles themselves. The deal was done. 'After this meeting,' wrote Sergei, 'father simply fell in love with Korolev.' And as proof of that love Khrushchev gave Korolev the rarest of privileges: a personal hotline to his office in the Kremlin.

Korolev had found his man. He would give Khrushchev the biggest missile on the planet with which to scare the Americans.

And, with Khrushchev now behind him, he would use that missile to fulfil the dreams he had nurtured ever since he had ridden as a child with his mother on a magic carpet.

BUT THERE WAS a price to all this. Korolev would also have to vanish.

His mighty new missile made him both too valuable to the state and too vulnerable to its enemies. Now his identity would have to be concealed. So too would the existence of his R-7, despite Khrushchev's boasts about missile numbers and the Soviet Union's capacity to strike at any corner of the world. So too would the vast new missile range in Kazakhstan codenamed NIIP-5 or 'Scientific and Testing Range No. 5', created in 1955 specifically to service the R-7 and which within five years would cover 6,700 square kilometres of flat Kazakh steppe, with its own assembly plants and closed dormitory town and four missile launch pads, including the one just down the road from Korolev's cottage.

In time the secrets of both the R-7 and the launch complex would be substantially penetrated by US intelligence. By January 1961, when Korolev was writing his letter to Nina, American aerial and satellite reconnaissance photographs had already yielded valuable information to the CIA about NIIP-5 and its missiles. But one astonishing operation on the ground also opened up a treasure trove of secrets about the R-7.

In 1960 the CIA actually succeeded in 'kidnapping' the upper propulsion stage of an R-7 rocket, minus its engine, of the type used to blast a Soviet Luna probe to the moon. This upper-stage vehicle had been placed on display in an exhibition in Mexico City. Incredibly it turned out to be a real one and not a mock-up, a case of monumental complacency on the part of the Soviets. On the last night of the exhibition, in a heist worthy of *Mission: Impossible*, US agents successfully intercepted the truck that was returning the crated space vehicle to the railway freight depot. They diverted the truck to a hidden location, opened the crate, disassembled its contents, measured its dimensions, photographed it from every angle and swabbed its insides for samples of propellant residue. The job proceeded without a hitch apart from one heart-stopping moment when an official Soviet seal was damaged by wire-cutters. Within hours the local CIA station was able to make a perfect replica. The whole unit was then reassembled, put back in its crate and returned to the depot before dawn. The Soviets never discovered what had happened. The CIA's findings helped to give American intelligence experts a more accurate indication of the power of the rocket that was capable of launching such a device. Such was the sensitivity surrounding the operation that the only CIA account published thirty-five years later in 1995 was heavily redacted, with words, photographs and sometimes complete paragraphs blacked out. It remains so to this day.

But even if the CIA learned many of the R-7's secrets, and even if by January 1961 they – and President Kennedy – knew the truth behind Khrushchev's exaggerations of Soviet missile numbers, there was one secret that they never penetrated, and that was Korolev himself.

On Khrushchev's own orders, Korolev's name was removed from all public discourse. New editions of histories of Soviet rocketry no longer included him after 1957, nor did they include the names of any of the other chief designers and senior engineers involved in the Soviet space and missile programmes. Occasionally Korolev was permitted to write articles for the government-controlled newspaper *Pravda* under a pseudonym, Professor K. Sergeev, in which he looked forward to the time when a Soviet comrade would conquer space; but these were his only disguised expeditions from the shadows. And it must have been hard for him too to see his name erased from his own achievements while his celebrated rival Wernher von Braun starred in a hugely successful Disney TV show about the future of space travel, ornamented the covers of *Life* and *Time* magazines, and even had a fawning Hollywood movie made about his life. None of this would happen to Korolev in his lifetime. 'Father never complained,' wrote his daughter Natalya many years later, 'but he surely felt hurt in the deep of his heart.'

As an additional smokescreen to cover Korolev's tracks, an inoffensive academician called Leonid Sedov was dangled before the CIA's spies, along with tantalising hints that here was the great mastermind behind all those Soviet space spectaculars. Sedov happened to be chairman of the grand-sounding Commission for Interplanetary Communications but in reality knew next to nothing about the Soviet space programme, which was precisely why he was wheeled out from time to time to talk about it. The ruse was brilliantly successful. 'At last,' claimed one top-secret CIA document confidently entitled *Here Are the Men and Formulas that Launched Sputnik*, 'we were successful in penetrating the secret of the mysterious Leonid Sedov.' But the mysterious Sergei Korolev remained unpenetrated. And just in case he was ever unmasked, he was protected on his travels around the USSR by KGB bodyguards to thwart any American attempt to kidnap or assassinate him. The man who had once been a public enemy of the Soviet state and who had dug for gold as a prisoner in a Siberian labour camp had now become one of its most precious and most secret assets.

WITH EACH NEW technological wonder unleashed by Korolev on the world, with each heavier satellite, orbiting canine, lunar or planetary probe, Khrushchev quickly discovered he possessed political weapons in his armoury far more effective than real ones. And just as quickly he discovered he could never get enough of them.

Each successive space 'spectacular' was exactly that: not so much part of a carefully structured progressive space programme but yet another glittering showpiece, preferably tied to an important political anniversary, and once again thrilling proof of Soviet ingenuity with which Khrushchev could bludgeon the Americans. The result was that Korolev rapidly became a prisoner of his own success. His several attempts to plan for the long term, to build a step-by-step, co-ordinated strategy that would ultimately take humans not only into orbit but onwards to the moon and even to Mars, building space stations and lunar colonies along the way – all such plans were rejected or simply ignored. And the pressure on him was unremitting. At one and the same time Korolev had to navigate a tortuous path between the jealousies of fellow rocket builders like his former denouncer Glushko, who loathed him, the seething resentment of military leaders who wanted him to stick to his proper job of making missiles, and Khrushchev who wanted *everything*. The demands on his schedule, his energy and his health, already compromised by his period in the Gulag, were almost crippling. 'Here I am,' he wrote to Nina, 'in a state of permanent anxiety and disquiet, and I so want to be home, in our dear little home, just the two of us. But I fear all these are just futile dreams.' Yet despite all these fears and pressures Korolev never lost sight of his great goal of human space travel. 'This is not some military toy,' a colleague, Viktor Kazansky, recalled him saying privately of the R-7. 'The purpose of this rocket is to get up there' – and he pointed a finger at the ceiling.

Once again Korolev's finely honed skills of manipulation and hard-headed pragmatism served his ambition. In August 1958, two months before NASA committed to its own manned space programme, he and his team prepared a secret document entitled *Materials on the Preliminary Work on the Problem of the Creation of an Earth Satellite With Humans on Board (Object OD-2)*. The title may not have exactly rolled off the tongue but the document itself was truly revolutionary, detailing a five-tonne manned orbiting spaceship that just *happened* to fit perfectly on top of his R-7 missile. But the real genius of the design, the core selling point, was that it could also double as an unmanned spy satellite. The proposed Object OD-2 would have two versions. In one of them, later to be called a Zenit, the generals would reap the benefits of photographic surveillance from a satellite that no known enemy weapon could intercept or destroy. And in the other, what would become the Vostok, Korolev would put his human cosmonaut. In a single hit, Korolev had pulled off a stunning coup. He could placate the generals – and he could also get his spaceship.

Like that first dazzling presentation to Khrushchev two years earlier, it was a superb piece of salesmanship, and it worked. By

January 1959, the Communist Party's Central Committee issued a secret decree calling for preparations for human space flight. A second secret decree followed in May 1959, a month after the Mercury Seven's first press conference, in which the spy satellite programme was formally placed under Korolev's control and included a commitment to 'a Sputnik for *human* flight'. Almost by sleight of hand, and aided by the helpful threat of this new American competition, Korolev had slipped his lifelong dream into the swirling cocktail of competing demands.

Now he just had to deliver it.

THIS, THEN, WAS the man who wrote to his wife on January 27, 1961, to tell her how much he believed in his cause – and that he was 'getting ready'. But since his deputy Konstantin Bushuyev had laid out the latest timetable for manned space flight just three weeks earlier, that timetable had slipped yet again, by at least another month. The two final Vostok dress rehearsal flights, each with its human-sized mannequin and accompanying dog, would now take place not in February but in March. Only after that, and only if everything went sufficiently well, would Korolev finally risk a human.

The six Vanguard cosmonauts were all now waiting. Yuri Gagarin was still tentatively at the top of the list. Ever since that first meeting in early 1960 three days after their induction, Korolev had been forcibly struck by Gagarin. At the time Korolev had not revealed his identity. But three months later he did, when the cosmonauts were invited to his headquarters at OKB-1 for their first glimpse of the Vostok. Along one length of wall was a row of shining silver spheres, like no flying machines the visitors had ever seen. As the men stood entranced, Korolev invited one of them to step inside the nearest sphere. Gagarin immediately volunteered; and the others saw how Korolev was visibly impressed by the way the fresh-faced young man first respectfully removed his shoes before stepping inside the immaculate Vostok. After Korolev left, Gagarin's friend Aleksey Leonov went up to him. 'You are the chosen one,' he said.

But Leonov was premature. Even as Korolev was writing to Nina the final choice had not been made. And time was slipping past. On the same day he wrote that letter, *Life* magazine published its latest issue which included a feature by John Glenn under the headline: 'We're Going Places No One Has Ever Traveled in a Craft No One's Flown.' Below were photographs of Glenn and his son watching a rocket launch at Cape Canaveral. Beside them, Alan Shepard's wife Louise also watched through binoculars. 'The announcement of schedules is a risky business,' Glenn wrote, but 'barring unlikely difficulties, one of us' – he did not reveal which one – would make the first space flight shortly after a chimpanzee had paved the way. And that chimpanzee flight, he added, would take place 'very soon.'

That chimpanzee flight was, in fact, exactly four days away when *Life* hit the stands on January 27, 1961; and for astronaut Alan Shepard, for NASA's Mercury programme and above all for America's hopes of finally beating the Soviets in the race to space, everything now hung on that flight going right.

NINE

SUBJECT 65

January 31, 1961: 02.00 hours
Cape Canaveral Air Force Station

THROUGHOUT THE NIGHT, the seven trailers parked in the fenced compound next to Hangar S at Cape Canaveral had been a fever of activity. They had been there for twenty-nine days now, and the smell and the noise coming from inside those trailers was enough to drive some of the hangar's regular occupants off the site. Shepard, Glenn and the other astronauts had their living quarters in Hangar S for use on their trips to the Cape, but they had since abandoned them for more salubrious comforts at the local Holiday Inn in Cocoa Beach. These days the place felt more like a zoo – which, in an almost literal sense, it was. For in those trailers were the six chimpanzees who, along with their twenty handlers and veterinary staff, had been flown to the Cape from the Holloman Aeromedical Research Laboratory in New Mexico at the beginning of January to prepare for Mercury-Redstone 2, NASA's upcoming chimpanzee flight into space. MR-2's launch was scheduled to take place later this morning at 09.30, and one of those chimpanzees would be on board.

The choice had narrowed down to two contenders: Subject 46, a female otherwise known as Minnie, was provisionally the back-up candidate. The front-runner was Subject 65, a thirty-seven-pound, approximately three-and-a-half-year-old male originally called Chang. Chang's name had very recently been changed to Ham after the initials of the laboratory that trained him, but for the moment that new name was being kept under wraps; it would only be released if Ham made it back from space alive. His selection as prime candidate had been based in part on his docility but largely because of his outstanding skills on the psychomotor, a training device that tested the chimpanzees' responses to flashing lights by making them push two levers within a specified number of seconds. If the animals failed, their feet were zapped by what a NASA press release described as a 'mild' electric shock.

By now, Ham had spent 219 hours on the psychomotor, first at Holloman and then in one of the dedicated trailers at Hangar S, lying strapped to a couch. Every time the left-hand blue light came on, he had exactly five seconds to press the left-hand lever; when the right-side white light was on, he had to press the right lever within a fifteen-second period. In either case if he failed he received that electric shock. A slightly later version of the psychomotor flung banana pellets at the chimpanzee as a reward for success; but in the version that Ham trained on there were no rewards, only punishments. It was this version that would be on the mission later this morning, flashing its blue and white lights and where necessary sending its electric shocks as Ham inside kept pressing those two levers while coping with the additional stress, if not trauma, of a rocket flight into space. Within the predicted fourteen-minute duration of today's flight, Ham could receive up to sixty shocks if he failed to respond to all the lighting cues in time.

The chimpanzee programme had started at Holloman back in April 1959, the same month that the Mercury Seven had begun their own programme, and certain elements of the animal training had closely mirrored those of the astronauts, a truth that none of those astronauts liked to dwell on, especially Shepard himself. The prospect that a chimpanzee would make exactly the same flight as his and with exactly the same profile and risks – and what was more, do it before he did – understandably did not sit well with the man who had more than three thousand piloting hours and could loop the loop in his jet around the Chesapeake Bay Bridge. Perhaps that was another reason, apart from the stench and the animal chatter, why Shepard had escaped from Hangar S to the Holiday Inn.

But the fact is that the Holloman chimpanzee training programme, much of it supervised by a bespectacled forty-two-year-old air force aeromedical technician called Edward Dittmer, adopted the same fundamental principle as the human programme, which is to say that it was based on *operant conditioning*. The point was to get used to all the novel sensations and stresses up there so that they were no longer novel, whether for a chimpanzee or for an astronaut. So the chimps went through the same centrifuge rides, pulling the same G-forces almost to the point of blackout and sometimes beyond, and both at Holloman and in two of those trailers at the Cape they even had a mock-up of the Mercury cabin where the animals could push the levers of their psychomotor in imitation of the levers the astronauts would manipulate at some point in their own flights; although in their case without being zapped in the feet if they made a mistake.

In its press release for this morning's flight, NASA included a statement from the American Medical Association, maintaining that 'only animals that are lawfully acquired shall be used and their retention and use shall in every case be in compliance with the law'. Ham himself might have disagreed with the spirit, and probably the letter, of that compliance given that he was first kidnapped from his mother at a very young age by animal trappers in the rainforests of the French Cameroons before later being sold to the US Air Force in 1959 for $457. At the Holloman laboratory, where experiments in connection with aviation and space research were being conducted on some four hundred different animals, Ham suddenly found himself joining a colony of fifty or so other 'lawfully acquired' chimps such as Little Jim, Duane, Bobby Joe and Elvis. Like Ham, they too were all very young. 'Usually,' said one of the laboratory's veterinarians, Captain Jim Cook, 'we try to get them as babies, one to three years of age.'

All of them began their training under Edward Dittmer. The first thing Dittmer did was to tie the chimpanzees to little metal chairs and 'set them about four or five feet apart so they couldn't reach each other and play'. They were made to sit like that for five minutes at the start, with the time gradually increasing over the following months until they were able to remain quietly in their chairs 'indefinitely'. Some of the chimps did not easily submit to this sort of conditioning, notably one called Enos who kept biting and attacking his handlers and who later at Hangar S would famously throw his faeces in the face of a visiting senator. But Ham proved more compliant. He became one of Dittmer's star pupils. 'I had a great relationship with Ham,' Dittmer said years later. 'I know he liked me. I'd hold him and he was just like a little kid.'

He was holding Ham now, in the early hours of that Tuesday morning, inside one of the trailers outside Hangar S, while getting on with his 'particular job' of implanting two electrocardiogram electrodes under the chimpanzee's skin to measure his every heartbeat during the flight. Dittmer also inserted a thermometer eight inches into Ham's rectum before taping it to his buttocks, where it would remain for the next sixteen hours. These and other sensors would send real-time physiological data from Ham inside his spacecraft to special panels both in the launch blockhouse and mission control. At just after 02.00, Dittmer moved on to the next phase of preparing Ham, which was to zip and lace him into a restraint suit reinforced by nylon tape, before attaching the 'psychomotor stimulus plates' – the electric shock plates – to his feet. These were then tested to confirm they were working properly. Ham was next strapped into his special couch. This formed one half of a container which, when closed, made a sealed box containing its own life-support system just like a spacesuit, and included the psychomotor itself at waist level. For the moment the top half was kept open allowing Ham to breathe normal air.

The whole operation took Dittmer thirty minutes. Now secured in his little container, Ham was moved from his trailer to a transfer van to wait out the hours before being driven the six miles to one of von Braun's Redstone missiles at Pad 5 – the same type of missile on the same pad where the 'Four-Inch Flight' had occurred two months earlier in November. It would turn out to be a long wait: two and a half hours in fact. But Ham was used to sitting for very long periods in tiny metal seats in confined spaces. He had been doing it most of his life.

HAM WOULD NOT be the first animal to be used for an American rocket test flight, although he would be the first chimpanzee. Since 1948, in a test called Project Albert, the US military had been strapping a variety of animals into the nose cones of ballistic missiles and firing them up into the sky. Originally, the reason was to test the impact of high-speed acceleration forces on living organisms in the expectation that humans would one day pilot advanced flying machines capable of a similar performance to rockets. That, at least, was part of the rationale behind Project Albert, in which a succession of nine-pound rhesus monkeys, tightly bound inside special aluminium capsules with their heads crimped painfully forward at an acute angle, were sealed into the nose cones of American-built German V-2 rockets and then blasted at over 3,000 mph to altitudes of forty miles – almost five times the height at which a modern airliner flies. At the apogee of the missile's ballistic arc the nose cone, with the animal still inside it, was supposed to separate from the missile and parachute down to the ground.

Unsurprisingly, the chances of any of those monkeys surviving such a mission were reckoned to be low. On the first one, on June 11, 1948, a photographer noticed that somebody had chalked *Alas, poor Yorick I knew him well* on the tail fin of the V-2 rocket carrying its first monkey, Albert. Albert was killed an hour later when the parachute carrying his nose cone shredded in the thickening atmosphere and he smashed at terrific speed into the desert. 'We got no information or data from the test,' said the project supervisor Jim Henry. But that did not stop Captain Henry from conducting three further V-2 rocket tests in which every animal – two more monkeys and a mouse – was also destroyed.

This was just the start of what would effectively become an animal killing operation over the next decade and beyond, as experimenters in both the US and USSR attempted to discover what flying on rockets and in space actually did to sentient beings. The list of unknowns was frighteningly long. Multiple unpleasant and largely fatal hazards, it was feared, awaited those who dared to trespass beyond earth's cradle. For 3.5 billion years since life had begun, all life had existed within the same gravitational environment. But nobody could be sure what would happen to that life in the weightless environment of space flight. There were fears that the cardio-vascular system might not work at all; fears of heart failure; fears of blindness, of difficulty in breathing or swallowing or even moving a single muscle; fears of uncontrollable nausea, ulcers or brain damage. And there were other fears once a spaceship was free of the planet's protective atmosphere: of horrific cancer-causing solar radiation or micro-meteorite bombardments that could suddenly tear that spaceship apart. Added to all these fears were yet further fears about the business of getting to and from space: the vibrations and the noise, the rocket accelerations that multiplied the weight of a body several times over, draining blood from the head, pooling it in the abdomen, squeezing and crushing flesh; or the searing heat of re-entry that could, if the spaceship were not properly insulated, burn that flesh to a crisp. The terrors were everywhere, even within the mind itself, terrors of isolation, panic, claustrophobia and ultimately of insanity.

At the beginning nobody knew. And so animals were sent up to find out.

Rhesus monkeys appeared to be the animal of choice in the US when it came to blasting complex living organisms to very high altitudes and eventually into space, although frogs and mice were also employed on occasion. The rhesus monkeys were more useful because of their obvious physical relationship to humans, although the animals could be excitable and difficult to train. By the time Ham was awaiting his own suborbital trip on the last day of January 1961, nine monkeys had already been killed in eleven flights, a tally that did not suggest favourable odds: among them a tiny, bushy-tailed squirrel monkey weighing just a pound called Gordo, barely bigger than a man's hand, who had disappeared along with the nose section of his rocket somewhere in the South Atlantic, possibly as a result of another parachute failure. Another victim was Able, a seven-pound female monkey who was recovered safely but died four days later when her electrodes were being removed. Photographs of these monkeys taken just before their flights are distressing in the extreme, their small size and hopeless vulnerability emphasised by the metal cylinders in which they are rigidly encased, their tightly restrained heads peeping out from a tangle of wires. Able's breathing rate soared to three times more than normal on her flight in May 1959 as she, along with an accompanying monkey called Miss Baker, plunged back into the atmosphere at 10,000 mph. A reporter afterwards asked Captain Ashton Graybiel from the navy's Aerospace Medical Institute in Pensacola how these animals behaved during their training. 'We didn't force a monkey to take a test if it objected to it,' he replied. 'These monkeys are almost volunteers.'

Until Ham, chimpanzees were not used for rocket flights, but – no doubt as 'volunteers' – they were extensively employed in ground acceleration tests at the Holloman Aeromedical Research Laboratory, well before Ham ever got there. These tests were initially designed to assess the impact of ejection or crash forces in high-speed jets or experimental rocket aeroplanes, and pigs were sometimes used as well as chimpanzees. Some of the tests were truly horrific, such as the Gee-Whizz machine in which animals were strapped to a rocket sled which accelerated to 169 mph before slamming to a stop in less than eighteen feet to see what would happen. What happened, according to Donald Barnes, an air force physiologist who joined the programme later, was that 'it pretty much smashed the brain inside the skull'. Another researcher described the animals' remains as typically 'a mess'. Eighty-eight chimpanzees were put through the Gee-Whizz; and in another variety of the *Alas, poor Yorick* facetiousness that seems to have characterised the experimental culture at the time, the pigs were occasionally photographed before their high-speed acceleration runs with signs hanging round their necks carrying the words 'Project Barbecue'. They were sometimes eaten after they had been killed.

Throughout the 1950s the variety and violence of these acceleration tests multiplied. The prospect of future manned rocket flights necessitated an even greater understanding of the impact of forces on different parts of the body in order to concoct the best means of protection. Holloman's chimpanzees – but less so with time, the pigs – now found themselves tearing along rails on their rocket-powered sleds at speeds even faster than jet planes before abruptly stopping or tumbling or being ejected off a precipice. Even the names of these tests evoke some nightmarish version of a fairground park: the Daisy Track, the Swing Drop, the Sonic Wind Number 1, the Bopper, the SNORT – which stood for the Supersonic Naval Ordnance Track at China Lake, California. Almost all the tests were filmed in ultra-slow motion, providing experimenters with a frame-by-frame record of the many gruesome ways in which the animals too often died. There was even a human volunteer brave, or insane, enough to undergo some of these tests too. His name was John Stapp, an air force officer and physician, and after somehow surviving twenty-nine rides, along with several retinal haemorrhages, cracked ribs and two broken wrists, he was finally grounded before he wound up killing himself. Stapp had a Hollywood movie made about his career entitled *On the Threshold of Space* and he featured on the cover of *Time* magazine, which was more than could be said of the animals who went through the same tests, or worse.

Meanwhile the Soviets used mostly dogs for their space experiments. An article from 1961 in the US journal *Aviation Week* discussed why they did this while the Americans used, first monkeys and then chimpanzees. A Soviet spokesman identified only as 'P. Varin' – which, in the closeted world of the Soviet space programme could mean anything or nothing – was reported as saying that while monkeys and chimpanzees 'have the same organ suspension and placement as man ... dogs remain calm'. Colonel Vladimir Yazdovsky, the doctor who devised the cosmonaut tests and also ran the space dog programme, once wrote that dogs 'respond well to training', unlike apes who 'are characterised by frequent breakdowns in behaviour and sometimes sharp aggressiveness'. Dogs, essentially, will do as they are told while chimpanzees might not.

On the very first page of NASA's seventeen-page press release before Ham's flight, the parallels between humans and chimpanzees are duly emphasised – not only do chimpanzees have very similar reaction times to humans but their 'high intelligence would make them outstanding performers in the conduct of complex tasks'. In contrast, lurking within the ethos of the Soviet cosmonaut programme was the notion that the best kind of cosmonauts did not need to be outstanding performers in the conduct of complex tasks; they just needed to be fit and obedient. Too much independence of mind was not actively encouraged, just as it was not encouraged in the larger Soviet society. Mark Gallai, a test pilot who supervised parts of the cosmonauts' training, later acknowledged that 'trust in the person flying was yet to be established'; like a dog, he could not be trusted to think. Instead he would be conditioned to respond to stimuli almost reflexively. It is not wholly coincidental that the nation that had given the world Ivan Pavlov would later train its own cosmonauts to behave somewhat like Pavlov's dogs.

The Soviet dog programme was initiated in 1951 – three years after Project Albert in the US – by Sergei Korolev himself as the first step towards his ultimate goal of putting a human in space. One of the lesser-known realities about this programme is just how many dogs actually flew. Today Laika is the most famous and perhaps the only dog's name most people remember, the story of her sad end still haunting as she died orbiting alone in her Sputnik in 1957 with no possibility of coming home. The four Vostok dog flights in 1960, in which Arvid Pallo served as leader of the rescue teams, are far less known, with the possible exception of Belka and Strelka's orbital trip in August of that year. But though the numbers are not certain, in part as a result of the secrecy surrounding the programme, one estimate suggests the Soviets sent up as many as forty-one dogs on rockets. Probably twenty-two of those dogs died. One, incredibly, ran away just as she was being brought out to her rocket and nothing would induce her to come back; she was replaced by a puppy that was unlucky enough to be found wandering about outside the staff canteen.

The dogs were all stray bitches – to ease the problem of urinating in a cramped space – and they were all mongrels, since it was thought they would be tougher than pedigrees. They needed to be light, not more than about seven kilos, or sixteen pounds, and small enough to fit inside their cramped containers. They also needed to be light-coloured so that the on-board film cameras could register them in the dim cockpit illumination. In fact the list of qualifications became so long that the dog-catchers who had the job of grabbing these animals off Moscow's streets were often exasperated. 'Perhaps,' one of them complained to the physicians, 'you'd also like them to have blue eyes and howl in C-major?' To help prepare the animals, initially circus trainers from Moscow's legendary Durov Animals Theatre were enlisted into the secret programme. Like Dittmer's chimpanzees, the dogs were slowly conditioned to cope with being restrained in increasingly tighter spaces for increasingly longer periods. At first they would bark and whine within a few minutes, but gradually they became used to the confinement until they could survive without protest inside tiny containers harnessed by metal chains allowing just a few inches of movement. By the end of their training they could do this for twenty days at a time.

As Ham waited in his parked trailer for his space flight, similarly restrained in his little container, two more of these dogs were in the final stages of their training at Dr Yazdovsky's Institute of Aviation and Space Medicine in Moscow before their own flights with a human mannequin in March – the two final dress rehearsals before a Soviet cosmonaut was sent to take his own chances in the cosmos. Ham too would need to fly before an American astronaut was risked in the same attempt. Both of these human flights, in the USSR and in the US, would be built on a long trail of terrified, trapped and helpless animals who were never 'volunteers' and who died in their numbers along the way.

But now, in the last hours before dawn, the lid of Ham's container was closed. Beneath its window his face could be seen. After one test run of his performance on the psychomotor, which he passed satisfactorily, his behaviour was deemed sufficiently stable for him to make today's flight. At 05.04, accompanied by Edward Dittmer and a second handler, he was driven in his trailer out to the launch pad, a twenty-minute ride that would take slightly longer than his trip into space. Lying on his back, strapped inside his little box, there was no mistaking the physiological parallels between him and the humans riding with him in his trailer. What nobody pointed out was that if he looked like they did, perhaps he might feel like they did too.

WAITING FOR HAM at the top of MR-2's gantry was Alan Shepard. As America's first astronaut-to-be, he was there to observe the final preparations and Ham's 'insertion' into the Mercury capsule. Five of the other astronauts were also at the Cape for this launch, either at mission control two miles north across the scrub or in the launch blockhouse at the edge of the pad. Shepard himself would shortly go down to the blockhouse, but first he was up on the gantry. In a NASA public relations film released later, Shepard is described as wanting to be there 'to wish Ham luck'. Perhaps so; but in truth it was hard for him to watch a chimpanzee take the flight before he did. Chris Kraft, mission control's flight director and a close colleague of Shepard's, saw how hard: 'Al would gladly have traded places with Ham on that January day. He had supreme confidence in what rigors the human body – especially his own – could handle. But the decision wasn't his to make.'

At the base of the gantry, Dittmer and an assistant removed Ham's container from the transfer van, placed it on a trolley and wheeled it into the lift. The rocket towered eighty-three feet above them against the black sky, its gleaming white form lit by arc lights. In the hours since midnight its two propellant tanks had separately been filled with ethyl alcohol and pale-blue liquid oxygen, the latter cooled to below its 'boiling point' of –183°C. At the moment of ignition these two propellants would slam together and combust, creating the blast of gas and fire that would thrust the Redstone on its way. But now, in the relatively warm night air, the cryogenic liquid oxygen steamed through the tank's vents, hissing across the pad and wrapping the rocket and its gantry and the slowly ascending lift in clouds of white vapour. 'It was quite a thrill taking Ham up there,' said Dittmer; but he also confessed to 'a twinge of anticipation', having seen too many missiles explode.

So far Ham's behaviour was, in the jargon of space flight, 'nominal'. Nor did Dittmer expect anything less. Five times over the past three weeks the chimpanzee had been taken through a dress rehearsal of this exact same procedure, the same battening down in his container, the same ride to the pad and then up in the lift. In yet another instance of operant conditioning, the experience was as far as possible normalised by repetition. But there was one experience that could never be normalised, the only one that really counted in the end, and that was the flight itself.

Once it reached the top of the gantry the lift door opened on to a short platform. At its end awaited the Mercury capsule. While Shepard stood by and watched, several engineers helped Dittmer ease Ham's container through the open hatch. Ham was now lying on his back in his sealed container inside the capsule. This particular capsule – Spacecraft Number 5 – had been delivered to the Cape back in October, since when it had been extensively checked and tested in Hangar S, the same hangar outside which the chimpanzee trailers were parked. In that time, one hundred and fifty separate flaws had been discovered that required rectification. With over seven miles of wiring and plumbing massed inside its slight, seven-and-a-half-foot-high cone, the engineers could only pray that nothing had been missed.

To anybody seeing it for the first time, Spacecraft Number 5 did not look much like a spacecraft at all. If anything, with its corrugated metal skin it looked more like an oversized grey trash can, although its shape was not cylindrical but more like a flask, with the astronaut – or chimpanzee – lying on his back at the flask's bottom. Compared to such futuristic, but never-built, winged wonders as the Dyna-Soar the whole assemblage looked alarmingly crude at first sight, almost as if somebody had knocked it together out of a few real trash cans over an afternoon. But there was nothing at all crude about it. If anything, it represented turn-of-the-sixties technology at its most cutting-edge, a product of engineering genius.

The core of that genius was a brilliant if eccentric aerodynamics engineer called Maxime Faget who understood that all those sexy, sharply streamlined, low-drag, winged spaceplanes swirling around in their inventors' minds would very rapidly burn up on re-entering the atmosphere, and that what was therefore needed was not streamlining but its exact *opposite*. The more *drag* that could be created, the slower the speed at which the spacecraft would meet the thickening air, thus giving its occupant at least a decent chance of survival. Which was why, in the teeth of aerodynamic orthodoxy, Faget devised and championed this ungainly-looking flask-shaped trash can now clamped to the top of the Redstone rocket with a chimpanzee inside it. With its big blunt bottom it would come home to earth more slowly and thus more *coolly*, providing far better protection for anything or anyone that happened to be inside. It took time for Faget to persuade the powers-that-be in the nascent US space programme that his way was the right way, and it probably did not help that he was addicted to Hawaiian shirts and doing headstands in technical meetings, but in the end he prevailed. Perhaps it was in his blood; his grandfather had once come up with a totally unorthodox cure that saved New Orleans from an outbreak of yellow fever. He called it 'Faget's Sign', an appellation that could just as well have applied to his grandson's phenomenal little space capsule.

In the Soviet Union, Sergei Korolev and his own capsule designer, Mikhail Tikhonravov – another eccentric, brilliant maverick and a sort of Russian version of Faget – had come up with the much simpler concept of a sphere, itself an excellent drag-creator; but they also had the massive R-7 rocket that could lift this heavier, bigger shape into space whereas Faget did not. His ugly corrugated flask was both smaller and much lighter than the Vostok, as it needed to be if it was ever to get anywhere near space aboard a relatively underpowered American rocket.

And for the same reason it was smaller and lighter on the inside too. In fact it was tiny, about the same size as a phone booth once you managed to squeeze aboard: 'You don't climb into the Mercury spacecraft,' joked John Glenn, 'you put it on.' The whole assembly was a miracle of miniaturisation, just to fit everything in. Apart from all those miles of wiring there were some ten thousand separate parts, almost every one of which, in a wondrous parody of *Gulliver's Travels*, had to be re-fashioned and re-sized to levels of littleness beyond anything recently believed possible. Every last inch and pound of weight counted. And the problem of course was that all this took time. Simply devising one lightweight, spherical tank the size of a volleyball containing a mere four pounds of oxygen took eighteen weeks to perfect; and that was just a single item. Behind the apparently endless delays and postponements that so frustrated the American public lay the simple truth that smaller took longer, even if this truth proved so difficult to communicate. The Soviets did not need to take such pains in order to be so advanced. They were ahead in the race, certainly. But for the same reasons, they were technologically behind.

In almost every respect Ham's Spacecraft Number 5 was identical to Spacecraft Number 7, the capsule assigned to carry Shepard himself into space and which was itself now in Hangar S undergoing its own checkout since its arrival there in December. Ham's flight would be entirely automatic, controlled by an on-board system that ensured his capsule oriented itself correctly in the various phases of flight – another complex system that was about to be trialled for the first time. But Shepard in addition would have the opportunity to 'hand-fly' his own capsule on his mission, or at least for a portion of it, using a controller like a small gaming stick that fired quick jets out into space and allowed his spacecraft to change its attitude in a crude approximation of normal flying. Unlike Ham, Shepard also had an abort lever he could hit if everything fell apart – the astronauts called it the 'chicken switch' – another piece of pilot autonomy they had fought the engineers for, even though aborts could also be controlled automatically and from the ground. In its own way the very design of the cockpit was proof of the astronauts' success in their battles for more piloting control. Almost every inch of it was littered with dials, knobs, indicators, lights and levers just like a 'real' aeroplane cockpit. There were twenty separate displays and fifty-five separate switches in there at least, far more than in the Soviet Vostok with its almost ludicrously bare central panel of just four dials looking for all the world like something out of a period Volkswagen or perhaps an East German Trabant, except that the drivers of those cars probably had more to do than the Soviet cosmonauts in their capsules.

What Ham's spacecraft did not have – and nor would Shepard's – was a proper window. The astronauts had fought for that too, arguing that no self-respecting pilot could fly his ship without a window, but the engineering challenges had proved too daunting to make one light enough, at least for now. A decent-sized window would come in later models. Shepard himself would have to make do with two tiny six-inch portholes and a periscope, not much from which America's first astronaut was supposed to view the glories and splendour of the planet from space, although viewing the planet was unlikely to be high on Shepard's list of priorities in the very short time he was going to be up there. Ham had the same two portholes in his capsule, but he was inside his container. The only view he was going to care about was of those two flashing lights on his psychomotor just inches away, and he already knew from painful experience the penalties for not heeding them properly. Edward Dittmer could see him in there now, his face peering through the transparent upper half of his container, his wide eyes blinking in the light of the film camera that would record his every expression throughout this mission, a strange anomaly of life in all the cold hard metal of his environment.

But then the hatch was closed and the last of its seventy bolts fastened. It was now 07.10, two hours and twenty minutes from the scheduled launch time. As Shepard headed to the lift and down to the blockhouse, he might have reflected that for all the extra controls he and his fellow astronauts had fought so hard for, in the end none of them mattered. A chimpanzee was about to make the same flight without any of them.

TEN

A HELL OF A RIDE

January 31, 1961: 07.30
Launch Complex 5/6
Cape Canaveral Air Force Station

FOR THE SEVENTY-ODD men in the launch blockhouse – and there were only men, since women were forbidden to enter – the long night's activities perceptibly shifted into their highest gear just as the sun began rising out over the Atlantic. Many of them had already been working round the clock for at least three weeks testing and preparing the Redstone's complex systems for this morning's launch, hardly seeing their wives and families and homes in that time. It was a familiar and exhausting ritual with every missile they fired into the sky, even if it also meant that the divorce rate among blockhouse personnel was just about the highest in Brevard County, an area that included not just the Cape but a chunk of eastern Florida. William 'Curly' Chandler worked one of the consoles at Pad 5 and knew his job was taking a toll on his marriage, but the job was too strong for him: 'It was bad for the family, real bad. But we were all dedicated. That was the biggest thing. I loved my job. Everybody did.' Ike Rigell, the Pacific war veteran who with the CBS reporter Ed Murrow had watched the Juno missile U-turn off the pad and blow apart in 1959, had also watched his own marriage to Kathy nearly go the same way in the years he worked here. 'I always cry thinking about those days,' said Kathy, years later. 'It was a very hard time.' Day after day she kept the household and their children going while Ike disappeared off to his missiles at the Cape. But there were real compensations too, not least the missiles themselves whose launches Kathy watched from her yard as the windows shook from the noise: 'They were so beautiful. It was so awesome to see one of those things go, and to know that I was part of it – because I fed him, and clothed him, and nursed him and because he was part of it.'

Like a band of brothers in wartime, these men in the Cape blockhouses formed the closest of bonds, looking out for one another, knowing and accepting one another's foibles, building the essential ligaments of trust in what they all knew was a very dangerous business. At just 750 feet and sometimes less from their rockets they were, literally, in the front trenches and every launch, as Rigell remembers, was 'a very fine line between total success and catastrophic failure'. Their trust in each other extended to the German rocket engineers who had come to America after the war. Von Braun, their supreme chief, was already a full-blown legend, a man who Rigell felt 'could talk to anybody … he could brief the President and thirty minutes later be out on the pad talking to a bunch of technicians'. Von Braun was in the bunker now for Ham's launch, crisp as ever in his white shirt and tie, his ID card clipped to his left breast pocket, a handsome, powerful, quietly reassuring presence moving among the engineers as they sat at their panels beneath the thick blockhouse windows.

And yet von Braun was essentially a visitor on this day. The real authority in the blockhouse this morning was the launch director, Dr Kurt Debus, in his mid-fifties and with a face distinguished by a duelling scar, another German who had worked with von Braun on Hitler's V-2 missiles and who had also been a member of both the Nazi Party and the SS. One of Debus's responsibilities was the countdown, an exhaustive point-by-point checklist that from start to finish took a very lengthy 640 hours, as Kathy Rigell and every one of the engineers' long-suffering wives were only too aware. And that was if everything went to plan. Already this morning, the count had had to be stopped several times as a result of an overheating electrical inverter in the capsule's automatic controls system. And it kept on overheating, so much so that at 09.08 – some twenty minutes before the planned launch time – the gantry had to be rolled back to the rocket and the capsule's hatch unbolted and re-opened to try to cool the thing down. But Ham was left inside his sealed container while Dittmer and his team took advantage of the extra time to test him out yet again on the psychomotor, just to make sure he was keeping on the ball in there.

LIKE A RELAY race, once the rocket lifted off the pad, responsibility for MR-2 would transfer from Debus and his team in the launch blockhouse to the Mercury Control Center two miles north. Here was the embryonic structure we know as mission control, today located at NASA's Johnson Space Center in Houston, Texas. But back in 1961 all the action was down at the Cape, concentrated in a large room with three rows of consoles facing a wall-sized illuminated map of the world, over which was superimposed a little two-dimensional model of a Mercury capsule.

Ham's ballistic flight, like Shepard's to follow, was not expected to extend beyond approximately 290 miles out into the Atlantic; but when the technology was ready, perhaps a year or so from now, one of the seven Mercury astronauts would get to fly the first fully orbital mission like the kind the Soviets were planning to fly very soon. That was why the map up on the wall compassed the whole globe. The Soviets possessed nothing this advanced. 'We literally wired the world for communications,' said Chris Kraft, Mercury Control's flight director, and the driving force behind the whole operation.

Sitting at his table in the second row Kraft – or 'Flight', as the mission's flight director was known – waited yet again for the countdown to restart. All these delays were getting on his nerves. The overheating inverter was still not behaving itself. The scheduled lift-off time of 09.30 had already been and gone. By now Ham had been on top of the Redstone for over three hours. He had been strapped in his container for more than seven.

At the age of thirty-six, Kraft was the oldest person in the room. A superb aeronautical engineer from Phoebus, Virginia, and famously unafraid to speak his mind, he was worshipped by his young team, most of whom he had picked personally. 'I'm Flight and Flight is God,' he once described his role. But he was not feeling so godly now. Like everybody in mission control, like Debus and von Braun and the engineers over in the blockhouse, like Alan Shepard and the other Mercury astronauts, and like the rest of the United States of America, he knew that this mission had to succeed.

Nobody had forgotten the fiasco of MR-1 and its four-inch ride off Pad 5 just two months earlier in November. In December, its successor MR-1A had performed pretty nearly flawlessly; but there was nothing alive aboard that time. Now there was. Once blasted into the skies, the Redstone was supposed to streak up to a height of thirty-five miles in less than two and a half minutes, at which point Ham's capsule would yank clear, ripping on further upwards in its ballistic trajectory to an altitude of 115 miles above the surface of the earth, above the atmosphere, above anywhere any human or chimpanzee had ever been; where finally, for four and a half weightless minutes at the apex of its beautiful arc, Max Faget's little miracle of a rubbish bin would sail along in space, with Ham inside pushing away at his psychomotor levers as fast as he possibly could.

Of course he would also then have to come back down. A complicated series of events had to happen in exactly the correct sequence and at exactly the correct moment for that to occur, any one of which could also go wrong. By this point the capsule would be sailing *backwards* through space, thanks to the automatic control system whose overheating inverter was currently causing so many headaches. This same system would pirouette his capsule blunt-end down as it reached the apex of its arc and three retro-rockets would fire to slow it down; then back Ham would plunge into the atmosphere, protected from the rapidly rising temperatures of re-entry by Max Faget's big blunt end as well as a heat shield, until the first of two parachutes popped open and Ham inside his capsule would drop, safely and slowly one hoped, into the cold waters of the Atlantic Ocean, ready to be fished out by a helicopter. The entire ride from start to finish was supposed to take just over fourteen minutes. To cushion his impact a pneumatic landing bag had been attached to Ham's capsule. Like a number of other systems on this mission it was also being flight-tested for the first time.

Little wonder then that Kraft was not feeling so godly as the clock ticked on another half-hour past 10.00. There were enough worries here to keep anyone's head spinning. He checked with the blockhouse physician, William Augerson, to see if Ham was still OK and not overheating along with that inverter. Augerson had indicators in front of him – heart rate, respiration rate, body temperature – that linked back to those electrodes and that rectal thermometer Dittmer and his team had implanted early that morning. He reported that Ham was fine. But as the count restarted for the umpteenth time at 10.15 Dittmer threw the chimpanzee three more psychomotor tests to make sure he could hit those levers without getting his feet zapped. Or maybe just to give him something to do.

Over his headset meantime, Kraft could hear the voice loops from various parts of the operation, from mission control, from Debus's blockhouse team, from the seven recovery ships steaming out in the Atlantic, a cacophony of back-and-forth radio chatter that he had become supremely skilled at disentangling, to the point of legend. Just as a conductor can keep multitudinous parts in his head at the same time, so too could Kraft; up to eight voice loops at once and one of them especially, which was the link to the Cape's range safety officer – or RSO – the man whose finger would be hovering over the abort button at the moment of launch. Just one press of that button would send a signal to the Redstone, igniting a strip of detonation cord placed inside its fuel tank and blowing up the rocket. It was an option again today if anything were to go *really* wrong, especially if the Redstone did something crazy and threatened to come down on a Florida suburb. But among all his other worries Kraft worried about that option too, because if the RSO could send that signal, then what was to stop the Russians who were sitting out there in their 'trawlers' monitoring this launch as they did every launch at the Cape: 'If I could send an abort signal, so could they,' said Kraft. 'Our frequencies were widely known'; which raised the alarming possibility that the Soviets might end up winning the race to space by simply blowing up America's own rockets.

There was of course the additional question of what would happen to Ham and his capsule if the Redstone *did* have to be aborted. This is where another of Mercury's unique systems came in – and another reason why the programme was so handicapped by delays, since the relevant hardware had first had to be developed and then repeatedly tested. Mounted at the very top of the capsule was a fifteen-foot red escape 'tower' or pylon, with its own rockets ready to fire and drag the capsule away from the stricken rocket – the same pylon that had shot off and slammed back down into the Florida scrub in the Four-Inch Flight. Despite the fiasco on that occasion the principle behind this escape system was sound, except that if the RSO ever did have to send his abort signal there were only three seconds before the detonation cord blew up the Redstone. And three seconds was not very long to pull Ham and his capsule clear, assuming the escape system even worked properly – which in a number of practice test flights at Wallops Island, Virginia, it most certainly had not.

But the time was now 11.52. The overheating inverter had been allowed to cool and the gantry had been rolled away, this time for good. The launch area had since been cleared of all personnel. Barring any more delays, there were two minutes to go. In Mercury Control Kraft watched his team at their panels, watched the giant wall map, watched the clock and watched a TV screen that gave him a live feed from the pad. Over his earphones he could hear Debus and the blockhouse engineers also checking their systems one last time, monitoring the Redstone's heartbeat to confirm that everything was functioning as it should. The physician reported that Ham's own heartbeat was also good. The RSO was standing by. The recovery ships and helicopters were all in position, 290 miles out in the

Atlantic. They were now reporting five-foot waves – not enough to scrub the mission but adding to the pressure to launch. Here at the Cape the skies were mostly powder blue and clear. There were no more holds. Everything was *go.*

In the blockhouse at his firing panel, Jack Humphrey, an engineer from Lexington, Kentucky, married with three children to whom he was devoted, pressed the firing command button on his panel, initiating a thirty-five-second automatic sequence of events inside the Redstone's guts that led all the way to lift-off. With a roar a blast of flame shot from its engine. Through the two-foot-thick blockhouse walls Humphrey and the seventy other personnel could feel its power. So could Alan Shepard, his eyes glued to the Redstone through the shock-proof windows. So too, no doubt, could Ham, strapped on his back in his little pressurised container at the rocket's apex. At exactly 11.54, nearly two and a half hours late, he was finally on his way. As one of the mission controllers, Gene Kranz, remembers, he was in for 'a hell of a ride'.

WITHIN THE FIRST few seconds things began to go wrong. At lightning speeds giant IBM computers up at the Goddard Space Flight Center in Maryland crunched the numbers transmitted instantaneously from multiple data points on the Redstone and flashed back a warning to mission control. The rocket was already climbing too steeply. It was off by one degree; not yet enough for the RSO to press that button, but the angle was increasing as the rocket clawed for altitude, its 78,000 pounds of thrust punching it faster and higher into the Florida skies – but punching it too fast and too high. And something else was wrong too. A thrust controller valve on the Redstone's engine appeared to have stuck in the fully open position, pouring too much precious liquid oxygen into the combustion chamber. The missile was tearing upwards with its foot hard down on the pedal, accelerating far beyond its predicted 4,400 mph to a much faster 5,857 mph. And as it did so it began to vibrate and to oscillate, twisting on its axis and shaking ever more violently. After just seventy seconds the vibrations had reached the point where the sensors inside Ham's capsule were reading off-scale.

On the physician's panel a needle flicked sharply upwards as Ham's heart rate shot from 94 beats per minute to 126. All the time the psychomotor kept on flashing its blue and white lights; Ham kept pushing his two levers just as he had been trained. The white light suddenly glowed – he had less than fifteen seconds to react – he pushed the right-hand lever correctly and on cue; but an electric shock stabbed his feet. The machine had been upset by the vibrations jolting the capsule. The two lights kept flashing; Ham kept pushing his levers – this time it seemed to be working. Chris Kraft never afterwards forgot the film footage of Ham's face: 'The onboard movie camera showed that he was desperately trying to do his job and stop getting that shock to the sole of his foot.' Freeze frames of that footage are deeply distressing sixty years later, each one an instant of trapped animal terror. But the rocket was ripping on upwards through 100,000 feet, vibrating and oscillating – and now it was even *bending*. The stuck-open fuel valve remained stuck. And the engine's turbopumps were also spinning too fast, another malfunction that added still further to the torrent of liquid oxygen entering the combustion chamber. The rocket was burning up fuel much too rapidly. And as Kraft knew, as von Braun and Debus and everyone else on that mission knew, right there was the very biggest problem of all.

From the rocket's launch to its fuel depletion was supposed to take 143 seconds. This was the point at which Ham's capsule would unclamp and continue alone on its ballistic arc up to a maximum 115 miles into space. But with the liquid oxygen pumping so quickly out of its tank the engine was clearly going to run out of fuel sooner. A clever timing device inside the Redstone devised by von Braun's rocket engineers had been set to abort the mission *automatically* – and thus launch the emergency capsule escape system – if any major parameter were breached up to 137.5 seconds, after which point the abort system would disarm itself. Kraft had argued with von Braun to set the timing earlier, *just in case* the rocket ran through its fuel too fast. Von Braun had stuck to his position – 'We've never had a

Redstone burn for less than 139 seconds,' he told Kraft. 'Never.' But now it looked like he was about to. Helplessly Kraft watched the seconds on the clock, wishing he had punched 'the damned arrogant German' and praying that the rapidly depleting fuel would keep pumping and the rocket's engine keep running just a fraction longer, until that magic number of 137.5 seconds was reached and the automatic abort switched itself off. The seconds ticked down. And then, at exactly 137 seconds, just *half a second* early, the rocket ran out of fuel – triggering an immediate abort.

The explosive bolts clamping Ham's capsule to the Redstone blew off precisely as they were designed to do in an abort. Within three seconds the red escape pylon fired its own rockets pulling Ham's Mercury capsule clear and punching it upwards at terrific speed, almost instantaneously increasing its rate of climb to 450,000 feet per minute – equivalent to getting from the ground to the cruising altitude of a modern airliner in five seconds – and giving Ham, as Kraft recalled, 'a 17-G kick in the ass – something worse than being blindsided hard by a pro football linebacker'. For a second or two Ham was seventeen times his normal weight, the sensation like a giant boulder pressing down on every part of his small body, knocking the breath out of him and almost causing him to black out. Even in their worst centrifuge rides none of the astronauts had experienced G-forces as violent as this. The weight was too much – the blood drained from Ham's head and he could barely see as his vision tunnelled down to just a tiny point. It was nearly impossible for him to move his arms – but still the psychomotor was flashing its blue and white lights and still he managed to push his levers in time to prevent those stabbing electric shocks. His heart raced up to 158 beats per minute as the escape tower was finally jettisoned and his capsule catapulted to its apogee, 156 miles up in space, 41 miles higher than it was ever meant to be, meaning that Ham would experience an extra two minutes of weightlessness. More than that, he would also land approximately one hundred and thirty miles further out into the Atlantic than any of the recovery ships were expecting him to be.

At about this point one of Kraft's team – known by his initials EECOM, the man responsible for the capsule's environmental system which was also going through its first flight test – spotted the cabin pressure beginning to drop from its nominal 5.5 pounds per square inch to just 1 pound per square inch. A snorkel designed to open and let in fresh air once the capsule was safely parachuting down had shaken loose with all the violent juddering and vibrations. The 100 per cent oxygen atmosphere inside the capsule was being sucked out into space. Ham's pressurised container was now his only barrier against the near-vacuum of his capsule. Had he not been inside it he would already be dead. 'I look around mission control,' recalled Kraft of that moment. 'Every face is sombre.' By this time Ham had been in flight for just five minutes. And he still had to get home.

Back on the ground, rapid calculations from Goddard's IBM computers were passed to the recovery ships and search planes, giving the new predicted splashdown zone in the ocean. Not only would the extra height add considerably to the range but the capsule no longer had any retro-rockets to slow it down on re-entry; these had been jettisoned, as they were designed to, along with the escape tower. Ham would be coming down fast, just as he had gone up fast. Once he swung over the top of the arc he would drop like a cannon-ball back to earth with only his drogue and main parachutes to slow him down towards the end. Meanwhile the six destroyers and one landing ship – the USS *Donner* – churned eastwards at full steam through the choppy Atlantic waves to the area where Ham would probably, but not certainly, be. At their best speed it would take another three hours to get there.

But Ham's ride was not over. As he fell back to earth the acceleration forces began building up again, reaching a massive 14.7 Gs. Once again his heart rate tore up. At 9 minutes, 35 seconds into the flight his heart was hammering at a terrific 173 beats per minute. And his respiration rate doubled – he was panting with fear. A minute later, as the worst of the G-forces crushed his limbs, drained the blood from his head and shoved his eyeballs back into their sockets, his heart rate careered to a very dangerous 204 beats per minute.

Missile Row, Cape Canaveral Air Force Station in the early 1960s, launch site for both US missiles and NASA's rockets.

A recent photograph of Gagarin's launch pad at what is now Baikonur Cosmodrome in Kazakhstan. The massive flame pit is clearly visible beneath the rocket. Very little has changed since 1961.

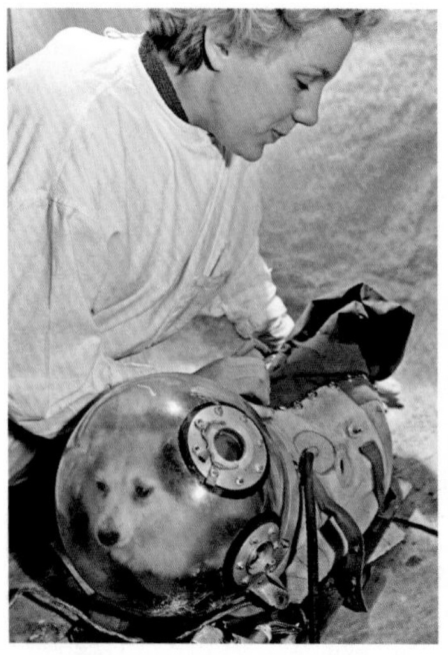

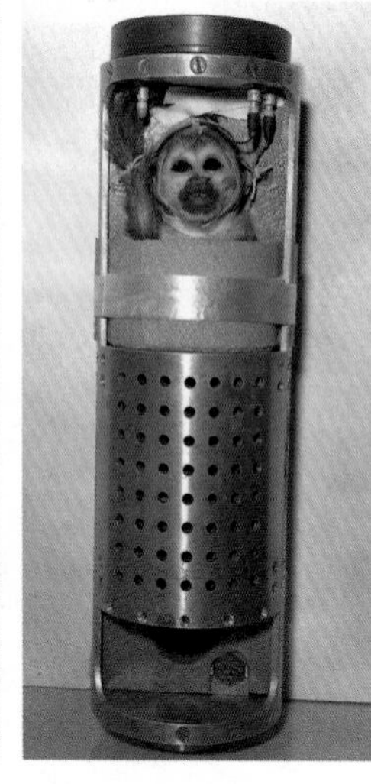

Miss Baker in her capsule before insertion into the nose cone of a missile in May 1959.

An early Soviet space dog is prepared for flight.

Ham in the lower half of his container before the lid is secured. On the left is his trainer Edward Dittmer.

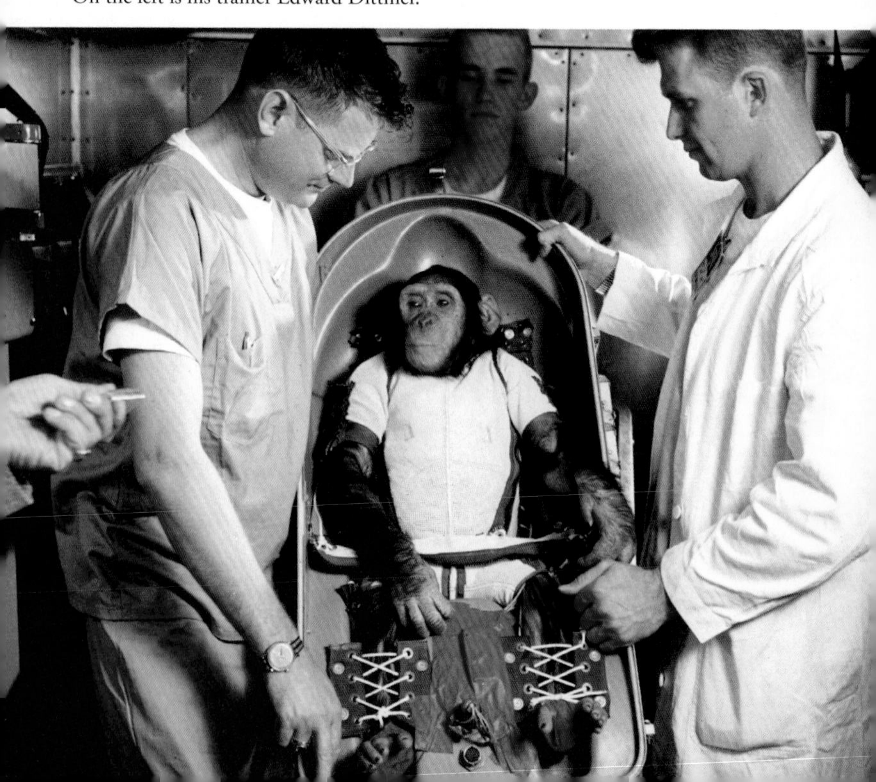

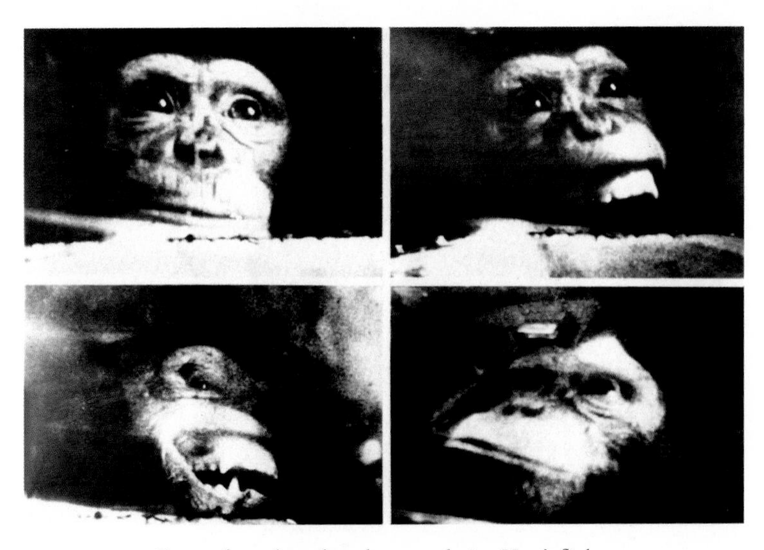

Frames from the onboard camera during Ham's flight.

'Boy those things were really scary.' A Juno II rocket explodes seconds after lift-off on July 16, 1959.

The Mercury Seven in full dress. Clockwise from top left: Shepard, Grissom, Cooper, Carpenter, Glenn, Slayton, Schirra.

The opposition. The Vanguard Six. Front row from left to right: Karpov (commander of the training centre), Gagarin, Nikolayev, Titov, Kamanin (chief of training and secret diarist), Popovich, Bykovsky, Anikeyev (cosmonaut but not one of the Vanguard Six), Nelyubov.

John Glenn, as American as apple pie and for many the favourite for first US astronaut.

The man who beat Glenn to the title, Alan Shepard.

Gherman Titov and Yuri Gagarin. Close friends, next-door neighbours and arch-rivals for first Soviet cosmonaut.

Wernher von Braun, America's top rocket engineer, at his periscope before a launch.

Von Braun with former associates in Nazi Germany.

Von Braun's great rival Sergei Korolev, here with Yuri Gagarin. 'Sergei Pavlovich loved Gagarin as a son,' said Korolev's daughter.

US astronaut training. The dreaded human centrifuge in Bucks County, Pennsylvania. A victim is about to step into the gondola.

Soviet cosmonaut training.
Titov demonstrates his acrobatic skills.

The isolation chamber or
'chamber of silence' today.

Mercury Control Center at Cape Canaveral. At the cutting edge of technology in 1961.

Launch bunker at the Soviet cosmodrome. Not at the cutting edge of technology in 1961.

Despite all the months of operant conditioning, despite all the restraint training and all the centrifuge rides and all the sitting tied in chairs for days at a time, Ham had never experienced anything as bad as this. One can only imagine the stark terror behind those clinical numbers on the physician's indicators. Traces of vomit containing blood would later be found inside his container; a self-inflicted abrasion would be found on his nose; but incredibly, throughout all this turmoil, Ham still kept pushing the two levers of his psychomotor without stopping. Perhaps his fear of those electric shocks was greater than every other fear. His left-hand lever score was perfect; his right-hand was near-perfect too. Apart from that one shock when the psychomotor malfunctioned after launch, he only received one more, moments before his capsule, now dangling from its main parachute, splashed into the ocean at exactly 12.12 p.m. The whole flight had taken just over sixteen minutes. He was 422 miles out in the Atlantic and alone. Now he had to wait for somebody to find him.

It took twenty-seven minutes for the first search aircraft to locate the capsule. By then water was beginning to seep inside. The splashdown had damaged the heatshield, snapping it back into the hull and breaching it in two places. The capsule was starting to list in the sea, allowing the stuck-open snorkel valve to let in even more water. Heavy Atlantic waves slammed against the side, ripping the pneumatic landing skirt around the capsule's base and eventually tearing the heatshield loose. A few minutes earlier Spacecraft Number 5 had been 156 miles above the earth. Now it was slowly sinking into the ocean, with Ham bolted in his container inside.

By the time the first helicopter arrived, Ham had been in there for well over two hours and the capsule had taken on eight hundred pounds of water. The pilots radioed that it was sitting alarmingly deep in the sea and tilting on its side in seven-foot swells. They moved fast. Frogmen jumped from the hovering helicopter into the water, attaching cables to the capsule's neck. Running the engine at maximum power, the pilots slowly hauled the heavy spacecraft out of the water, then flew with it to the USS *Donner* which was still racing towards the landing site. There it was lowered carefully to the ship's deck. Nine minutes later, with most of the ship's complement looking on, Ham's container was opened up and he was released from his couch. He was given an apple.

A photographer on board caught the moment he was given that apple and it went round the world. The New York *Daily News* splashed it on its front page the next day, as did many other newspapers across the country: 'Looking far less perturbed than a subway rider, the proudest little guy in the animal kingdom gleefully reaches for his reward, an apple polished to a sparkle.' Everybody had fallen in love with the 'cuddly astronaut' – or 'chimponaut' – and *Life* duly put him on the cover of its February 10 issue, headlining the inside feature, 'A Happy End for Ham's First Flight'.

There was surely no doubting that big grin on the chimpanzee's face, despite the fact that it took four men to try and put Ham back in his special container to get a picture for the *Life* shoot and even then they could not do it. He went berserk, baring his teeth and screaming at the photographers. And at the press conference in the Cape the day after his flight, he 'balked and screeched' when two vets attempted to do the same for the twenty-five cameramen and reporters who had turned up to see the 'grinning hero'; he also bit the finger of his own handler, Joe Pace. He had to be taken back to his trailer just to quieten him down. But that huge grin on that first photograph with the apple kept alive the myth that an animal who had undergone what the space journalist Jay Barbree, who also witnessed the launch, called 'a disaster' should have ended up smiling about it. In fact, Jane Goodall, possibly the world's leading expert on primates for the past fifty years, described Ham's beaming face as 'the most extreme fear I've seen on any chimpanzee'.

Before the end of the week the *El Paso Times* reported that 'preliminary examination at Cape Canaveral showed Ham to be in excellent condition', although it did not report the fact that he had also lost over 5 per cent of his bodyweight during his sixteen-minute flight. He was utterly exhausted and dehydrated, not least because he had not been allowed any water or liquids from the time he was first placed in his couch until the time he was examined by a veterinarian on board the *Donner* – a period of nearly sixteen hours. But the message out there was that 'physiologically, an Astronaut would have been perfectly all right on this trip', a sentiment with which Alan Shepard would have agreed, since he was eager to go next and be that 'Astronaut' – with a capital 'A' – before a Soviet Cosmonaut got there before him. Ready too were Chris Kraft and his people at mission control, along with NASA's Space Task Group, the organisation running Project Mercury. They all wanted this next flight, despite the problems with the Redstone and its stuck fuel valve, despite its vibrations and oscillations and its half-second early abort, and despite the problems with the capsule, all of which could surely be remedied before the end of March, in time for Alan Shepard to step onto the stage and rescue America's reputation in space, and much else besides, before the world.

But others were less sure. Von Braun, for one, was not yet prepared to commit another of his Redstones to a manned shot until a full investigation had been made into what, precisely, had gone wrong – and why. Nor were most of his own engineers at what was now called the Marshall Space Flight Center in Huntsville, and for the same reasons. It was enough to drive Kraft mad. He was furious about that abort timing decision and frankly he did not care much for those German rocket engineers anyway. 'The damn Germans still haven't learned who they work for,' he once said, and nobody in mission control had forgotten how he had once marched over to one of them, Joachim Kuettner, and yanked his microphone cable out of its socket for speaking German and not English. But right now, Kraft's word did not appear to carry much sway. The official word out there may have been that Ham's flight was, with one or two minor caveats, a rip-roaring success, a mantra repeated in all those lush headlines and chimp-with-apple photographs. But at NASA's press conference immediately after the flight nobody on the platform would be drawn on the only question anybody in the press corps, or the astronaut corps, or frankly the rest of America, really cared about which was – would a human still go next?

ACT II

DECISION

FEBRUARY–EARLY APRIL, 1961

What would happen to anyone who disclosed a state secret? Everyone knew what would happen. And it wouldn't be a pat on the head.

Boris Volynov, cosmonaut

Sure, I knew all seven of the Mercury astronauts. And four of 'em still owe me money.

Jay Barbree, veteran space reporter

ELEVEN

THE RISK EQUATION

AS THE DAYS slid past after Ham's space flight the silence from NASA was deafening. There was neither confirmation that the manned flight would soon follow, nor that it would not soon follow. There was nothing. Meanwhile the Soviet space juggernaut rolled forward with yet another of its space spectaculars. On February 4, less than a week after Ham's trip, the official news agency TASS claimed that the world's biggest satellite had just been launched into earth orbit, this one a monster weighing almost seven tonnes. The claim was a lie. The satellite was not a satellite at all. It was in fact Venera 1, the Venus probe that Korolev had masterminded and which had launched early that morning from the pad near his little cottage in Tyuratam. But things had swiftly gone wrong: the R-7's fourth stage carrying the probe had failed to ignite, leaving it stuck in low orbit instead of striking out millions of kilometres across interplanetary space towards Venus. As always, however, failures could be dressed up to look like successes: the Soviets simply pretended it was meant to be an orbiting satellite. And in a closed society where were the independent witnesses to suggest otherwise?

But that did not stop rumours spreading that something was being covered up. There was even a wild claim by two amateur radio hams in Italy, the brothers Achille and Gian Battista Judica-Cordiglia, that the Soviets had actually succeeded in putting a human in space but that some sort of catastrophe had occurred leaving the man – or possibly even two men – to suffocate and die. The brothers announced that they had even picked up the sound of heartbeats and human breathing, sounds which had then abruptly stopped. Their claim was never verified, and nobody else heard anyone breathing up there; but in the end that was beside the point. Whether it was the world's heaviest satellite, or even these sensational rumours of a lost cosmonaut, the Soviets were still manifestly *ahead*. Meanwhile NASA's strange silence about its own immediate plans persisted. On February 8 President Kennedy himself added fuel to the flames of confusion when he was asked in a press conference if he had ordered an acceleration of the American manned space programme – and if he also considered the US to be in a race with the Russians. No, he replied. 'We are very concerned that we do not put a man in space in order to gain some additional prestige, and have a man take a disproportionate risk.'

Disproportionate risk? Was Kennedy already throwing in the towel? Here, surely, was his science advisor Jerome Wiesner's backstage signature, counselling caution in the face of a potential – who knew, perhaps even likely? – catastrophe, one that would seriously damage the president's standing in the world. Inexperienced and untested as he was, still less than a month into office, Kennedy could ill afford such a disaster. And especially not when the Soviets were meddling more than ever in Fidel Castro's Communist Cuba or in the hotly divided South-east Asian country of Laos, threatening to extend their Red menace across new swathes of the world's geography. 'So,' the president told the massed reporters at the press conference, 'we are going to be extremely careful in our work.'

And he clearly meant it. On February 11, just three days after that press conference, Professor Wiesner, in his capacity as Chairman of the President's Science Advisory Committee, appointed his panel of inquiry to examine the worth of NASA's manned space programme just as he had recommended to Kennedy in January. The panel would be led by Donald Hornig, a Princeton chemist and explosives expert who had once worked on the atomic bomb. Hornig's brief basically boiled down to one question: 'Was Mercury ready to fly?' Before it ever did, the ten members and four specialists on his team would have to spend time finding out.

To large sections of the American public, let alone the American press, none of this procrastination made sense. Everybody had seen that picture of Ham grinning away with his post-flight apple. Things could not have been *that* bad. Yet here was their president appearing to back off from a key, perhaps the key, contest with the Russians. Then on February 12, one day after Wiesner's panel was appointed, the Soviets launched a second Venera probe from Tyuratam. This time TASS did not have to lie. The launch went perfectly and the probe set off on its astonishing journey across the solar system. Even the *New York Times* joined in the applause, gushing about Venera and its bold adventure in a front-page headline and seven further columns. And not just the *Times*: President Kennedy himself made a statement at his next press conference, fulsomely congratulating Premier Khrushchev on the USSR's latest 'impressive scientific achievement'.

On the same day as the *New York Times* ran its story, the respected American technical journal *Aviation Week* also ran a double-page article entitled, 'Mercury Chimp Test Might Be Repeated'. While Venera was already on its way to another planet and Russian citizens, according to the *Times*' Moscow correspondent, were celebrating in the streets, *Aviation Week* meticulously spelled out NASA's apparently stalled ambitions: instead of sending the first astronaut to space there was every possibility it might next be sending a second chimponaut. No doubt the Soviets were busy taking note. All they had to do was buy *Aviation Week*. Two weeks had now passed since Ham had ridden his Mercury capsule into the heavens, and still NASA had not made any official announcement. Just what was going *on*?

BEHIND THE SCENES what was going on was this: a raging, bitter and prolonged debate that was tearing NASA's manned space flight programme in two. At stake was the prestige not just of the United States' space programme but of the United States itself. On one side of the argument were the rocket builders at the Marshall Space Flight Center in Huntsville, effectively the Wernher von Braun team. As both Marshall's director and a hugely admired national celebrity, von Braun's authority in the rocket business was overwhelming. Since Ham's flight he had continued to have serious misgivings about the wisdom of putting an astronaut on top of the next Redstone and blasting him into space. Too many things had gone wrong, not just with Ham's flight but with the 'Four-Inch Flight' before it; too many things to justify a manned shot before all the problems could be ironed out. Most of his team agreed with him, many of them fiercely loyal comrades von Braun had personally led across the shattered landscape of Germany in 1945 to surrender to the Americans. Among them was his Cape launch director Kurt Debus. 'At least one unmanned shot,' Debus urged in his journal on February 6, 'must be obtained with flawless performance of the Mercury-Redstone mission booster flight.'

That was one side of the argument. But others in the space programme were appalled. Their centre of opposition was Bob Gilruth and his Space Task Group at Langley, Virginia, the people primarily responsible for Project Mercury and the seven astronauts. Gilruth and his spacecraft engineers were fully abreast of the issues with the Mercury capsule in Ham's flight – the sudden loss of cabin pressure with the failed snorkel valve, the problems with the ripped landing bag, the subsequent leakage of seawater through the hull – but they were convinced these were all easily and quickly fixable problems; certainly not ones that merited yet another delaying test flight. As for the Redstone rocket, they argued, surely its problems could also be quickly fixed? Ham had not *died* after all. He had survived, albeit with a bump on his nose. And time was moving on. Everybody could see that the Russians were stepping up the pace.

Even the president had just been telling Khrushchev how well they were doing.

On February 13, the day *Aviation Week* published its piece about another possible chimpanzee flight, the two sides finally sat down at a meeting in Huntsville to thrash out the problem face to face; and to decide, as the record put it, 'man or no man'. Attending were von Braun and senior members of his team, Gilruth and members of his Space Task Group, and leading figures from the McDonnell Corporation in St Louis, the company responsible for building the capsule. This was a round-the-table conference, with all the principal decision-makers participating but with the battle-lines on either side unmistakably drawn.

Joachim Kuettner, the man whose microphone cable Chris Kraft had once yanked out, set the tone for the von Braun side by presenting a pre-prepared list of ten weak spots on the Redstone, prioritising those that needed urgent attention before he believed it was safe for a human to fly. Among them were the stuck thrust controller that had allowed the rocket's fuel to burn up too soon, the issues with the Redstone's vibrations and its 'bending', and of course the critical complication with the abort timing that had led to Ham's capsule aborting half a second before the system disarmed, flinging him far too high and too fast in his ballistic leap into space.

Then von Braun stepped in. He reminded the Space Task Group team of their own longstanding rule for mission reliability: *all* the parties involved had to agree they were ready to launch before that launch could happen. A very sharp exchange of words followed between the two sides. Gilruth, the quiet, balding Minnesotan, could be as forceful as any ex-SS German rocket engineer when he needed to be. Referring explicitly to Gilruth and his Space Task Group, von Braun afterwards told Don Ostrander, an air force general and director of the Office of Launch Vehicles at NASA, that 'their feelings are quite bitter and feel that we are letting them down and … chickening out, etc.' That was a major understatement. In reality, Gilruth and his team were more than 'quite bitter' – they were *furious* that von Braun's gang were 'chickening out', and their fury flooded the ranks beneath them. 'Some of those people were just awful,' said Gilruth in an interview twenty-five years later in 1986. And even forty years later mission control's Kraft was still fuming with von Braun who, as he wrote in his autobiography:

> brushed aside our comments, and over the strenuous objections of Bob Gilruth … ordered a series of engineering changes and demanded an additional unmanned launch … Now we had a timid German fouling our plans from the inside.

Despite its intentions nothing was resolved at that Huntsville meeting, but it deepened the rift that already existed between the two sides. 'We were never a team,' remembers Terry Greenfield, one of Debus's launch engineers in the Cape blockhouse. The bitterness was sometimes accentuated by memories of who was on which side in the war, with Gilruth – echoing one of Kraft's phrases – bluntly asserting that von Braun 'doesn't care what flag he fights for'. That may well have been true but it was also not the point. The point was that if anything *did* go wrong with the first manned mission, it could very well go wrong on live television in front of an audience of tens of millions. As one eminent scientist, George Kistiakowsky, put it at the time, it could therefore end up as the most expensive public funeral in history.

FOR THE ASTRONAUTS, and for Alan Shepard especially, all this uncertainty was becoming unnerving. Shepard was keen to get going. Along with his back-up John Glenn and the designated astronaut for the second manned flight, Gus Grissom, his training had shifted into the highest gear. The three men now embarked on an intensive course of sessions on the two Mercury simulators, one at Cape Canaveral and the other at Langley, operating them between fifty-five and sixty hours a week. Back and forth the astronauts shuttled between the two locations and other training centres, flying themselves in sleek supersonic F-106 Delta Dart jets helpfully placed at their disposal by NASA to avoid being delayed – or perhaps mobbed – on ordinary commercial airliners.

The simulations were gruelling. Everything was thrown into them. Their purpose was in part to teach the astronauts to handle just about every emergency the supervisors could dream up, however far-fetched, nightmarish or insane. There were 275 separate malfunctions the system could spring on them, so many that it sometimes felt that a normal flight was one that consisted of non-stop emergencies. Often Shepard and the other astronauts, each playing different roles, were looped electronically into Kraft's team at Mercury Control, and using those room-sized IBM computers at the Goddard Space Flight Center in Maryland, they would run full-scale simulations of entire missions from start to finish to find out if every bit, part and cog in the machine was working just as it should and if everybody was on top of their jobs; and if not, how to put it *right*. It was a sophisticated operation, at the very edge of technological know-how for the time; and it was also exhausting, especially with the long hours of debriefing that came afterwards.

There were compensations for all this hard work away from home, and not just in all the money provided by the highly lucrative *Life* contract. For Glenn in particular one of those compensations was exercise. He *loved* jogging, which was just as well because at nearly forty he was the oldest of the Mercury astronauts and he also had a weight problem. From a paunchy 195 pounds he had jogged and dieted himself down to 165 pounds which was as close to the limit that the little Mercury capsule could bear; and like a jockey it was essential for him to keep off the pounds if he wanted to stay in the race at all, even as Shepard's back-up. At the Cape he was often awake early pounding the surf, eating up mile after mile of beach as the sun burned over the Atlantic. In his quest to be both physically and morally fit – the faithful husband, the regular churchgoer – Glenn was doing no more or less than Glenn always did, which is to say doing it 100 per cent.

None of the Mercury astronauts were formally required to do any exercise at all. 'They are all big boys,' said their physician, Dr William

Douglas. 'They will take care of themselves.' But Shepard had also taken up jogging in advance of his flight, despite not knowing when, exactly, that would be. And while he was training there were other, perhaps less salubrious, compensations too, many of them to be enjoyed just a few miles south of the Cape on the hot sticky strip of land called Cocoa Beach. Here was a boomtown that in ten years had sprung up out of nothing next to the missile site, a heady mix of glamour, tackiness, commercial hard-headedness and neon-lit fantasy where there was plenty of fun to be had. There was, for instance, the fun of racing cars, especially the gleaming brand-new Corvettes the celebrity astronauts had been gifted by a local Chevrolet dealer to 'executive drive' – an ingenious publicity stunt for the company. These races were as cut-throat competitive as everything else the astronauts did, with Shepard often leading the pack over Grissom and Cooper – he was, remembers his younger daughter, Julie, 'a terrifying driver'. Sometimes the gang raced their Corvettes on the actual beach itself, even hooking up a tow-rope to the back bumper so that one of them could water-ski in the surf behind. And sometimes there were accidents. But Glenn chose not to take part in any of these capers. As someone who drove a small, slow and very sensible German-built Prinz that did fifty miles to the gallon – a distinct source of pride – this was hardly surprising. 'Definition of a sports car', he once chalked on the astronauts' classroom blackboard. 'A hedge against the male menopause.'

Nor was it surprising that Glenn played the moral card when it came to another kind of compensation, the 'Cape cookies' who miraculously seemed to materialise around the Cocoa Beach Holiday Inn pool or bar whenever the superstar astronauts showed up in town: among them the air hostesses on layover, the secretaries, cocktail waitresses or female contenders for the yearly Miss Orbit award at the Mousetrap Steakhouse. The reporter Jay Barbree quickly won the friendship of the astronauts, or at least the hardcore speedster element, when he went drag-racing with them, but he won their *trust* when he bumped into Gus Grissom with a woman who was not his wife at the Mousetrap and never breathed a word. There were rumours that Shepard especially, with his impish blue eyes, played away; but they remained just that, rumours. Some of the other astronaut wives, perhaps not so secure in their own marriages, even took it upon themselves to break the news to Shepard's graceful, dignified wife, Louise, whom they nicknamed, with a mixture of irony and awe, *Saint* Louise. But she deflected their concerns. Her husband would never be unfaithful, she told them coolly, 'because I'm the one he really loves'. Bill Dana, a popular comedian whom Shepard idolised, was equally scornful of the rumours. If Shepard had really fooled around as much as that, Dana said, 'his dick would have fallen off'.

Still, the rumours could cling, even if none of them ever appeared in the many airbrushed articles in *Life* magazine. But Shepard and Glenn had clashed more than once over the 'proper' way for America's chosen heroes to behave, or to be seen behaving, not least because of their high public profile. After one particular incident involving a call girl and an unnamed astronaut, Glenn had angrily warned them all to keep their 'pants zipped' and not jeopardise the programme and thereby cede moral leadership to the 'godless communists'; but Shepard instantly shot him down with one of his icy stares, the kind that once seen was never forgotten according to his niece Alice: 'He'd give you that famous stare and you knew you were in trouble.' To Shepard Glenn was meddling in matters he had no right to meddle in. It was bad enough that he drove around in that ridiculous Prinz but this pants-zipping lecture was worse than any of it. Meanwhile Glenn himself could never quite escape the belief that in this battle against those godless communists he was surely the better choice for America's First Astronaut, a belief widely shared by the press and public too. In fact Glenn believed it so much he had even tried to get the original choice overturned, appealing in writing to the chief of the Space Task Group Bob Gilruth himself. But Gilruth did not reply to his letter. So there he was, stuck as Shepard's understudy. And unless Shepard broke a leg or got the flu or crashed his Corvette, nothing was going to change that.

At least while the ranking remained secret it was bearable. Then on February 22, the astronauts gave a press conference at the Cape where NASA's press officer John 'Shorty' Powers introduced Glenn, Grissom and Shepard as the first three who would get to fly in space – only without saying in what order they would do it. It was a galling experience for Glenn, especially when one reporter asked him directly if 'he would like to go in the next Redstone', to which Glenn could manage only a strained, 'Absolutely', and then go on to mutter something about the whole operation being 'a team effort'. When the same question was asked of Shepard, he pulled out every stop: 'I think the answer is Yes, an overwhelming Yes, a resounding Yes.' And those capitals are right there in the official NASA transcript.

But for all Shepard's resounding yesses the question of *when* he might get on that Redstone remained outstanding. More than a week had passed since the acrimonious meeting at Huntsville, more than three weeks since Ham's flight, and still nothing definitive had happened. Just the day before, on February 21, Gilruth had called his own press conference to announce the successful result of an unmanned test flight of the Atlas rocket, the launch vehicle that in a year or so was intended to carry an astronaut into earth orbit. But when Gilruth was asked about the timing of the upcoming Mercury-Redstone manned flight, he swatted the question like a fly: 'I don't think it will be fruitful to talk about this,' he said, 'because, frankly, I don't know.' Then he made a couple of opaque remarks about various 'tests' currently being conducted, before deftly changing the subject. It was truly mystifying. While the Russians were busy lobbing enormous satellites and planetary probes into space here was the chief of NASA's Space Task Group certainly saying, and apparently knowing, nothing about his own manned programme. Meanwhile the Russians were keeping quiet. And if past experience was anything to go by, that usually meant they were up to something.

TWELVE

AT OUR HEELS

ON THE DAY the second, successful, Venera probe began its three-month journey to the planet Venus, Lieutenant General Nikolai Kamanin wrote in his secret diary about the man behind this and so many other space successes. Sergei Korolev's energy, knowledge, engineering and organisational talents were 'undoubted', Kamanin declared in his February 12 entry, even if he could also be rude and despotic. But 'he has not yet spoken his last word. Among the pioneers of space his name will always be one of the first.' Leaving aside the point that Korolev's name was then quite unknown to the outside world, Kamanin's assessment of Korolev's place in space history would undoubtedly prove valid, as would his conviction that Korolev's last word was yet to come. But it was arguably a prophetic statement to make in the middle of February 1961, because in his determination finally to achieve his dream of human space flight Korolev was about to gamble his position, his reputation and his genius on one, hugely risky, roll of the dice.

By this time the chief designer was exhausted. The two Venera probes had consumed much of his energies, as well as everybody else's – 'I really want to sleep, this is the second night without sleep in a row,' complained Kamanin after the failed first launch on February 4, and he spoke for them all – but as always Korolev's net went far wider than the Venera programme. His hand was everywhere: in probes to the moon, two of which had launched the previous year, in probes to Mars, in the spy satellite version of the Vostok he had promised to the generals, in the development of his R-7 rocket both as a weapon of war and as a space vehicle, in other ballistic nuclear missiles, in the Vostok dog programme – and of course in this business of putting the first human being in space.

The list of responsibilities and accomplishments was long, far longer than it was for his rival Wernher von Braun. But with that list came not only exhaustion and corrosive frustrations, but also fear. In the five years since Korolev had first stunned Khrushchev with his R-7 missile, the premier's love affair with his chief designer, as his son Sergei had put it, had noticeably cooled. Even if Korolev's brilliant talents had given Khrushchev a host of impressive victories over the west, even if all those Sputniks and orbital dog flights had allowed Khrushchev to gloat and glory, the fact remained that Korolev's passion for space, in Khrushchev's mind, was also leading him astray. His primary job was to develop and build missiles. But instead, as Khrushchev told Sergei, Korolev 'was apparently more interested in setting records in space than working on defence'.

And knives were now out for him. If Khrushchev's patience was wearing thin, the patience of others was wearing even thinner, some of them formidable figures like Leonid Brezhnev, Chairman of the Presidium of the Supreme Soviet, who actively questioned the enormous cost of all these pointless space ambitions. A secret ministry of defence report, delivered to Brezhnev on February 15 just three days after the second Venera's successful launch, decried the spending of 'distracting' and 'unbearable' funds on space, funds that should have been used to build new missiles to defend the motherland against 'the enemy'. Meanwhile Khrushchev was increasingly showing favour to Korolev's former denouncer and arch-rival Valentin Glushko who was himself developing newer missiles with much quicker turnarounds than the R-7, using alternative 'storable' fuels. Khrushchev had repeatedly asked Korolev to co-operate with Glushko on these next-generation missiles, but Korolev had repeatedly equivocated. He hated these storable fuels, which he regarded as extremely hazardous, the 'devil's venom'. More than that he hated Glushko. In one heated exchange with Khrushchev on the subject the Soviet premier even had to remind Korolev sharply who was in charge. Proud, arrogant, obstinate and often infuriatingly self-righteous, Korolev was in real danger of overstepping himself and losing the support of the most powerful man in the USSR, a man he had previously cultivated with such deftness and skill. More than ever he needed a really big win.

The Americans now gave him that chance.

NASA's caution was Korolev's opportunity. He had only to ask his own wife Nina, once his English-language translator, to translate the American papers or *Aviation Week* for him. It was all there: the delay over the weeks since Ham's flight, the absence of any official confirmation that one of America's astronauts would fly next. Here was Korolev's chance to beat the Americans even more comprehensively than when his beeping Sputnik had first sent waves of panic across their country and to win back the premier's favour. One of his engineers, Oleg Ivanovsky – the man who had first encountered Korolev driving his car at manic speed – had personal experience of this consuming obsession of his chief:

> He would tell us that 'the Americans are at our heels, and the Americans are serious people'. He wouldn't use the word '*Amerikantsi*' but '*Amerikan-ye*' as if these weren't just American residents but the entire American culture we were competing with.

That determination to win this race against the Americans was far from new. But what was new were the risks Korolev would now take to win it just when NASA was hesitating. In one master stroke he would snub the *Amerikan-ye* and achieve his dream while at the same time feeding Khrushchev's vanity and regaining the premier's confidence; and perhaps even his love. The Americans had given him a gap; one that Korolev would try to slip through.

And so with each week in February that NASA continued to announce nothing, the Soviet manned space programme rapidly, and secretly, accelerated its pace. In his diary Kamanin recorded the sudden change. By February 20, Korolev was urging the construction of another *ten to fifteen* Vostoks as part of a sustained manned flight series far more ambitious than Mercury's mere seven flights. By February 22, on the same day Shepard, Grissom and Glenn were introduced at their press conference as America's first three astronauts, a meeting of Korolev's designers was resolving to launch the first human mannequin flight *without* even ground-testing many of the Vostok's on-board systems. By February 24, just two days later, new tentative flight schedules were set down: the first mannequin flight on March 2–3, the second on March 20–25. And by late March or early April – the first human flight.

By this last week of February, in other words, Korolev and the members of his state commission responsible for the Vostok programme had effectively committed themselves to three major orbital flights in the totally untried, unflown manned version of the Vostok – the Vostok 3A – and all within the space of a month. And they had just committed to do this despite the abysmal record of four Vostok 1 dog flights in 1960, every one of which had failed in one way or another apart from the Belka and Strelka mission, which had itself run into difficulties. Everything would now be thrown into this gamble. Perhaps Kamanin best caught the mood when he confided in his diary on February 24: 'My personal opinion is this: without risk, space is not to be mastered, but to be afraid of risk and possible sacrifices is to slow down space exploration.' While President Kennedy was worrying about disproportionate risks, Korolev was embracing those risks to get ahead and be first. Even if the greatest of those risks was a man's life.

PERHAPS THE MOST exciting news for Yuri Gagarin that February was that Valentina would very soon be giving birth to their second child, a brother or sister for little Elena. The baby was expected at the beginning of March. Gagarin had already written to his mother Anna on February 13 asking her to come and visit them in their Chkalovsky apartment outside Moscow. 'Valya is feeling well,' he wrote. 'Not long to go now.' He was busier than ever and apologised that his letter was brief. 'I am still at work from morning till night.' But what, precisely, he was busy at he did not say. He never did. And of course his parents never asked.

Nor did his own wife, at least not for the first six months of his training. The wives of all the other cosmonauts were rather less in the dark, not because they were officially supposed to know anything but because they had either guessed or their husbands had broken the rules and told them. Tamara Titova had known ever since Gherman had written all those illicit letters while being tested in hospital. Grigory Nelyubov's wife, Zinaida, knew because she was a typist at the Cosmonaut Training Centre. And Zinaida's glamorous best friend, Marina Popovich, knew because she was a top-notch pilot herself and was also, in her daughter Natalya's words, 'hot-blooded' and sometimes downright scary; definitely not the sort of person who was going to put up with any kind of nonsense from her husband, Pavel.

But Valentina Gagarina did not know, not until it was decided by the authorities in the summer of 1960 that the time had finally come to tell the wives just what exactly their husbands had signed up to. Colonel Yevgeny Karpov, the chief of the Cosmonaut Training Centre, was called in to break the news every one of them except Valentina knew already. 'He invited all the women, all the wives of the future cosmonauts to a kind of meeting,' says Gagarin's daughter, Elena, 'and he told them what kind of work their husbands were doing there. And my mother was the only one who had no idea.' Tamara Titova remembers that meeting well. After revealing the truth, Karpov went on to give them a lecture about their duties as Soviet wives. 'You must subordinate your routine to their routines,' he urged. 'You must create the conditions they need to rest, and not let them worry about everyday domestic life. The fate of the nation is in your hands.'

It was quite a demand to have the fate of the nation in your hands, not least because none of these women had any clue what flying in space really meant – not that their husbands knew much more. But they all knew about fear. They were military wives who had long worried about becoming military widows in a visceral instant of tangled debris and smashed bodies, left to raise their children alone on a meagre Soviet widow's pension. It was, says Tamara Titova of her husband Gherman, 'frightening'

> to see him off and you do not know if you will ever see him again or not. Although I drove those thoughts away from me. I was young. Perhaps I didn't fully understand what risks my husband was taking each day. On summer days I would sit by the open window, listening to them take off and fly … But sometimes you heard this silence … and then the sirens went off and God forbid, it had happened … Some kind of accident when landing or a plane had spun in and swerved off the runway. And then everything just stopped.

Such fears however were rarely openly discussed in the tight little community the cosmonauts and their wives now inhabited. That world was 'all Soviet', as Tamara Titova recalls, 'with no privileges, nothing', where the cosmonauts lived cheek by jowl in apartments that by most western standards would be regarded as tiny and drab – with no TVs, no refrigerators, not even telephones – but by Soviet standards were relatively comfortable. Compared to the homes many of the cosmonauts had grown up in, they were almost luxurious, especially if, like Gagarin, that home had once been a dugout with a dirt floor.

Money was also tight. Existing in their controlled, secret world there were no equivalent *Life* deals here; the families had to survive on a junior officer's pay of between 50 and 60 roubles a month: $12–$15 in 1961 money (or perhaps $100–$130 in 2021); approximately one twentieth of the Mercury astronauts' salaries along with their *Life* monies. At one point Tamara Titova and two other wives found themselves polishing other people's floors just to make ends meet. And while Alan Shepard and some of his fellow astronauts were racing their latest-model Corvettes up and down Cocoa Beach, Gagarin and his fellow cosmonauts were taking the bus or train in Chkalovsky. None of them could afford a car.

By the time the Gagarins were expecting their second child at the end of February 1961, most of the twenty cosmonauts had been training for almost a year. Much of the training would have been immediately familiar to their Mercury rivals. Both employed 'Vomit Comets', the rollercoaster aircraft rides designed to simulate the experiences of weightlessness for thirty-odd seconds – over five days in May 1960 Gagarin did this seventy-five times and loved every minute, describing it with boyish glee as 'fun' in his official report. Both included centrifuge rides and simulator sessions, although the Vostok simulator was far more primitive than the Mercury version and there was only one of them; not that it mattered much since the cosmonauts were never expected to 'fly' their spacecraft anyway. Both training regimes also included an arduous timetable of academic studies, with hundreds of classes in mathematics, biomedicine, flight dynamics, life-support systems, astrophysics and, exclusively in the cosmonauts' case, Marxism-Leninism.

While Glenn and Shepard chose to get fit, the cosmonauts had no choice. A substantial, perhaps *the* substantial, element of their training was devoted to transforming them into even more perfect physical specimens than they were when they were first selected. Every morning began with a light group gymnastics session, typically followed over the course of the day by other, less light, forms of physical exercise, until the working weeks sometimes felt like a never-ending schedule of running, jumping, squatting, crunching, trampolining, acrobatics, diving, and team games like ice hockey and volleyball. Gagarin stood out as a star performer, just as he had at college in Saratov. Titov, at least in the early days, rebelled. 'He was not afraid to speak his mind,' says his wife. 'He was not afraid of anything.' But even Titov toed the line over time, and by now could do handstands balancing on the palms of his hands. Colonel Karpov noted with satisfaction how far his pupils had come: 'The men were doing some very complicated exercises on these trampolines, not up to circus standards, perhaps, but their performance was neat, bold and certain.' It is hard to imagine Bob Gilruth, as chief of NASA's Space Task Group, noting approvingly that the Mercury astronauts were almost up to circus standards in their training.

The cosmonauts also learned to parachute. They needed to since, unlike the Mercury capsule, the Vostok was too heavy to land safely beneath its own parachute with a human inside. The only way for them to get home was therefore to eject from the Vostok in the final minutes of descent and parachute independently to the ground. The training took place in batches at Engels, a town some eight hundred kilometres south of Moscow and, coincidentally, just across the Volga river from Saratov. A KGB officer accompanied the men. They were introduced as a sports team. 'No one was supposed to know who we were, no one,' remembers Boris Volynov. The lie was strictly enforced: 'What would happen to anyone who disclosed a state secret? Everyone knew what would happen. And it wouldn't be a pat on the head.' The training was tough, even brutal. With up to sixty-six jumps in five weeks, sometimes at night and over water, things occasionally went wrong. One cosmonaut smashed his leg in a bad landing. Another ended up in a deep spin and was lucky to survive. Titov in particular scored high marks when his main parachute refused to open at all on one jump and he was forced at the last moment to deploy his reserve. It was noted that he did not lose his nerve.

Gagarin went there in April 1960. It was his first prolonged absence from home since he had joined the cosmonaut corps the previous month. His second absence was in late July the same year. This time he was away for ten days, but he did not spend it out of doors. He spent it instead inside a sixteen-inch-thick, steel-walled, soundproofed room, totally cut off from all human contact to see if he could cope without going mad. Along with the parachuting, the 'isolation chamber', more graphically known as the chamber of silence, was another unique dimension of the Soviet training programme. The chamber was in the Institute of Aviation and Space Medicine in Moscow where the dogs were also trained for their space flights. 'Sometimes,' recalls Boris Volynov, 'people didn't get through it':

> On about the third or fourth day, sometimes on the second day, they started to pound the walls with their fists and legs, stopping the experiment and begging to be let out ... Maybe they scared us on purpose.

They did. Since the dog flights had told the doctors nothing about the impact on the *human* mind of flying alone in space, cut off from the world – and, should anything go wrong, cut off absolutely, possibly without radio contact, for days or weeks – the experience needed to be simulated first. Isolation was only part of the exercise. The men were also required to perform complex number puzzles at intervals while observers watched through tiny portholes or on closed-circuit TV screens. Lights might flash or alarms shriek at any time without warning, and again each cosmonaut's reaction was carefully recorded. Even day and night were deliberately confused. Volynov spent his twenty-sixth birthday in the chamber: 'I wanted so much for someone to be with me ... to hear some good kind words ... A human word, only a single word, what would I not have given for a single word?'

None of the cosmonauts were told in advance how long they would be in there. Gagarin's ten days was the minimum; others spent longer. But Gagarin handled his loneliness in part by imagining he was flying around the planet and looking down on the cities and oceans below: 'Although I had never been abroad,' he wrote later, 'in my mind I flew over Peking, London, Rome and Paris and over my native Gzhatsk.' Apart from the touching juxtaposition of his provincial home town with these great capital cities, there is also a poignancy in Gagarin's imaginings of places beyond the USSR's borders, places he would not expect to see as an ordinary Soviet citizen – as if the sealed chamber were itself a metaphor for the world he and 216 million of his fellow citizens inhabited. At any rate, he survived the ordeal; as did Titov who spent most of his fifteen days inside the chamber reciting Pushkin, and Popovich who spent his ten days singing Ukrainian folk songs. Should any of them ever find themselves marooned in orbit, presumably they could be relied on to cope.

All six of the Vanguard front-runners had completed their parachute training and isolation chamber sessions by the time they took their examinations in January 1961. But there was still one last element of their training left. None of them had yet visited the secret cosmodrome in Kazakhstan and none of them had witnessed a rocket launch. They would get their chance soon enough. Sometime around the middle of March the six men were expecting to depart for Tyuratam. There they would witness the second of the two human mannequin space flights Korolev had timetabled in rapid succession before the actual human flight. After a year of intensive training they would finally get to see the mighty R-7 rocket that one of them was hoping to ride into space within the next few weeks.

For Gagarin the upcoming trip was both a blessing and a burden. It meant he would be leaving within a few days of Valentina giving birth. Of all his absences from home so far this would surely be the hardest. But everybody was now at the mercy of Korolev's relentless new schedule as he galloped to get ahead of the stumbling Americans.

IT TOOK NASA a month to make its decision. It was made as February slipped into March, and exactly as von Braun had wanted. The next Mercury-Redstone launch would not be carrying Alan Shepard or even another chimpanzee. It would be an unmanned test flight. A date was provisionally set for the last week of March. If everything went to plan, and if all the requisite fixes to the rocket and capsule actually worked, Shepard might get to fly in the last week of

April: three weeks, that is, after Korolev's provisional, and secret, date for the first Soviet manned flight.

The decision had not been reached easily. Gilruth had taken his case all the way up to James Webb, NASA's chief. But Webb, a tough ex-marine aviator, lawyer and oil executive in his mid-fifties, had only just been sworn into the job on February 15, two days after the bitter Huntsville meeting. A shrewd operator, he was very shortly to face a taxing round of hearings on NASA's budget before an exacting House Committee, and he was not about to jeopardise his case by making enemies. After more than two years of glitches and sometimes humiliating failures, NASA had become a weakened organisation. Its very existence might be imperilled by another public failure. Nor, no doubt, had Webb forgotten the president's own recent concerns about racing ahead too fast.

So Webb took von Braun's side and turned Gilruth down. Aiding his decision were no fewer than three separate detailed probability studies commissioned by von Braun in February to calculate the statistical likelihood of rocket failure, a process that was either rational or reeked darkly of obsessive Teutonic thoroughness depending on which side of the argument you happened to be on. The result was a 98 per cent probability that Shepard would survive the mission. Better than reasonable odds, perhaps; but in the end even a 2 per cent chance of Shepard getting killed was still 2 per cent too much. Yet such was the anxiety about the American public's likely reaction to yet more dither and delay that the date for this next test flight was for the moment kept secret, as was the fact that an astronaut would not be going on it.

Meanwhile, behind the scenes, Alan Shepard himself moved mountains to try to get his flight back, telling anyone who would listen: 'We're ready to go. Let's go!' He even intercepted von Braun himself. 'For God's sake,' Shepard told him, 'let's fly now!' But von Braun refused to budge. He and his Huntsville team had won their case for caution. They had chosen to reduce the risk of failure and slow down just when the Soviets had chosen to accept that risk and speed up. But that did not mean that the Americans were out of the race. Far from it in fact – because in order to win, everything the Soviets now did had to go right, beginning with the first dress rehearsal flight of the curious human mannequin they called Ivan.

THIRTEEN

IVAN IVANOVICH

March 9, 1961
Tyuratam Cosmodrome
Soviet Republic of Kazakhstan

VLADIMIR SUVOROV WOULD never forget the first time he filmed Ivan Ivanovich. In a spotlessly clean room within the rocket complex's huge assembly building in Site 2, three men in white laboratory coats brought in a large, sealed box. As Suvorov watched through the eyepiece of his beloved 35mm Konvas movie camera, the men opened the box and carefully lifted Ivan out. Suvorov followed them across the room, still filming as they eased the cosmonaut into his spacecraft couch and strapped him down. Ivan lay there on his back, compliant, unmoving, lifeless, the strangest of sights, as Suvorov wrote afterwards:

> He was dressed extraordinarily. Bright-orange spacesuit, white helmet, thick gloves and high-laced boots. All this made him look like a cosmic alien. His head, the 'skin' of his body, arms and legs were made from synthetic material ... mimicking that of the human skin. His neck, arms and legs had gimbal joints so they could be moved. The spaceship flights with a human

> being on board were obviously not far away and 'Ivan Ivanovich' was the last stage before them. Dressed in a complete cosmonaut spacesuit he looked somewhat unpleasant with his fixed false eyes and a mask for a face.

Suvorov had been filming the space programme for two years and some of the things he had witnessed and recorded over that time often made him feel like he was 'in some fantastic movie. But it wasn't a fantastic movie. It was *reality*.' Ivan Ivanovich was another such reality. Here was the life-sized mannequin that would be launched into orbit aboard the first 'man-rated' Vostok 3A capsule on top of an R-7 rocket.

That at least was the plan; except that with so many systems on this spacecraft not yet fully tested – or in some cases not tested at all – anything and everything could still go wrong. Suvorov already had experience of what going wrong looked like, having filmed an R-7 rocket a few months previously that blew up less than two hundred metres from his camera crew. He and his team had been lucky to get away with their lives that time, especially as they kept on shooting while chunks of burning metal fell all around them. The two dogs inside the rocket on that occasion had been less lucky.

But the job was addictive. It had begun in 1959 when Suvorov was asked to join a special unit that had just been formed to film the USSR's burgeoning space programme. At the time he was still in his early thirties working as a cinematographer at Moscow's prestigious Studio for Science Films. Several directors, sound recordists and cameramen were also asked to join. 'It was a top-secret assignment,' wrote Suvorov in his memoirs. 'For several days that followed we waited for security clearances. At last we got them. Did we know then what an unforgettable and fascinating life lay ahead?' Presumably they did not; and even if they did, they were not allowed to tell anybody about it. A handsome, well-built figure with thick black hair and a shoulder strong enough to bear the weight of the superb, triple-turreted Konvas, Suvorov quickly fell in love with his new posting. The filming, he was told, had two purposes: to help the spacecraft engineers by recording their mistakes as well as successes and, when appropriate, to advertise to the Soviet public and the world the glorious achievements of Soviet technology, most of it in glorious colour. The mistakes would not be advertised.

'What a time!' wrote Suvorov. 'We were making history.' He felt this so strongly that one day he even dared ask Korolev – a man he describes as a 'mysterious personality' known to them all only as 'S.P.' – if he might be permitted to make daily notes about their work. Korolev agreed, so long as Suvorov refrained from writing about *him*. And so, like the cosmonaut chief Nikolai Kamanin, the cinematographer also ended up keeping a diary, except that his was not illicit. He wrote his daily impressions in a school exercise book. Each evening the exercise book was locked away in a specially guarded place; and whenever Suvorov went home on leave, it was duly deposited with officers of the KGB.

Parts of this diary have survived and, like Kamanin's again, they offer a revelatory glimpse into the exotic, secret world that Suvorov was filming. The pace of work had always been relentless, so much so that as a Christmas present in 1960 his wife had given him a calendar for that year with 250 days marked off in black crosses – for all the days he had been away. But by the first week of March 1961, as the two final dress rehearsal mannequin flights were being prepared, that pace had stepped up dramatically for everybody. Suvorov and his crew were daily, and even nightly, shooting off thousands of feet of film covering such critical tests – ejection-seat tests, water landing tests, capsule drop tests – as there was still barely time for the engineers to perform. Sometimes they would discover Korolev watching them. 'Keep shooting!' he once yelled, when a trial capsule's parachute failed to open and the Vostok came hurtling down, smashing into the ground. Sometimes Korolev would even come into the projection rooms to watch and comment on their rushes, like some Hollywood movie mogul.

In that spring of 1961, Korolev seemed to be everywhere at once, 'his slightly jutting chin betraying a strong will and uncompromising character', as Suvorov described him, and he once complained sadly to Suvorov's cameraman that he was no longer a scientist but an organiser. As always, his rule inspired both fear and something like love: 'By his mood, gesture, bent of his head, his subordinates could tell straight away what to expect: a reprimanding or a praising.' At times the praises *were* unexpected. When one of his engineers accidentally started a fire inside the spacecraft during a test but owned up to the mistake, the man was given an engraved watch as a reward. Korolev needed the truth, not time-servers.

FOR THIS FIRST dress rehearsal flight, Ivan Ivanovich would have company. Under his orange spacesuit, stuffed into cavities in his chest and thighs, was an assortment of animals that would accompany him on his ride into space: mice, guinea pigs, microbes and various biological specimens. Other animals and specimens would ride with him in a separate sealed unit inside the Vostok, among them another eighty mice – forty white and forty black – several more guinea pigs, reptiles, plant seeds, human blood samples, human cancer cells, bacteria and fermentation samples. The dog would also be in this separate unit, another mongrel stray that had undergone the same restraint conditioning as all the other mongrel strays. Her name was Chernushka, or Blackie, and she had been flown to the cosmodrome from the Institute of Aviation and Space Medicine in Moscow with her trainer Dr Adilya Kotovskaya, the same physician who supervised the human centrifuge rides; a sort of multi-tasking that was common practice at the institute, where the partition between human and canine subjects was easily blurred.

Dr Kotovskaya had brought along a second dog too, Udacha, or Lucky, a lively black and white mongrel. Udacha would be going on the second mannequin flight later in March. Arriving with both dogs at the cosmodrome's airport on a characteristically freezing February day, Kotovskaya was astonished to find Korolev himself there to greet her:

> I was dressed in some flimsy beret, and I even remember my coat – it was very thin. He said, please, wait here for me. And he went to his cottage and he brought me his helmet, his own fur-lined helmet made of soft leather, his personal helmet. It was of great value to him. He said, there's no need to be freezing, put it on and wear it. That was our first meeting and it amazed me.

Like Suvorov and everyone else, Kotovskaya was instantly spooked by Ivan Ivanovich's creepily lifelike features when she first saw him, his rubberised flesh-coloured skin, his full mouth and lips, and 'the lashes on his eyes'. Even Korolev was not immune to the effect the dummy had on those who saw him – including on himself, as one of his engineers, Vladimir Yaropolov, remembered:

> We all knew that it was categorically forbidden to smoke on the testing site and that it would be a strange sight to see someone sitting there reading a novel. In defiance of these rules, Ivan Ivanovich could often be seen sitting in an armchair with a book on his knees and a cigarette between his fingers, elegantly placed on the armrest. Ivan Ivanovich sat so naturally in this position that at one point Sergei Pavlovich Korolev mistook him for a real test engineer and tore strips off the engineering chief for breaking formal procedure. Once he realised his mistake, he burst out laughing along with everyone else.

There was one final addition to Ivan's person before he was ready. A major objective of this flight was to test the reliability of radio communications as Ivan set off on his voyage around the planet. And so, along with all the mice, the guinea pigs and the microbes, a tape recorder was tucked underneath Ivan's now-bulging spacesuit inside another cavity approximately where his stomach was. Its one purpose was to broadcast traditional Russian folk songs from space sung by the world-famous Pyatnitsky Choir, as a Vostok engineer reveals:

> The main purpose was to ensure reception of voice transmissions from the spacecraft. We rejected a numerical countdown, fearing western radio stations would monitor the human voice and raise a clamour throughout the world alleging Russia had secretly put a man into orbit. A song also aroused objections because it would be said in the west that the 'Russian' cosmonaut had lost his head and started singing! It was then decided to tape the popular Pyatnitsky Russian Choir, and when the dummy ... suddenly sang like a choir, it was very funny.

Now fully stuffed, dressed and ready, Ivan and his substantial menagerie, a veritable 'Noah's Ark' as one senior medical officer put it, were transported to the waiting R-7 at Site 1, placed inside the Vostok capsule at its top and, at exactly 09.29 Moscow time on March 9, launched into the clear, cold skies of the Kazakh steppe. Within several minutes the Vostok and its myriad occupants had successfully separated from the rocket to begin their single orbit of the earth, while back on the ground Korolev and his team waited to hear Ivan sing.

A THOUSAND MILES SOUTH of the Bering Strait, halfway between Alaska and the far-eastern USSR, lies the tiny island of Shemya. Barely more than a speck of rock in the ocean, this lonely piece of American territory was – and still is – America's secret eyes and ears. A major electronic intelligence, or ELINT, station had been based here since the 1950s, to watch, monitor and track Soviet missile activity in the heart of the cold war.

And not just missile activity. After Sputnik, the ELINT operators had also been secretly monitoring Soviet space activities. But when the two dogs Belka and Strelka made their eighteen-orbit flight in August 1960, these same operators noticed something new. Embedded in the transmissions from their capsule was what looked like a television signal. The primitive eighty-three-megacycle transmission would require special decoding equipment to see the actual pictures, a process that took some time, but after the images were finally extracted by CIA technical analysts they could clearly see the two dogs inside the orbiting capsule. And when two more Soviet dogs flew into orbit on December 1 of that year, the CIA analysts were able to process and extract those images too. Like the earlier TV pictures, they were crude and blurred, but the face and forelegs of one canine passenger could be distinctly made out. Judging by its black and white markings in the very few stills declassified by the CIA thirty-four years later, the analysts were almost certainly looking at Pchelka who, along with her companion Mushka, would shortly be blown up by the Vostok's on-board bomb to prevent them returning to earth outside the USSR.

By the time of Ivan's first launch in March 1961, and with the Soviets appearing to be close to placing a human in space, the CIA had upgraded their monitoring equipment to allow their ELINT operators to 'demodulate' the TV signal and watch the images very nearly in real time. Two sets had been built by the National Security Agency and delivered to the monitoring stations at Shemya and in Hawaii. And it was at Shemya, on the evening of what was still March 8, 1961, thirteen hours behind Moscow time, that the ELINT operators intercepted a new TV signal on their oscilloscopes coming from a spacecraft in orbit somewhere west of Alaska. This time they could watch it within minutes of receiving it. There again was another dog, a black one this time. They were looking at Chernushka. But where the declassified CIA document on the subject is strangely silent, even decades after the event, is whether they could also see what looked like a rather substantial spaceman singing traditional Russian folk songs with a hundred voices. And with his mouth shut.

NIKOLAI KAMANIN HAD already left the cosmodrome in the early hours of that morning to fly the thirteen hundred kilometres north to Kuybyshev in Russia – today called Samara – a town nestling in a loop of the mighty Volga and in the approximate area where Ivan Ivanovich and Chernushka were expected to land. Along with running the cosmonauts' training regime, assessing their examination results, participating in the state commission managing the Vostok programme and keeping an illegal diary, Kamanin was now also co-ordinating the search and rescue teams to find and recover the capsule, its mannequin and its ark of animals; another instance of extreme Soviet multi-tasking. Vladimir Suvorov joined Kamanin with his camera crew. With him too were Dr Vladimir Yazdovsky, the fearsome director of the Institute of Aviation and Space Medicine, and Arvid Pallo, still fresh from his New Year adventures rescuing the dogs Zhulka and Alfa from their Siberian wilderness.

For Pallo, this latest adventure must have unravelled with a powerful sense of déjà vu. Signals from the Vostok's sphere appeared to indicate that it had landed successfully and reasonably close to the planned target, some 260 kilometres north-east of Kuybyshev. Ivan, who should have automatically ejected from the sphere in the last phase of descent, ought to be somewhere nearby. But freezing wintry conditions and raging snowstorms were once again going to make the business of getting to them both a serious trial.

With the snow falling thick and fast, it was impossible to fly all the way by helicopter. Kamanin and his rescue team tried to travel the last kilometres by horse, but the horses got stuck in heavy snowdrifts. They then tried walking. By 4 p.m., fourteen hours after they had left the cosmodrome and almost five hours after Ivan had landed, they found him lying on his back on a snowy plain beside his orange parachute. One of his legs was badly bent. He was staring up at the sky and once again spooking everybody out with those eyes. Suvorov filmed him.

> 'Ivan Ivanovich' is not injured but it feels creepy to look at him lying motionless in the snow with that fixed glance of those fake eyes and a deadly mask of a face. He looked precisely as if a real cosmonaut had been killed during the landing.

Ivan was loaded onto a sleigh and carted off to the nearby village of Tokmak while Kamanin trudged on across the plain to the Vostok sphere. The correct procedure at this point was to wait for the specialist bomb disposal officer to turn up in his helicopter and disarm the capsule's on-board bomb before removing Chernushka's container. But with the appalling weather there was no hope of any helicopter getting to them until at least the next day. Kamanin worried that Chernushka, her eighty mice, guinea pigs and all the other animals would freeze to death; and so, as the commander of the rescue operation, this tough, war-seasoned aviator and Hero of the Soviet Union broke the rules and ordered the animals to be brought out to safety. The team then waded back through the snow to Tokmak where, as Kamanin wrote that night:

> a large crowd of collective farmers and children gathered near the village council – everyone was impatient to see a dog that flew around the entire planet in an hour and a half at an altitude of more than two hundred kilometres. While I was on the phone with Moscow, Vladimir Ivanovich Yazdovsky managed to show the villagers Chernushka and give them the shortest, but a very convincing, lecture on space exploration.

What the villagers themselves made of the forbidding doctor's lecture we will never know. Later that evening the Soviet news agency TASS broke the news to the world, hailing the flight as a triumph. The next morning *Pravda* reported that it had achieved its 'main objective', which was, explicitly, 'to prepare for manned flight'. Kamanin took up the same theme in the diary: 'Yes, another brilliant victory was won,' he wrote. 'Everything is ready for a manned flight into space.' And so it seemed: as if this first of Ivan's two dress rehearsal flights had indeed gone right despite all the things that could have gone wrong. So far Korolev's gamble appeared to be paying off. There was still one more practice flight – and that flight was only two weeks away.

But when Boris Chertok, one of Korolev's most senior electronics experts, afterwards examined the post-flight data he noticed something alarming. The Vostok was actually made up of two sections, an 'instrument module' containing, among other systems, the retro-rocket – the Soviets called this a braking engine, a better description of its function – and the sphere itself or 'descent module' containing, essentially, the cosmonaut. Once the braking engine had slowed the spacecraft down sufficiently before re-entry the two sections were supposed to separate automatically, with the sphere descending alone until its cosmonaut finally ejected.

But in Ivan's case the two Vostok sections had not properly separated. Instead they had plunged into the atmosphere attached by a thick cable, like two boots tied by a shoelace, spinning and threatening at any moment to collide. 'Final separation,' wrote Chertok many years later, 'took place only after the cable … burned up in the atmosphere.' Had it not done so, Chernushka and all the other animals – or indeed a human cosmonaut – might never have made it back alive. But 'in the frenzy of preparation for the next launch', nobody took much notice of the failure. Nor did they do anything about it. They were all hurrying too much.

FOURTEEN

THE BIGGEST MISSILE SITE IN THE WORLD

ON THE SAME day as Ivan orbited the earth, Yuri Gagarin turned twenty-seven. Only two days earlier on March 7, slightly later than expected, Valentina gave birth to their second child, another daughter. They called her Galina – or Galya for short. Gagarin's parents were thrilled to learn the news. With his wife and now two children Gagarin appeared to be settling down as a responsible family man with a respectable career in the air force. On March 14, while Valentina was still recovering in hospital, Khrushchev made a speech in which he declared that 'the time is not far off when the first Soviet spaceship with a man on board will soar into space'. Of course Valentina knew that already. What she did not know was who would be inside that spaceship. Nor did her husband, yet. Meanwhile he visited his wife and tiny Galya for the week they remained in hospital and played 'crazy, noisy games' at home with Elena, sometimes chasing her around the tiny apartment.

He was not home for long. A week after his birthday, on March 16, he and his fellow five Vanguard cosmonauts left Moscow for the first leg of the long flight south to the cosmodrome at Tyuratam, stopping for the night in Kuybyshev halfway. They would remain at the cosmodrome for the next ten days while once again Ivan Ivanovich, accompanied this time by the other dog, Udacha, flew what everybody hoped would be the final dress rehearsal flight.

With them, as ever, was Nikolai Kamanin. Those ten days would also be an opportunity for Kamanin to study the cosmonauts closely, and three of them in particular. Gagarin, Titov and Nelyubov were still the front-runners to fly first, in that tentative order. Gagarin remained the favourite. But on the flight to Kazakhstan he appeared to have lost some of his natural exuberance. Kamanin kept an eye on him:

> Yuri Gagarin is the first candidate for the flight – but for some reason he looks paler and more silent than others. His not quite usual state could be explained by the fact that on March 7 he had a second daughter, and he had only yesterday brought his wife home from hospital. Perhaps the farewell with the family was not easy and this is weighing heavily on him.

The cosmonauts landed at the cosmodrome's airport on March 17. The moment they stepped off the plane Korolev was there to greet them and Vladimir Suvorov to film them. Suvorov was not impressed. 'They are all rather short,' he wrote in his approved diary, 'and do not look extraordinary. One of them will be the first to go to space soon.' Korolev gathered them together on the following day for a friendly chat, asking each of them – in Kamanin's words – just 'one to two technical questions' about the Vostok capsule. After answering these one to two technical questions 'satisfactorily', Korolev told them something of the chequered history of the Vostok dog flights so far. It is revealing – and a stark contrast with the Mercury astronauts – that the cosmonauts were learning such details just a month before one of them was expected to fly. After having told them, Korolev was delighted at the 'willingness' of the cosmonauts 'to fly today if need be'. Despite the catalogue of failures they had just been hearing about, their obedience and devotion to duty, if necessary to the point of self-sacrifice, remained unquestioned.

The six men were then conducted to a gigantic structure known as the *Montazhno-Ispytatel'nyj Korpus*, the Assembly and Test Facility, another Soviet mouthful thankfully abbreviated to MIK. Here they got their first look at the R-7 rocket and the Vostok capsule that, barring any last-minute mishaps, Ivan and the dog Udacha would be flying into space within a week. Suvorov's film footage of that encounter still exists: the cosmonauts in their uniforms walk into the MIK, where the silver rocket lies on its side, its length taking up almost the entire building. They are dwarfed by its size. Andriyan Nikolayev reverently touches one of its huge nozzles and then the camera slowly pans up and across all thirty-two of those nozzles – thirty-one more than on von Braun's Redstone. For the briefest moment the camera settles on Gagarin, standing next to Popovich. His face is etched with wonder as he gazes at the enormous rocket.

While in the MIK the cosmonauts had a discussion with senior engineers about the instruction manual for the Vostok. This manual, entitled *Instructions from the Central Party Committee to the Cosmonaut on Use and Control of the Spacecraft Vostok 3A*, had been compiled two months earlier in January and in its original version was all of thirteen pages. But Korolev had insisted that this was too long and so the version the cosmonauts discussed was shortened to make things even simpler. Nevertheless, as Kamanin notes without apparent irony in his diary, the men 'made several significant amendments' to those instructions during this meeting, one of which was that it would be much better if they were to put their gloves on fifteen minutes before launch and only after the hatch had been closed. Another 'significant amendment' was that it might be a good idea to carry a logbook on board the Vostok. We are a very long way from the years of vigorous, high-level discussions between the Mercury astronauts and NASA's engineers in which the astronauts' voices were at least genuinely heard, if not always acted upon. There were no prolonged dialogues with Korolev's designers about additional windows or cockpit displays, or in fact any such dialogues at all. The clear impression is that even in these final weeks the cosmonauts were not so much participants in this programme; they were sightseers.

And as it started so it continued. Two days later on March 20, the cosmonauts were given an opportunity to practise putting on their spacesuits for what, even at this late stage, appears to be the first time. The spacesuit would be the cosmonaut's last line of defence against a cabin depressurisation failure or oxygen leak. Yet in the rush to get a man up in space before the Americans the original intention had been to dispense with them altogether. They were only developed as an afterthought over just seven months from the middle of 1960. Each spacesuit, a complex, multi-layered, personalised life-support system had to be tailor-made to fit its owner. But thanks to a bureaucratic or engineering glitch, only three of them were ready by the time the cosmonauts now got to try them on. Gagarin, Titov and Nelyubov all had their own suits. Popovich, Bykovsky and Nikolayev did not. While the first three got dressed the other three just stood there and watched.

On the following day, March 21, the cosmonauts were invited to a meeting to discuss some of the issues of search and rescue, and in particular the issue if, as a result of some malfunction, the Vostok landed in the sea instead of on Soviet soil. Kamanin mentions that the 'cosmonauts were pleased' with everything they were hearing, before also going on to write that 'they do not know the serious shortcomings in equipping the search ships with search facilities'. Nor did they know that if they *did* land in the sea 'unfortunately … the rescue of a cosmonaut is not at all assured and a lot of work will have to be done to solve this problem.' Kamanin did not write – at least not in this entry – that in sea trials the emergency dinghy had proved to be unstable and leaky and would almost certainly sink. Nor did he remind himself that the 'lot of work' that needed to be done had to be done within three or at most four weeks, if they were ever to meet Korolev's timetable.

Two days after that, on March 23, the cosmonauts were given classes on how to use the Globus device, a navigation instrument with a colourful little globe of the earth – rather like a miniature version of what one might find in a schoolroom – which sat in the centre of the Vostok's limited instrument panel. The Globus was supposed to show the cosmonaut where exactly he was over the earth at any time and also where he was predicted to land from the moment the Vostok's braking engine was fired. It stretches credulity to suppose that the six men, even with their rudimentary simulator back in Moscow, would not already know how to use this critically important navigational device, but if Kamanin is to be believed they still found the lessons very helpful. Sounding increasingly like a school inspector, he commends the classes for being 'fairly well prepared' and 'certainly useful for the cosmonauts'.

This, then, was how the cosmonauts were introduced to Tyuratam. Everything they saw in the eight days before Ivan Ivanovich's second flight was new. Everything about the cosmodrome was a revelation, even if they had been told some of it before. Conducted from place to place like a party of goggle-eyed tourists, they found themselves – just as the cameraman Vladimir Suvorov had when he first went there in 1960 – in a 'fantastic movie'. The stern Dr Yazdovsky watched them as they soaked up the drama:

> The cosmodrome astounded the cosmonauts. The huge installations, the Cyclopean launch complex ... it all seemed grandiose and fantastic, but at the same time made their future flight more real. Everyone understood that it was not months any more, or weeks or even days before the first manned flight in a spacecraft into the expanses of the universe.

In the area called Site 10 they saw the conurbation called Leninsky with its now more than ten thousand inhabitants where only six years before had been the isolated railway halt at Tyuratam. They saw the new road with its accompanying high-voltage pylons cutting like an arrow thirty kilometres north across the monotonous steppe to Site 2 with its gargantuan MIK assembly building, its military barracks, its arsenal of nuclear warheads – which one can assume they did *not* see – and Korolev's incongruous little cottage, one of two there. But of all the astounding tours by far the most spectacular was the one they were given on March 22, from ten in the morning until lunchtime according to the very precise Kamanin, of the cosmodrome's launch pad at Site 1 by its chief designer, Vladimir Barmin.

According to one of the pad engineers, Vladimir Solodukhin, the purpose of Barmin's tour was, baldly, 'to get the cosmonauts acquainted with it'. This was their first opportunity to see the pad from which Ivan and Udacha were expected to launch into space three days later on March 25, and from which one of them would launch soon after that. As the six young men stood blinking in the brilliant spring sunshine, Barmin proudly showed them around the structure that Kamanin, with perhaps a touch of facetiousness, called 'his extensive establishment'; but which was, in fact, one of the secret wonders of the modern world.

FOR ONE THING, it was enormous. The launch pad was dominated by a massive concrete platform bristling with a complex array of steel-lattice service and supporting towers that would clasp the R-7 rocket – still being checked over in the MIK – until the final moments before lift-off. These were impressive in themselves; but more impressive still was the feature over which the platform jutted, a huge artificial crater shaped like a peaked military cap, 250 metres long, 100 metres wide, 45 metres deep. This was the R-7's flame pit, an excavation of a million cubic metres of earth, surfaced with thirty thousand tonnes of concrete, and designed for the single purpose of containing the mass of flame, fire and smoke from the biggest rocket yet built at the instant of its launch. Nothing like it existed anywhere on the planet. Mark Gallai, one of the cosmonaut instructors, was awed when he first set eyes on it. 'No photograph, no piece of cinema could possibly convey the scale of this construction,' he wrote in his memoirs. 'Imagine this. A huge mountain is cut off at its base, turned upside down and pressed into the ground ... and then removed.' He compared it to the Grand Canyon. Like the canyon it was 'a magnificent abyss'; but unlike the canyon it was entirely man-made.

The work on the launch pad and its flame pit had begun in the late summer of 1955, just seven months after the first contingent of what would shortly become more than ten thousand workers broke ground near the Tyuratam railway outpost to start building the site codenamed NIIP-5: the world's largest missile complex in one of the world's remotest places. The total area of the complex in this Kyzyl-Orda region of Soviet Kazakhstan would be approximately a hundred times the size of Cape Canaveral. And that was just the complex itself. The area isolated from human habitation to contain the fall of missiles or spent rocket stages would be the size of Kentucky.

In part it was this very remoteness that had recommended the site to Korolev in 1955 when he needed somewhere to test his revolutionary R-7 missile. The tests here could be conducted largely in secret and far from built-up areas, and where the skies were clear three hundred days a year. There was another advantage too: it was closer to the equator than any existing Soviet missile site. Thanks to the rotational velocity of the earth, this would give Korolev's missiles – and later his space-bound rockets – a welcome extra kick of speed to help them on their way. And that tiny railway halt was also a benefit, a convenient stopping point on the Moscow to Tashkent line that allowed everything required for the construction and maintenance of the missile complex, 'every stone, brick, board and nail' as one construction worker put it, to be delivered thousands of miles from points all across the USSR by train.

But if the site had much to recommend it, it also had serious disadvantages. There were good reasons why almost the only humans living there were the Kazakh nomads who for millennia had wandered with the seasons and their animals across the semi-arid steppe – and all of whom were ruthlessly expelled when the builders arrived. This was a hostile, wild and dangerous land of sudden sand storms and thick, choking dust, of ferocious mosquitoes, venomous snakes and vicious scorpions. And despite those three hundred days of clear skies the climate here was extreme, with scorching temperatures of 46°C in the summer plunging to minus 45°C in winter, and often much lower in the bitter winds. Only for one brief magical month in April did the wilderness spring into life and colour as tulips flowered in profusion across the steppe; but for the other eleven months the landscape was a flat, treeless, empty expanse of sand and scrub from horizon to horizon. 'Our first impressions were depressing,' wrote General Aleksey Nesterenko, the cosmodrome's first commander who arrived in March 1955. 'You look around at all this, and a hopeless sadness takes hold of you.'

The choice of Nesterenko as commander was astute. Not only was he one of the few personalities big enough to confront Korolev when necessary but he was also seasoned in combat, having commanded batteries of Katyusha rocket launchers against the Germans in some of the worst fighting of the war. Building this colossal cosmodrome in such a remote, inhospitable place meant fighting another sort of war, one that also demanded heroic reserves of courage and sacrifice along with those of energy, and as a result Nesterenko's thousands of workers were drawn largely from the ranks of men who had survived the hell of battle too. They lived in tents, in railway carriages, in dugouts, with no baths, poor food and filthy drinking water, and they worked in shifts round the clock. Back and forth, day and night, an endless stream of trucks crawled across the steppe, in dust so thick their drivers had to leave their headlights on in the midday sun, often grinding to a halt in the sand. The dust was everywhere. Like a cloud, it 'hung all over the area', as one veteran remembers, 'filling noses, eyes and ears … and permeated food, bread and petrol'. But if the dust was bad, the climate was terrible, with those raw, freezing winters and the blazing heat in summer that almost felled Nesterenko himself: 'We were sweating and greedily sucking in the air,' he wrote later, 'like fish thrown on the shore.'

In these conditions, dysentery spread like wildfire. The filthy drinking water left hundreds of workers sick and sometimes killed them, as did the plagues of disease-ridden rodents, reminding Nesterenko why this desolate region of the world was described in Soviet encyclopaedias as 'the Home of the Black Death'. But still the work went on, building the long straight road from Tyuratam to the launch pad at Site 1, building the settlement of Leninsky and the MIK assembly complex at Site 2, building the launch pad itself with its flame pit so deep that workers often joked they were digging the biggest hole in the world. But they completed the launch pad in just nine months and the bulk of the cosmodrome in a little over two years, ready for the first test of the R-7 missile in May 1957. It was one of the largest civil engineering projects in history and even Korolev, no slouch himself when it came to engineering challenges, was astonished at the speed in which it was done. 'Did they really do it?' he is said to have exclaimed when told. 'How is it possible that they could do it in such a short time?'

The same energy, sense of duty, toughness and sheer resilience that drove the construction of the missile complex and its new town drove those who came to work there too, the thousands of engineers, support personnel, technicians, soldiers and their families, some of whom were sent but many of whom volunteered, often moving far from their homes to live in this strange new world out in the semi-desert. Most were young, like Vladimir and Khionia Kraskin, a couple who arrived from Leningrad in the autumn of 1956 to begin work as telemetry specialists. They found a place surrounded by guards and barbed wire, where photography was forbidden, and where any contact with foreigners on their occasional leaves back home had to be reported immediately to the KGB. They were also forbidden to tell their wider families or friends where they were, giving out their address only as Number 10, Tashkent 50.

Life in the spartan town was tough – this was no Cocoa Beach with its louche bars, topless waitresses and Holiday Inns – and for the first three years the Kraskins had to live with their infant son in army barracks, with barely any privacy and only outside toilets. The work was backbreaking and relentless – 'we had no days off' – but it was exhilarating too, and even decades later in her eighties Khionia describes the atmosphere as 'romantic', as if 'we were living a sort of student life there'. The mood is curiously reminiscent of 1940s Los Alamos, New Mexico, where talented young people came together to build the atomic bomb. But conditions in Tyuratam, certainly in those pioneering years, were far worse. 'We had gone through the privations of war and the horrors of the Leningrad siege,' said Khionia. 'That's why we were not hard to please.' Boris Chertok, the rocket engineer who worked under Korolev, perhaps best captured this sentiment and its impact on the entire Soviet space programme in this period:

> After the devastating war, we could not in our wildest dreams have imagined the possibility of enjoying the achievements of civilisation that were firmly entrenched in the American way of life. But perhaps this was to our advantage over the Americans. We didn't think about having a normal roof over our heads, while for them fundamental comfort was as essential as air.

The Americans meanwhile were quick to uncover the site. Intelligence hints from the mid-1950s had already suggested that the Soviets might be building something very big and potentially threatening somewhere south of the Aral Sea in the USSR. The only maps CIA analysts could find of this remote and inaccessible region were German military maps from the war. From these they spotted an old railway line built in the previous century by a British mining company, and this led them to suspect an area close to the tiny railway halt at a place on the German map called Tjur-Tam. High-flying U-2 spy plane missions were authorised by the then-President Eisenhower to begin photographing the terrain from 70,000 feet. On August 5, 1957, three months after the first – unsuccessful – test of the R-7, the pilot of one of these U-2s flying towards Tyuratam noticed something unusual slightly off his planned course and turned to photograph it. He had just stumbled upon what the U-2 programme's CIA director, Richard Bissell Jr, later described as the 'crown jewel of Soviet space technology, whose existence had not even been suspected'. The pilot had found the R-7's launch pad, its colossal flame pit visible from thirteen miles up in the sky.

A second U-2 mission was flown on August 28, only a week after the first successful test launch of the R-7. This time the U-2 returned with astonishingly detailed photographs of the launch site shot from directly overhead. The images stunned the CIA interpreters. A missile using a site that big must, they estimated, be twice the size of the heaviest American missile. In fact they underestimated: it was almost three times as heavy. Within a week of the mission, technicians at the Naval Photographic Intelligence Center had constructed a scale model of the facility based on the photographs. Further secret U-2 flights over Tyuratam were authorised, until on May 1, 1960, one of them was shot down by a Soviet surface-to-air missile. The CIA pilot, Francis Gary Powers, had been overflying the USSR and photographing, among other places, the missile complex.

Powers ejected to safety, but he was captured. Initially Eisenhower denied all knowledge of the spy plane. A cover-up story was concocted in which the U-2 was described as a NASA weather plane that had strayed off course after its pilot fell unconscious. Another U-2 was quickly painted up in NASA colours and its photo released to the press to corroborate the story. But within ten days of the shooting, and under pressure from Khrushchev who revealed that Powers was in Soviet custody – 'and now just look at how many silly things the Americans have said!' – Eisenhower was finally, and humiliatingly, forced to come clean. Powers was put on public trial in Moscow, convicted of espionage and sentenced to ten years' imprisonment on August 19, 1960, the same day that the two dogs Belka and Strelka launched from the very facility he had been photographing to begin their sensational eighteen orbits of the earth. For Khrushchev it was a double victory over the United States.

The U-2 flights were stopped. But the surveillance continued. When, in the early hours of October 24, 1960, an R-16 Soviet missile blew up on its pad at Tyuratam incinerating the commander of the USSR's strategic missile forces, Marshal Nedelin, along with at least another seventy-three people in the worst rocket accident in history, the Soviets kept silent. Their TASS agency reported only that Nedelin had died in a plane crash. The CIA was immediately suspicious. As a result of its suspicions, the agency authorised a mission of its latest and most secret surveillance device of all, a remarkable spy satellite codenamed Corona whose existence was known only within a tiny circle. On December 7, 1960, Corona photographed the missile complex at Tyuratam while passing one hundred miles above it in one of its orbits. The film then detached from the satellite in a special capsule – nicknamed a 'bucket' – and fell back to earth. At 60,000 feet the bucket's parachutes deployed, rapidly slowing its descent, until it was 'caught' in mid-air by a special aircraft towing an airborne claw that was then winched aboard. After the recovered film was developed, interpreters were amazed to observe a huge black scar on one of the Tyuratam missile pads. They also noticed a substantial gravesite in nearby Leninsky that had not been there before.

'It was horrible to see the funerals,' said Khionia Kraskina. 'That was a terrible moment in my life.' She and Vladimir lost many of their friends in the disaster. Vladimir arrived at the accident site shortly after it had happened and the memory of what he saw there never left him, 'the smell of burned flesh ... the torn bodies ... lying around the rocket'. He was overcome with nausea and was sick. When he finally got home hours later, Khionia remembered he was 'sobbing', and she had to wipe the blood off his clothes. At the burials in the Leninsky cemetery that the Corona satellite spied from orbit, the long rows of coffins reminded her of the German siege of Leningrad during the war: 'I have no words to describe what I felt. It was complete depression. It was a full emotional breakdown.' The tragedy shocked the tight-knit community in Leninsky. 'A dead silence,' said Khionia, 'came over the town.'

But nothing could be allowed to slow the pace of expansion in the months following the accident. By the time the six cosmonauts were marvelling at the R-7's massive flame pit and the rest of Vladimir Barmin's 'extensive establishment' on that bright March morning in 1961, the pad where the accident had occurred had already been repaired. There were now four fully functioning missile launch sites at Tyuratam and there would soon be others. The work of construction never stopped and the rockets kept launching, more with every month. And the message was always the same. Khionia Kraskina heard it everywhere: it was always 'hurry, hurry, hurry ... faster,

faster, faster, because the Americans had to be overtaken!' – even if that also meant that people would die along the way. Such was the price of progress and competition. Death was always there, always close and often sudden, as indeed the young cosmonauts themselves were just about to find out.

FIFTEEN

THE PRICE OF PROGRESS

March 23, 1961
Botkin Hospital
Moscow

DR VLADIMIR GOLYAKHOVSKY received the phone call in the early hours of the morning. For the chief surgeon-traumatologist at Moscow's sprawling Botkin Hospital in the north-western section of the city, it had been another long and busy night. Although still only thirty, Golyakhovsky was well used to handling the difficult and often distressing cases he faced every day, but something about this one sounded different. The anxious voice on the other end of the line identified himself as a Colonel Ivanov. He told the surgeon that in the next few minutes a patient would be arriving with severe and extensive burns. 'Stand by to administer emergency aid,' said Ivanov. 'I will be accompanying the patient.'

Within minutes a military ambulance, escorted by a cortège of five or six black official Volga cars, the type usually driven by party bigwigs, screeched up to the hospital entrance. A gaggle of military officers jumped out of the cars, some of them wearing medical insignia. They raced to the back of the ambulance where a patient was being unloaded on a stretcher. The stretcher was carried up a short flight of stairs to the emergency department where Dr Golyakhovsky was waiting. Even this seasoned trauma specialist was shocked by what he saw next:

> The patient on the stretcher exuded an odour of singed tissue. Silently I helped them carry him into the shock treatment room. A nurse and an intern helped me take off the blanket and the sheet covering him and I couldn't help shuddering. The whole of him was burnt. The body was totally denuded of skin, the head of hair. There were no eyes in his face; everything had been burned away. It was a total burn of the severest degree. But the patient was alive and even tried to say something through burned lips. I bent down close to his awful face and managed to decipher what he was trying to say: 'Too much pain … do something, please … to kill the pain.'

Golyakhovsky tried to give the patient intravenous injections of sodium chloride, glucose and morphine but he could not find a single vein on the man's body. The only part of his skin that had not been burned away was on his feet, and so Golyakhovsky injected him there. With the shot of morphine the pain seemed to ease a little but Golyakhovsky knew that the man could not survive much longer. He asked Colonel Ivanov his name and was told, briefly, it was 'Sergeyev'. But something about this Sergeyev must have been very special, judging by the number of high-ranking military officers filling the room.

IN FACT THE patient's name was not Sergeyev but Valentin Bondarenko, and at the age of twenty-four he was the youngest of all twenty Soviet cosmonauts. He was also one of the most well liked. 'He was a very good-natured, merry fellow,' recalled Pavel Popovich, who remembered how Bondarenko used to run up and down the stairs of the Chkalovsky apartment block knocking on doors and trying to coax his cosmonaut friends into a game of soccer. Like

Popovich he was Ukrainian and also loved to sing Ukrainian songs. Every account agrees that Bondarenko was quiet, perhaps even shy, but with an appealing ability to laugh at himself and never one to brag. He was passionate about his soccer and by far the best table-tennis player in the group. In the very few photographs of him that have survived – or that have ever been released – he emerges as a clear-eyed young man with full lips and neatly combed dark hair. The others gave him a nickname – '*Zvonochek*' or 'Tinkerbell' – and it was meant affectionately. And he was brave too. One cosmonaut, Georgy Shonin, describes how Bondarenko once rescued a little boy who was standing precariously on the window sill of a fifth-floor apartment in Chkalovsky: 'I still tremble even now when I recall Valentin climbing up that drainpipe. Each second he could fall down with the creaking pipe.' But he did not fall down and he saved the boy, accepting the applause of the crowd below with his characteristic diffidence. If Bondarenko showed any pride at all it was not for himself but for his family: for his father, a much-decorated partisan who had fought the Nazis in the war, and for his wife, Anna, and their four-year-old son, Sasha.

On March 13, three days before his Vanguard Six comrades left for the cosmodrome, Bondarenko entered the isolation chamber at the Institute of Aviation and Space Medicine in Moscow. He was the seventeenth cosmonaut to take his turn behind the chamber's soundproofed steel walls. Medical sensors were attached to his body, to his head, waist, arms, wrists and feet. The double doors were sealed behind him like submarine hatches and the chamber slowly depressurised to the equivalent altitude of four and a half kilometres, almost the height of Mont Blanc in the Alps. This would simulate the cabin pressure of the Vostok. To allow Bondarenko to breathe properly in this thin air, it was enriched with oxygen – up to 40 per cent of the total atmosphere, almost twice the normal proportion. Again, this was intended to simulate conditions inside the spacecraft. Bondarenko settled down in the one chair in the narrow, gunmetal-grey chamber with its single monitoring television camera and four tiny portholes, its desk, its lamp, its curtained toilet and its electric hotplate on which to cook his meals, the cramped claustrophobic space that would now be his home for an indefinite period. Like all the cosmonauts he was not told how long that would be. But like all of them too he knew that for each minute of that time every aspect of his behaviour would be monitored, assessed and recorded.

For the next ten days Bondarenko lived alone in his little cell, eating, sleeping, sometimes solving complex number puzzles, sometimes responding to abrupt, frightening alarms, mostly doing nothing at all. On March 23, early on the morning of the tenth day, he was following a standard daily routine in which he was required to remove his medical sensors and clean them before re-sealing them on his body. What happened next is still not entirely clear but it seems that on this occasion he ripped off the sensors and wiped away the sticky residue on his skin with a cotton-wool pad soaked in alcohol. Without looking, he tossed the pad away. It landed on the electric hotplate, which he had left switched on to warm up some food. There was an explosion – and in the oxygen-rich atmosphere the whole chamber was suddenly on fire.

In the first few seconds Bondarenko desperately tried to put out the flames. But they were spreading too quickly and within moments his woollen tracksuit which, like everything else in the chamber, had been soaking in the oxygen-rich environment for the last ten days, caught fire too. He tried to beat out the flames with his hands but that only helped them to spread further until his tracksuit began melting into his skin. He was now screaming in pain. With the fire alarm shrieking in the laboratory, the supervisor on duty, Dr Mikhail Novikov, and a team of nurses rushed to the double doors and their two big hand-wheels. But it was impossible to open them yet – the pressure differential between the chamber and outside was too great. It would take all of several minutes to equalise it before they could reach Bondarenko and get him out. They could only watch in horror through the chamber's portholes, or turn away, as the conflagration raged unchecked inside. When at last they were able to open the doors and drag Bondarenko's smouldering body clear, he was still conscious. Over and over he kept repeating, 'It's my fault … no one else is to blame … I'm so sorry.'

It took Bondarenko sixteen hours to die. At the Botkin Hospital, the closest medical facility to the institute, Dr Golyakhovsky did everything he could to soothe the dying man he knew only as Sergeyev. At one point he was astonished to receive a personal phone call from the chief surgeon of the Soviet Army himself, General Alexander Vishnevsky, who told the doctor that he was under intense pressure to have Sergeyev moved to a specialised burns unit. The general said he was resisting that pressure because he feared the patient would die from shock if he were moved. But he agreed to send specialists to the hospital to help Golyakhovsky. While this Sergeyev, now unconscious, lay dying on his bed, Golyakhovsky could not help noticing a young army officer sitting by the only telephone in the emergency department quietly reporting on the deteriorating condition of the patient to what were clearly a series of very senior ranking officers. Years later, he convinced himself that this young man must have been the subsequently famous Yuri Gagarin. But Gagarin was not in Moscow. He was two and a half thousand kilometres away at the cosmodrome in Tyuratam.

We do not know exactly when Gagarin and his fellow Vanguard cosmonauts heard what had happened, although Kamanin, who was with them, records receiving 'the horrible news from Moscow' in his diary the evening Bondarenko died. He also says that the accident was 'senseless'. It is likely Kamanin told the cosmonauts straight away. Their colleagues back in Moscow were certainly told that night, including Boris Volynov, who had miserably celebrated his twenty-sixth birthday in the chamber. Volynov was despatched directly to the institute to try to find out how the fire had started: 'I got there and the first thing I saw was the seat on which he had tried to put out the flames, burned in the shape of his body.' The tragedy sent shockwaves through the small cosmonaut community, not least because Bondarenko was so young and so well liked. 'We were speechless,' remembers Boris Smirnov, a photographer in the space programme. 'Everybody loved him.' The single dissenting voice was

Dr Yazdovsky, the much-disliked director of the institute. 'Valentin violated fire safety instructions,' he wrote briefly in his memoirs. 'It cost me a delay in promotion.'

But beyond the tightly controlled circle of those who were allowed to know, the accident, like all the other accidents, was hushed up. 'Everything was taken away, everything,' says Irina Ponomareva, a nurse who worked in the isolation chamber programme but was off-duty on that March day. 'Everything is classified even today, somewhere in the archives.' Bondarenko was quietly buried on a cold spring morning a few days later in Kharkov, the city in Ukraine where he was born. His wife, Anna, was present. He was not buried as a cosmonaut but as an ordinary air force officer. Anna was given a widow's pension and returned with her four-year-old son to live in Moscow. Meanwhile the story of her husband's death would remain secret for another quarter of a century until the state newspaper *Izvestia* released the first officially sanctioned details of the accident in 1986. In time the inscription on Bondarenko's grave was changed, from 'With fond memories from your pilot friends' to 'With fond memories from your pilot and cosmonaut friends'. But there are some who believe that if the Soviets had only moderated their policy of silence and released the details of the story when it happened another, hideously similar, tragedy might well have been prevented. Six years later in January 1967, a fire broke out in the 100 per cent oxygen atmosphere of the Apollo 1 capsule during a launch rehearsal test while it sat on the pad at Cape Canaveral. Before the hatch could finally be opened all three American astronauts inside, Gus Grissom, Ed White and Roger Chaffee, were killed in the flames.

None of the Vanguard Six cosmonauts were able to attend Bondarenko's funeral. They were still at the cosmodrome awaiting the imminent launch of the second Ivan Ivanovich flight. Yet their friend's death must have been a shocking reminder of the multiple and gruesome ways any of them might die too, and one of them possibly very soon. And that if they did, the world might never know about it.

BUT NOTHING COULD be allowed to hold up the schedule. On the morning of March 24, less than twenty-four hours after Bondarenko's death, a meeting of the state commission responsible for the Vostok programme was held at the cosmodrome. Its purpose was to update the principal players on the status of the programme and, more urgently, on what everyone hoped would be the last dress rehearsal flight. Korolev was present, along with some powerful figures adding weight to the occasion, including Mstislav Keldysh, Vice President of the Soviet Academy of Sciences, Sergei Rudenko, First Deputy Commander-in-Chief of the Soviet Air Force, and a number of other chief designers whose departments were building key systems for the Vostok. Kamanin, helpfully for posterity, was also there to record the discussions that night in his diary. None of the cosmonauts were invited.

The commission heard first from Semyon Alekseyev, the chief designer at Plant 918, the organisation that had failed to make more than three spacesuits for the six cosmonauts. Alekseyev now had more bad news. Essential seat ejection tests that were supposed to have been carried out in the previous weeks had still not happened. On top of that the emergency survival package, including the leaky dinghy, was still not ready. Further water tests would be needed and these would take another seven to ten days. Of course everybody in the room already knew that if the cosmonaut landed in the sea his chances of being rescued quickly were slim at best. The only people who did not know this were the cosmonauts themselves. But they were not present.

The bad news did not end with Plant 918. Next the deputy chief designer of OKB-124, the engineering bureau responsible for the Vostok's life-support system, stepped in to report that tests of an essential piece of equipment devised to absorb the cosmonaut's sweat and keep the cabin air dry had failed totally. After a ten-day trial, records Kamanin, 'a whole puddle of saline solution formed inside the cabin of the ship'. OKB-124 planned to test the apparatus again using a different absorbent chemical, but this would require yet more time – another two weeks. And there were no guarantees it would work either.

Since the first manned flight of a single orbit was only supposed to last approximately 100 minutes, the cabin's humidity levels should still remain safe even with this flawed air-drying system. But everything depended on the Vostok's single braking engine. As long as it fired correctly and on cue, slowing the capsule before re-entry into the atmosphere, the cosmonaut would get home.

But what if the braking engine failed even to start?

There was already an emergency plan for this. The Vostok's orbit was not predicted to be very high – approximately 217 kilometres at its apogee. At this relatively low altitude an alternative route home, if the braking engine failed, was to allow the capsule's orbits to decay *naturally* with the pull of gravity until it finally dropped on its own account back to earth. There was only one catch: to do this would take approximately ten days, ten days in which the friction of the atmosphere would also cause the capsule to heat up as it gradually descended with each orbit. The temperature inside would climb. The cosmonaut would sweat. The sweat would raise the cabin's humidity levels. And with no functioning system to dry that air, the cosmonaut might well be dead by the time he landed – killed by his own sweat.

The flawed dehumidifier, in other words, could well be a death sentence to any future cosmonaut. And yet, in the very next sentence after he describes this potential catastrophe, Kamanin writes that 'the launch of the Vostok ... with the mannequin will take place on March 25 at 08.54 Moscow time'. Despite all the difficulties and setbacks they had just been talking about, the final dress rehearsal flight was still set to go ahead the very next day.

None of Kamanin's other concerns are mentioned in this entry, nor do we know if they were even discussed. But they pepper his diary elsewhere. By now his worries were keeping him awake at night. He was worried about the R-7's troubled third-stage engine that had already failed at least twice before. He was worried about the capsule's unreliable automatic attitude-orientation system that was required to position the spacecraft *precisely* before re-entry, without either burning up in the atmosphere or skipping back higher into space. He was worried about communication black spots: 'There is,' he fretted, 'no confidence in the reliability of the connection.' And perhaps there were other worries not even mentioned in his diary, among them Boris Chertok's own worry about the separation failure he had spotted between the two sections of the Vostok on the last mannequin flight. But if Chertok is to be believed, in the hurry to press on this was never discussed at the meeting either.

Korolev's schedule was merciless. 'In that spring,' wrote the cosmonauts' instructor Mark Gallai, 'everybody involved in preparing for manned space flight wanted – really wanted – to be first.' Kamanin echoed the sentiment and the fear lurking behind it: 'All the time,' he wrote, 'I am haunted by the idea that we act with slow and spread fingers.' Like his chief, the greatest of all his worries was 'that we do not fall behind America'. There it was again: the '*Amerikantsi*' are coming, 'the Americans are at our heels!' just as Khionia Kraskina kept hearing and Korolev himself had declared.

And the Americans were on their heels – they were right on them. Because at exactly the same time as the state commission was meeting in Tyuratam to approve the final dress rehearsal flight, an American rocket was sitting and steaming on a launch pad seven thousand miles away at Cape Canaveral ready for its own test flight. The unmanned Mercury-Redstone Wernher von Braun had insisted on was about to launch into space. It would lift off from Pad 5 at the Cape just after midday on March 24 – less than thirteen hours before Ivan Ivanovich and his dog lifted off from Site 1 at Tyuratam on March 25.

SIXTEEN

BOOSTERS AND DUMMIES

March 24, 1961
Launch Complex 5/6
Cape Canaveral, Florida

THEY HAD FIXED things fast. Once they had won their battle to fly a further test of the Redstone, von Braun and his rocket engineers at the Marshall Space Flight Center in Huntsville got to grips with the work at once. And they came up with solutions within two weeks. In the end the fixes boiled down to just nine. Changes were made to that troublesome thrust controller that had stuck wide open on Ham's flight like a jammed accelerator pedal causing most of the problems. To damp down the vibrations that had jarred every bone in Ham's body, 210 pounds of additional ballast were added to the rocket's instrument compartment. The hydrogen-peroxide system that had spun the rocket's turbo-pumps too rapidly, pouring too much propellant into the engine's combustion chamber, was also adjusted to stop that from happening again. And then there was the major abort timing foul-up, the system set in Ham's flight to disarm at 137.5 seconds – exactly half a second *after* the last of the propellant ran out, triggering the abort that had flung Ham's capsule too high and too far out into the Atlantic. Now the timing was advanced several seconds so that the abort system would disarm well *before* the propellants ran out – preventing any future unwanted abort. And there were five other fixes too, some of them made at Huntsville and some at the Cape where the Redstone arrived at the end of the first week of March. By the end of the second week it had left its hangar on the back of a low-loader and was standing on Pad 5, ready to be mated to the Mercury capsule it would carry into space.

And throughout this same period the people at NASA's Space Task Group, who had so furiously opposed von Braun in February over this extra flight, lifted barely a finger to help him.

The capsule that would sit on top of the rocket was not even a 'real' capsule. As Mercury Control's flight director Chris Kraft put it – and he was one of the most furious out there – 'we refused to waste a Mercury capsule on von Braun's unnecessary Redstone test'. Instead von Braun's team had to make do with a 'boilerplate' version, essentially a dummy that looked like the real thing but was not the real thing. This particular boilerplate had no instrumentation of any kind and was inert, literally so because it had no retro-rockets nor even a working escape system in case of an actual abort. It was not much more than a Mercury-sized and -shaped bucket filled with ballast to make it the same weight as the genuine article. It was not even new either, having been used before in a very early – and failed – test on another rocket back in 1959. Bob Gilruth, the Space Task Group's chief, was basically giving von Braun used goods from the back of his shop, and he was able to do so because he did not want them back. The boilerplate capsule would remain stuck to the Redstone all the way up and all the way down into the Atlantic and then further on down right to the bottom of the sea. It was not even worth retrieving.

And in case von Braun failed to get the point, Gilruth made it another way just to ram it home, this time over the flight's name. Coming as it did after MR-2, Gilruth's people categorically refused to call it what it in fact was – MR-2A or perhaps even MR-3, the designation that had been reserved for the first manned flight. Instead they insisted it be called 'MR-BD' – the Mercury-Redstone

Booster Development flight – as if deliberately distancing themselves from everything it represented, namely a pointless exercise in Germanic timidity and political self-preservation that could, and probably would, cost America its best chance to win the race to space. Meantime at the Cape, Kraft's mission control would still be obligated to provide technical support for MR-BD, but that support would be frankly minimal, stripped back to a skeleton team at most. Glynn Lunney, one of Kraft's men, remembers the mood among the mission controllers all too well: 'It was, c'mon – how many times have you gotta run these damn things? We're telling you, it's ready to *go*!'

Such were NASA's sensitivities by now that official publicity of this upcoming Redstone test flight – yet *another* test flight – had been deliberately withheld until the very last moment. An internal NASA memo entitled 'Information Plan: Redstone Development Test MR-BD' dated March 21, just three days before the flight, is a public relations exercise in paranoia almost worthy of the Soviets, with its carefully detailed timetable of when, precisely, advance news of the test could be released. The answer was on page two: 'No advance public announcement will be made regarding this test,' it said – that is, there would be no advance public announcement *at all*. An off-the-record briefing to selected reporters would be permitted forty-eight hours in advance of the flight 'for planning purposes only'. Every piece of press kit material was to be marked in capital letters, '*HOLD FOR RELEASE UNTIL LAUNCHED*.' It was a total media blackout.

And yet, for all NASA's squirming embarrassment about this test flight and for all von Braun's caution, his engineers had nevertheless moved like lightning. There was even a last-minute opportunity to check how noisy the Redstone would be for the rescue crews come the day a human astronaut did finally fly, by planting a team of fire-men in an armoured personnel carrier a thousand feet away from it to see what the blast did to their ears. This was all part of a bizarre escape procedure that someone had dreamed up in case the said astronaut ever needed to get out of his capsule in a hurry while it was still on the pad: essentially it involved him first ejecting the hatch and then leaping (while in his spacesuit) onto a remotely controlled cherry-picker platform sixty feet up and stationed close by, which was then supposed to swing him over to the rescue team in their armoured personnel carrier for a quick getaway before, say, the rocket blew up.

With the fire crew hunkering down in their carrier and after a countdown with almost no holds this time, MR-BD launched into sunny skies at just before 12.30 on the afternoon of March 24. It was the start of what would be an almost-perfect flight. The rocket hit every one of its planned trajectory points almost precisely, climbing at over 5,000 mph to an altitude of 115 miles before plunging back down into the Atlantic 311 miles downrange – just 1.7 miles beyond and less than 3 miles to the right of its expected target. Along with its boilerplate capsule the entire assembly then sank to the bottom of the ocean. The flight had lasted exactly 8 minutes and 23 seconds. Sixty-five data points transmitted from the Redstone as it flew told the same successful story. Perhaps the vibrations had been a little on the high side but they were still within limits. A confidential NASA memo written the next day summed up the whole story in its final sentence: 'Conclusion: All booster corrections were satisfactory.' Even the ears of the fire crew had survived.

In mission control Chris Kraft had observed the whole thing: 'It worked,' he wrote many years later in his autobiography. 'Whatever the Germans had done to it was enough for them to spend a lot of time patting themselves on the back. I was just glad to be rid of the damned thing.' But for Alan Shepard the news was devastating. 'The March 24 Redstone flight was an absolute beauty,' he wrote. 'I should have been on that flight. I could have led the world into space.' And there was an additional twist of the knife too, since that same Redstone now lying on the floor of the Atlantic Ocean was the very one that was supposed to have launched him into history.

As for all those carefully assembled press kits held back for release until after the launch – when they were finally sent out nobody much gave a damn. The story had all the qualities of a damp squib. The

New York Times gave it a tiny paragraph on page 11 of the next morning's edition. In the New York *Daily News* it managed to make it onto page 7 squeezed next to a liquor sale advertisement, under the uninspiring – and, rather more to the point, uninspired – headline, 'The Redstone Clicks'. Yet despite all that Germanic patting on the back the same report included this disquieting comment from Walt Williams, NASA's deputy associate administrator and one of its most senior chiefs, when asked if there was now finally a date for an American manned flight: 'It is still too early,' he said, 'to set a definite date.' And he was not the only one saying that. Kurt Debus, the launch operations director, was also heavily still playing the caution card in spite of the fact that this latest Redstone was supposed to have 'clicked': 'However,' he told the *Schenectady Gazette* on March 25, 'a close look at the tapes might reveal a slight flaw which could necessitate another Redstone test launching.'

Another test launching? Whichever way one looked at it Project Mercury seemed to be mired in a quicksand of apprehension, if not downright dread, of something going horribly wrong once a human being ever sat on that rocket and actually lit the fuse. No doubt Donald Hornig's ongoing panel of inquiry into the programme, commissioned by the president's science advisor Dr Wiesner, had something to do with that dread too. Having begun his investigations on February 11, Professor Hornig and his panel members were now deep into the process, moving about the country from NASA facility to NASA facility, from contractors to subcontractors, asking lots of questions. And the more questions they asked the more the rumour mills kept grinding. There was even a story going round that another thirty chimpanzee flights might have to be flown before it was deemed safe for a human to fly. *Thirty* chimpanzee flights – that would take years. The Russians could be on the moon by then. They could be on Mars.

And even if those thirty chimps were just a rumour, it was pretty clear that President Kennedy, perhaps influenced as he was by Wiesner, was not exactly falling over himself to assist NASA's cause. At a meeting in the White House cabinet room only two days before this latest MR-BD flight, James Webb, NASA's brand-new chief, had argued with some force that 'the Soviets have demonstrated how effective space exploration can be as a symbol of scientific progress and as an adjunct of foreign policy'. But Kennedy was not listening. The budget he approved the very next day for NASA's fiscal year 1962 was a hulking 20 per cent less than NASA had been hoping for. More tellingly, it contained no funds at all for the Apollo spacecraft and therefore America's long-term programme to put a man on the moon. Essentially the president had just made it clear that he was uninterested in committing to *any* sort of human space flight after Project Mercury. Mercury was it; Mercury was the end of the line.

And right now he had other things on his mind. Some of those international trouble-spots he had hinted at in his inaugural address back in January had suddenly become a lot more troublesome by March. One of them was in South-east Asia, where the issue of Soviet air supplies to pro-Soviet Pathet Lao rebels fighting the US-backed government in Laos had very recently mushroomed into a major crisis. On March 23, the same day that Bondarenko died, the president openly challenged the Soviets in a press conference to stop supplying arms to the rebels. If they did not, he warned, 'those who support a genuinely neutral Laos will have to consider their response'. To back up that challenge he sent the US Seventh Fleet to the South China Sea along with helicopters and a further two hundred marines who happened to have been in Japan performing as extras for the movie *Marines, Let's Go!*, and who abruptly disappeared from the set.

Nor was Laos the only flashpoint. There was Cuba too, at just ninety miles from Key West, Florida, a truly next-door danger with its Communist leader Fidel Castro. In November 1960, only nine days after his election, Kennedy had been initiated into one of the Eisenhower administration's great, if not very well-hidden, secrets: an ambitious and highly risky CIA plot to invade Cuba and remove Castro using CIA-trained anti-Castro Cuban exiles. The rebel force had been preparing for months in the Guatemalan hills and the Guatemalan president wanted them out. Here was another momentous decision that Kennedy was being pressed to make, and make soon. On March 14 he had already authorised the CIA to continue secretly planning the invasion as long as US forces were not seen to intervene. But like brickbats, all these issues were coming at him from every angle and from opposite sides of the globe, testing his resolve and his experience, which after only two months in the job was minimal; and the consequences could be catastrophic not just for the United States but for the human race, if he got his decisions wrong.

In such a cocktail of competing demands, sending an American into space inevitably took a back seat; especially when there was even the slightest chance that astronaut might blow up on live television. Little wonder then that although NASA's deputy administrator, Hugh Dryden, made repeated requests to the president's office for a meeting to appeal the original budget decision, all his appeals were flatly turned down. The president, Dryden was told by David Bell, director of the Bureau of the Budget, was too busy. He simply did not have the time to spare. To which Dryden, with exquisite prescience, responded: 'You may not feel he has the time, but whether he likes it or not he is going to have to consider it. Events will force this.'

ONCE AGAIN IVAN Ivanovich's spacesuit, chest and thighs were stuffed with a variety of rodents. Once again another forty white and forty black mice, along with all those guinea pigs, reptiles, plant seeds, human blood samples, human cancer cells, bacteria and fermentation samples, would also be travelling, this time with the dog Udacha in her separate sealed container inside the spacecraft. Once again there would be a tape recorder to broadcast the Pyatnitsky Choir singing Russian folk songs in full-throated voice as the whole ensemble circumnavigated the globe at five miles per second. And not just the Pyatnitsky Choir this time: although the tape recording itself has not survived, there are indications that it might also have contained a spoken recipe for cabbage soup. Whatever the reasons for this extra piece of whimsy, there could be no doubting to any CIA eavesdropping post or ELINT operator listening in that this would be, in every sense, a truly Russian flight.

There were other changes too. After an urgent request to the Moscow Prosthetic Appliances Works, Ivan was sporting a brand-new head. His previous one had begun to look a little bashed-about after so much manhandling, space travel and cigarette pranks. But with his new head came a new problem: maybe he was now a bit *too* lifelike. 'We worried that if anybody who didn't know saw him after he landed, they would think he was a dead man,' recalls Dr Kotovskaya, who herself had been startled by Ivan's creepy eyelashes when she had first seen him. 'And that would be very bad.' With all the furore the previous year over the shot-down spy pilot, Gary Powers, there was genuine concern that Ivan might be attacked by hordes of patriotic Soviet peasants if they ever got to him first. Something had to be done, and quickly. A last-minute solution was to slip a hand-written sign with the word '*Maket*' – 'Dummy' – inside the visor of his helmet to forestall any acts of violence, however well intended, let alone further accusations or conspiracies from Italian radio hams or anybody else about dying or dead Soviet cosmonauts. As we shall see, it was a tactic that would prove only partially successful.

Another last-minute change was the name of the dog. Udacha – Lucky – was all well and good, but like Ivan's head this too was replaced. She became Zvezdochka – Little Starlet. Nobody has fully established quite why this happened but most accounts suggest that the idea came from Gagarin. He is said to have asked Korolev himself for the name change because, one day very soon, 'we are going to need that luck for ourselves.' At least it makes a nice story.

Despite an anxious moment when one of the sensors on the R-7's troubled third-stage engine failed and had to be disabled, Ivan Ivanovich along with his ark lifted off the pad on schedule at 08.54 Moscow time on March 25. They left with a roar that lit up the sky and shook the ground like an earthquake. For all six cosmonauts, none of whom had witnessed a rocket launch before, the experience was breathtaking. One of them, Pavel Popovich, never got to see it but he heard and felt it even though he was underground. Along with Kamanin and the rest of the launch crew, he was in the launch bunker buried beneath several metres of reinforced concrete and solid Kazakh earth just a hundred metres from the pad. A rocket the size and power of the R-7 would probably have flattened the sort of surface blockhouse near the little Redstone in Cape Canaveral. Only an underground bunker could survive this inferno.

The other five cosmonauts observed the launch from a tracking station less than two kilometres from the pad. An engineer, Svyatoslav Gavrilov, was also there. He watched the men as the minutes ticked down to zero and Yuri Gagarin in particular:

> He went up to the railings of the viewing platform and he did not take his eyes for a single second off the rocket standing in the distance. Then I saw his face drew in and his eyes darkened. It occurred to me that he was imagining himself sitting in the spaceship's cabin … 'Ignition!' … An instant later a cloud of smoke mixed with steam and dust came pouring out of the exhaust evacuation duct and roaring thunder began shaking the earth and tearing the air … Yuri was not able to hold himself back. A broad smile lit up on his face and he started to clap wildly. His comrades joined in. So did we. Then everybody was clapping.

Eleven minutes later the Vostok capsule separated from the rocket's spent third stage and entered earth's orbit. Racing at nearly 18,000 mph, it sailed over the USSR, over the Pacific and on around the planet, while the tape recorder underneath Ivan's spacesuit broadcast its traditional Russian folk songs and perhaps its cabbage soup recipe to the unrolling world below. Meanwhile back at the cosmodrome, Kamanin collected his cosmonauts and told them it was time to go home to Moscow. Long before they had even packed their bags, Ivan's Vostok had already completed its single lap of the earth. After a flight of approximately a hundred minutes, it landed back on irreproachably Russian soil somewhere between Votinsk, the small town in central Russia where Tchaikovsky had been born, and Izhevsk, the home of Mikhail Kalashnikov, inventor of the eponymous submachine gun. And just in case all that was not Russian enough, it also landed in another snowstorm.

In a script that must have been depressingly familiar by now, the snowstorm prevented any access to both the capsule and Ivan for the next twenty-four hours. Arvid Pallo led the search and rescue team. Once again, like some fairy-tale prince, he found himself facing yet another succession of near-impossible obstacles before he could get to Ivan and Zvezdochka. First, his rescue aircraft crashed on take-off from Izhevsk Airport. Having survived that, the team decided to use a helicopter instead, wisely making sure a landing site was prepared in advance so that, in Pallo's words, 'the helicopter would not capsize on landing'. After touching down in heavy snow somewhere near Ivan and his capsule, the team had to corral a group of peasants into taking them by horse-drawn sled to the site. At this point there are two versions of the story. In one, the villagers, on seeing Ivan in his orange spacesuit lying apparently dead in the snow, became very angry with the rescuers over their apparent indifference to his fate. The 'Dummy' sign inside Ivan's helmet did not calm them down. Before things got badly out of hand the village elder was persuaded to go up to Ivan and touch, as one rescuer put it, 'the rubbery, cold face of the dummy cosmonaut' to confirm he was not real.

In another version of the story, as the cameraman Vladimir Suvorov has it in his diary, the villagers thought Ivan was another American spy just as everybody had feared they might:

> Probably already anticipating the governmental awards the peasants, with a local militia man, surrounded the motionless and silent figure and tried to grab him. At that particular moment the search group arrived and saved Ivan Ivanovich. They said that the fellows were so frustrated by the revelation that they smashed their fists into the face of the dummy. But Ivan Ivanovich was too tough for them.

Presumably Ivan would need a new face yet again – if the story is true. But not for the press photographers. Like every Vostok flight, this one had been launched in secret in case it blew up or had to be blown up, but by mid-afternoon on the same day Soviet state radio was broadcasting news of its sensational success. The dog was mentioned but not the dummy. This was followed by a triumphant press release from TASS, proclaiming that 'preliminary investigation of the landed spacecraft showed that the test animal feels normal', a remarkable claim given that Pallo and his rescue team had not actually reached the test animal or the spacecraft yet. For those who still needed reminding, the TASS release also declared that after two successful dog flights in quick succession, 'the time is not far off when a man will be put into orbit and returned to Earth.'

The same mantra was repeated three days later on March 28 at a packed press conference of Soviet and foreign reporters in the grand conference hall of Moscow's Academy of Sciences, where the two dogs Chernushka and Zvezdochka were paraded before the cameras like celebrities. Gagarin, Titov and the other four cosmonauts also attended incognito. The Academy's portly vice president, Aleksandr Topchiyev, another of those figures purposely wheeled out before the press because he knew almost nothing about the Soviet space programme, waxed eloquent about the successful 'Korabl-Sputnik' dog-carrying flights so far, by which he meant the Vostok dog-carrying flights – the word Vostok still being a secret. He also took the opportunity to belittle American efforts to fly a *suborbital* manned mission. 'We don't consider [it] to be interesting – or even space flight, in the true sense,' he sniffed. As for the future Soviet spaceman himself, any suggestion that he would be merely a passenger on such a flight was indignantly dismissed: 'There will be a great deal of work for him to do. There will be a control panel. He won't have time for conversation or card-playing.' While the reporters took notes and the photographers snapped hundreds of pictures of the dogs, nobody took the slightest notice of the six young men in uniforms sitting in the front row.

In the US, the nation's media appeared to be almost as ecstatic as the Soviets, with the story making it to the front page of Sunday's

New York Times on March 26 and winding on through another stack of gobsmacked columns inside – a whole lot more newsprint than the puny paragraph on page 11 the *Times* had devoted to America's own Mercury-Redstone test of the day before. Likewise the New York *Daily News* made far more fuss than its Saturday 'The Redstone Clicks' piece, this time shouting the manifestly obvious in capital letters to the rooftops of all five boroughs and beyond: '*REDS ORBIT, LAND DOG. GET READY FOR SPACEMAN*' – by which the paper clearly did not mean an American spaceman. It all looked like a done deal, surely just a matter of time now before the race was won and the Soviets took the cup of glory. But behind the scenes, in places where neither TASS nor any other reporters ever trod, the story was not quite so simple; and perhaps the deal not quite so done.

SEVENTEEN

THE MILITARY-INDUSTRIAL COMMISSION

March 27, 1961
OKB-1
Kaliningrad, near Moscow

KAMANIN'S INSOMNIA WAS getting worse. Sometimes he lay awake in bed all night endlessly turning over the problems in his mind. While TASS and the *New York Times* were trumpeting this latest success of Zvezdochka's flight, the tough middle-aged war veteran and cosmonaut chief was thinking about all its failures and spelling them out in his secret diary. Even after this second dress rehearsal there were still too many of them.

Within twenty-four hours of Ivan and Zvezdochka's recovery, Kamanin was already noting how the dehumidifier system was still a mess, and that yet more tests would be needed. But as matters stood, 'there is,' he noted, 'no confidence in the cosmonaut's life support equipment in the event of the ship's descent due to natural deceleration' – the descent that would take ten days. In other words, if the Vostok's single braking engine failed, the cosmonaut inside would still likely die. Nothing there had improved.

Meanwhile those water tests of the emergency survival equipment that had been promised days ago had not yet happened: the equipment container, recorded Kamanin, 'is still leaky and poorly stable; its contents will be flooded and its radio transmitters would quickly fail'. Nothing there had improved either. The best that Kamanin and everyone else could hope for was that the cosmonaut never had to face a water landing. If he did, he was as good as dead.

And there were other failures Kamanin did not even write down. TASS had made a big fuss of the capsule landing in the precisely 'predetermined area', but in truth it had overshot that predetermined area by at least 660 kilometres. More alarmingly, once again the two sections of the Vostok had failed to separate seconds after the braking engine had stopped, as they were supposed to. Instead they had plummeted together into the upper layers of the atmosphere. Like the last time only the heat of re-entry had finally burned off the cable joining them. But neither of these two issues, both of them patently major, were even discussed by the engineers after Ivan's second flight. As Boris Chertok recalled in later years: 'Neither I nor any of the individuals still living who were involved in those historical launches can recall why such serious glitches … were not the subject of a report before the State Commission or even a discussion with Korolev.' His own amazement at himself is still palpable, decades after the event.

One by one, the risks were piling up, just as time was fast running out to fix them. But Sergei Korolev kept driving the juggernaut forward, whipping himself and everybody else towards his promised land of human space flight. By now he was working fifteen hours a day, usually seven days a week – he even had his engineers design a specially shaped key to his office at OKB-1 so that he could find it quickly in his pocket. But often he slept in that office, briefly breakfasting on sausage, black bread and black tea before facing another day and half the night at his desk, with its old grandfather clock ticking away behind him, and its telephone with the personal hotline to Khrushchev to remind him of whose favour he ultimately needed if he were to realise those vaulting ambitions that filled his dreams. Apart from his wife Nina, he now saw almost nobody outside his circle of work. He had already missed his daughter Natalya's wedding in January and she saw him only once in the whole of that spring when she paid a surprise visit to his house in Moscow. They walked for an hour or two in the garden, a rare and precious few moments with her father that she would cherish for the rest of her life. 'I told him immediately that I loved him very much ... We kissed and he hugged me tightly.' She thought he looked exhausted.

ON MARCH 30, five days after Ivan's flight, another meeting of the Vostok state commission took place in Moscow. The meeting lasted two hours. It was officially chaired by Konstantin Rudnev, a forty-nine-year-old industrial manager who had replaced Marshal Nedelin in the job after the marshal had been vaporised in his deck-chair in the missile explosion the previous October. But as always the real man in charge was Korolev. After Korolev had reported on the results of the previous flights – without saying anything about the separation failures and duly emphasising all the successes – a vote was cast on the simple question, as Kamanin later noted it down: 'Who is for a human flight next?'

The answer was everybody.

That meeting ended at 6 p.m. By 6.30 p.m., Korolev was sitting down with the Military-Industrial Commission in the Kremlin – the next tier up the official decision-making chain in the Soviet hierarchy. The chairman this time was Dmitry Ustinov, Deputy Premier of the Soviet Union and a Hero of Socialist Labour, the nation's highest civilian honour, awarded to him personally by Stalin. Also present was Pyotr Ivashutin, former deputy head of the Soviet counter-intelligence department SMERSH and now deputy chairman of the KGB, an imposing figure in his early fifties with a face like a slab of meat and whose presence in the room demonstrated the importance with which the Communist Party bosses viewed this project.

Korolev presented his case for manned flight with his characteristic combination of authority, energy, deception and shameless cunning – beginning with the latter, especially given who was in the meeting, by first passing round a couple of albums of photographs taken from orbit in the last two Ivan flights. The photos revealed excellent ground details, especially one of them, a shot of the city of Alexandretta in Turkey, which 'clearly' – according to Kamanin who was at the meeting too – revealed its air base with its 'concrete runway'. Turkey was a NATO member. The military potential did not need further stressing.

Once that was out of the way, Korolev produced a comprehensive document, carefully prepared in advance, containing a detailed report on the status of the Vostok programme and ending with a formal request to the Central Committee of the Communist Party, effectively the USSR's supreme governing body, to authorise the first human flight of a cosmonaut in space. Not one of the issues that were keeping Kamanin awake at night found their way into this report. A single brief sentence – placed in brackets – makes a passing reference to the Vostok's '(air regeneration system, 10-day supply of food and water etc.)', without also mentioning that if that air-regeneration, or air-drying, system ever had to last ten days the cosmonaut would probably be dead. Elsewhere, the report confidently asserts that 'in the event that the satellite ship does not enter orbit due to lack of speed, it can descend into the ocean', which was also certainly true but not the whole truth, given that the 'satellite ship' would then rapidly sink, as might the cosmonaut in his unstable life raft, who was anyway severely handicapped by a 'lack of means', as Kamanin had baldly put it, 'to indicate his location'.

None of these nor any of the other trouble-spots were anywhere in the report. Korolev was a superb manipulator – he had proved it again and again – and now he was both under pressure and determined to deliver the most spectacular of all his space spectaculars. And so the discussions in the Kremlin that evening continued, centred around a document that was necessarily and very largely economical with the truth, if its ambition were ever to be realised.

One by one Korolev ticked off the points that needed to be agreed. First, the name of the spacecraft. To the outside world all these Vostoks were still satellite ships – *korabl-sputniks* – an uninspiring, and by now overused, appellation for such a dramatic and world-changing mission. With this first manned flight the time had come at last to release the secret name – Vostok – 'East'. Intriguingly, an undated page exists in Korolev's handwriting listing eighteen alternative names for Vostok, also all beginning with 'V': *Vozrozhedenie* – Resurrection; *Volya* – Mighty; *Voinstvenny* – Bellicose. But apart from half of these names being totally unpronounceable to foreigners, the public presentation of this flight would need to stress the peaceful, not the martial, purposes of the USSR, whatever NATO bases or runways the cosmonaut might happen to spot out of the window as he passed over the top of them. So Vostok it remained.

As for the public presentation itself, three pre-prepared TASS bulletins were agreed upon. The first would be released after the cosmonaut reached orbit – but before he landed back on earth. The reason was logical, given the current state of the world: should anything go wrong and the cosmonaut landed on foreign soil, 'it would exclude any foreign government from declaring the cosmonaut a spy'. Nobody wanted a Gary Powers scenario in reverse, in which the cosmonaut played the part of the downed spy pilot. No doubt memories of what those villagers might have done to Ivan Ivanovich's face were also fresh.

The second TASS bulletin would be released only after the cosmonaut had landed safely. A third bulletin would be released if he landed in the sea, so that other nations could assist 'in the cosmonaut's rescue', a wise precaution given that the chances of any Soviet sea rescue would be extremely patchy at best, as Korolev well knew. There was no fourth bulletin should the R-7 and its cosmonaut never make it into orbit in the first place. If the rocket blew up after launch the world would not be told anything. Like Bondarenko in his isolation chamber, the cosmonaut would die unknown.

These issues were all soon decided. But another generated more heated discussion in that meeting than all the others, and that was the on-board Emergency Object Destruction device, or bomb. This had been fitted to all seven Vostok flights so far. It had been used once, when the braking engine on Pchelka and Mushka's flight the previous December had gone wrong, threatening to send their capsule beyond the USSR's borders. Before it ever got there the bomb had instantly destroyed it along with the two dogs. Now the KGB deputy chairman, Ivashutin, argued that it should also be fitted to the human flight. Like the dogs the cosmonaut would have to be destroyed rather than risk revealing any secrets about the spaceship. Incredibly, Ivashutin was opposed by every other person present. In the end, faced by this unanimous front, even this stern, powerful, mistrustful figure was forced to back down.

With that resolved, the report was finally ready. The last line stated its request to the supreme power in the USSR 'to authorise the launch of the first Soviet satellite-ship with a human on board', with the date for that launch set between April 10 and 20. Without wasting any more time, Korolev sent the report and its request the next morning to the Central Committee of the Communist Party. The Central Committee was a large and cumbersome institution, formally controlled by a much smaller governing body, the Presidium. But its decisions ultimately came down to just one man: Nikita Khrushchev. Korolev had set the clock ticking. Now he had to wait for a reply.

BUT AS HE waited, there was of course still one outstanding issue the report did not address: who would be that 'human on board'? As the head of cosmonaut training, Kamanin had been observing the six men while they were recently together at the cosmodrome. Over the ten months since he had first met them he had got to know them fairly well; but on this latest trip he had got to know them much better. In a diary entry on March 20, five days before Ivan's flight, he breaks off from a discussion about the Venus probe to 'write a few lines about the six brave cosmonauts':

> In November last year … my meetings with the cosmonauts became more frequent. I was the chairman of the examination committee, I knew their qualifications for the flight, their personal data, but I did not really notice the differences between

> them – they were all cosmonauts for me and nothing more. Today is the fifth day we have been together all the time. I sit in classes with them, we play sports together, we eat at the same table, we play chess, we watch movies. They all trust and respect me, and I'm beginning to notice their purely individual traits and interests. Yesterday, for example, when we all went to the movies after dinner, Titov asked permission not to go with us, but to read Pushkin – it turns out he is interested in poetry and reads a lot. Popovich, Nikolayev, Bykovsky and Nelyubov play chess decently and sometimes cards ... Yura Gagarin is indifferent to cards and chess, but he is keen on sports, and he doesn't miss an opportunity to enjoy a witty anecdote or a funny joke.

Once again Gherman Titov's individualism, his addiction to Pushkin and his occasional preference for his own, rather than for the group's, company is noteworthy – certainly to Kamanin, whose reading of these personalities would play a central part in the final decision. Notable too is Gagarin's humour, the quality that stands out in innumerable contemporary accounts and one that perhaps helped to leaven all that relentless drive and energy. The fleeting reference to Nelyubov is revealing. Since the cosmonauts' examinations in January he was still tentatively ranked third after Gagarin and Titov. But his sometimes blatant egotism meant his chances of climbing into first place were fast slipping away.

Now, back in Moscow, all three of these men recorded solemn 'personal' addresses for future broadcast designed to appear as if they were spoken on the day of their flight. They were also filmed for future purposes walking about Moscow in the melting spring snow with the trees still bare, visiting the Lenin Hills, Red Square and Lenin's Tomb, all the sacred Communist sites, chatting together, laughing, pointing things out while passers-by barely gave them a second glance: just three young officers in uniform enjoying a day's leave in the capital. But Titov, ever alert, noticed that 'the cameramen filmed Yura more than us. And I thought, Ah – it's Yura. Although nothing had yet been decided and of course I still hoped that they would give the first flight to me.'

That walkabout happened on April 3. By then Korolev had been waiting for the Presidium's response to his report and its request for three days. Just before four o'clock that afternoon it finally arrived:

> Presidium of the Central Committee of the Communist Party of the Soviet Union
>
> 3rd April 1961
>
> Top secret
>
> 1. (The Presidium) authorises the proposal ... on the launch of the space satellite ship 'Vostok-3A' with a cosmonaut on board.
>
> 2. (The Presidium) authorises the TASS bulletin schedules on the launch of the spaceship with a cosmonaut on board ... and, regarding the launch, grants the Launch Commission the power, if required, to make adjustments based on the results of the launch, and grants the Commission of the Presidium of the Council of Ministers of the Soviet Union Military-Industrial Commission the power to publish them.

In Soviet-speak that was a yes. After telephoning Kamanin and telling him to 'speed up the departure' of the six cosmonauts back to the cosmodrome, Korolev departed himself. That evening he was on a plane heading for Tyuratam. The dream he had nurtured since he had first ridden with his mother on a magic carpet around the world lay directly ahead. He had promised the supreme power in the land that he would achieve it within the next seventeen days – and possibly in just seven.

ACT III

FINAL COUNTDOWN

APRIL 4–11, 1961

All the time I am haunted by one thought … Who should I send, Gagarin or Titov? Who should be sent to their certain death?

Nikolai Kamanin, diary entry April 5, 1961

EIGHTEEN

BRINKMANSHIP

April 4, 1961
State Department
Washington, DC

AT SIX O'CLOCK in the evening in Washington, as Sergei Korolev was on his way to the cosmodrome in Tyuratam, President Kennedy convened a secret meeting at the State Department to decide what to do about Cuba. The Laos issue was still simmering in South-east Asia but Cuba was now the priority. The question of whether to sanction an imminent CIA-backed invasion of the country and remove its leader Fidel Castro could not be postponed much longer. Over the next two hours, as the capital's dull, overcast skies darkened outside the windows of a seventh-floor conference room next to Secretary of State Dean Rusk's office, some seventeen men around the table hammered out the options before the president.

Leading the case for the CIA was Richard Bissell Jr, its fifty-one-year-old deputy director for plans. This seemingly innocuous title in fact gave Bissell great powers. Under his control were at least fifty CIA stations around the world along with hundreds of covert operations, including the U-2 spy plane programme that had provided the first astonishing aerial photographs of the Tyuratam missile site.

Yale-educated, charming and highly sociable, Bissell was also, in his own words, 'a man-eating shark' and a hawk of hawks. Now he urged the president to back the invasion. One of Bissell's CIA advisors, Colonel Jack Hawkins, confirmed that Brigade 2506, the thirteen hundred Cuban anti-Castro exiles secretly funded and trained by the CIA in Guatemala, were waiting and ready to go. Bissell had also taken on board Kennedy's repeated concerns that the US should not be seen to be behind this invasion by altering a key element of the plan. The original landing location was switched to three beaches at a much quieter, less obtrusive site on the southern coast of Cuba, a hundred miles from Havana: a place called Bahia de Cochinos, or the Bay of Pigs.

Kennedy's sensitivities to accusations of US involvement in the invasion had forced a change in the choice of air cover too. There would be no US fighter jets committed to the invasion. Instead the Cuban brigade would have to rely for air support on a small number of obsolete B-26 propeller-driven bombers dating back to the Second World War, painted up in Cuban Air Force colours and flown by Cuban exiles to maintain the lie that this was an *internal* Cuban insurrection. An advantage of the new invasion site at the Bay of Pigs was that it also had a runway nearby long enough to support the fiction that this was the base from where these 'rebel' B-26s had taken off – when in fact they would be taking off from Nicaragua.

But time was short. The Soviets were busily arming Castro's forces at a pace that might doom any delayed invasion to failure. Crates containing Soviet-built MiG-15 jet fighters had already arrived at Havana's docks, while the Cuban pilots who would soon fly them were completing their training in Communist Czechoslovakia. And the Cubans were daily becoming more aggressive. Less than a week earlier one of their gunboats had intercepted a US vessel owned by Western Union that had been laying telephone cables in international waters between Miami and Barbados. The gunboat turned tail only when an American destroyer arrived on the scene and chased it off. Tensions were escalating. On the very morning of this State

Department meeting the *New York Times* had carried a stark headline on its front page: 'Cuba is Warned'.

Most of the men in the State Department meeting openly supported Bissell's case for invasion, including three members of the Joint Chiefs of Staff, and Secretary of Defense Robert McNamara, the professorial-looking, bespectacled Harvard Business School graduate who appeared to approach every political issue in terms of the balance of profit and loss. The removal of the Soviet-backed Castro in an invasion by non-American Cuban exiles, one in which no American lives would be at stake and no American involvement proven, tipped that balance towards the profit side, and it was this coolly analytical conclusion that McNamara promoted. A few in the room were less sure, including Secretary of State Dean Rusk, a passive, almost faceless figure sometimes disparaged as 'the silent secretary'. Rusk had a reputation for sitting on fences and he appeared to be sitting on one right now; whatever his private reservations, he did not speak up. Only a couple of people disassociated themselves from the invasion altogether. One was the Arkansas Democrat Senator William Fulbright, later to be a passionate opponent of the Vietnam war, who argued that the CIA plan was a frankly morally dubious adventure and who made, as the president's special assistant Arthur Schlesinger wrote afterwards, 'a brave, old-fashioned American speech, honorable, sensible and strong'. The other opposing figure was Schlesinger himself. But he said nothing.

Kennedy ended the meeting by taking a poll of the room. Almost everybody, whether enthusiastically or reluctantly, voted in favour of the operation. Adolph Berle, a Latin-American specialist at the State Department, began to make a long-winded speech when Kennedy interrupted and told him just to vote. Berle did: 'I'd say, let 'er rip!' With the result of the poll clear, the president reminded everybody how essential it was that the invasion be made to look like an internal uprising and that American hands were seen to be clean. Then he stood up. As Senator Fulbright left the room Kennedy turned to him: 'You're the only one in the room who can say, "I told you so."'

The die was cast. But Kennedy nevertheless reserved the right to cancel the attack at any time up to twenty-four hours beforehand. Despite the poll that had just been taken, he was not yet fully committing himself. He appeared both to want and not want this invasion. He was caught between his desire to replace Castro with someone more 'democratic', and friendlier to US interests, and his fear of a questionable and potentially dangerous entanglement, especially with the USSR. But he had, wrote Schlesinger, 'enormous confidence in his luck'. So far that luck had held. The question was, would it still?

Meanwhile a tentative date for the operation was set. Originally it had been scheduled for April 5, then April 10, now it was April 17, thirteen days from the date of this meeting. It was, in other words, within the same time frame in which Korolev had just promised the Presidium of the Central Committee of the Communist Party that he would launch a man into space.

IN MOSCOW, EIGHT hours ahead of Washington, it was early morning the next day. A heavy snow was falling on the city, mocking the late spring thaw that had cheered the capital over the past week. It fell thick and fast through the night, blanketing the streets, transforming the grey city into a pristine winter landscape. Kamanin was there to see it. He was up well before dawn, finishing the last of his preparations before leaving for the long flight back to Tyuratam with the six cosmonauts as Korolev had just ordered. His wife, Musya, was also up to make him breakfast. Parting from Musya was always difficult and it had happened too often recently. But this time things felt different. Perhaps it was because of what lay ahead: 'The cause for which I am parting,' wrote Kamanin in his diary that day, 'has captured me entirely.'

As dawn broke he was on his way to the military air base at Chkalovsky, close to where the cosmonauts lived and where they had undergone their zero-G training in aerobatic planes. Still gripped by the momentousness of the occasion, the grizzled soldier betrays a poetic sensibility while his car sweeps through the still, white city:

> Early in the morning Moscow is deserted and perhaps at its most attractive. The Stone Bridge, Manezhnaya Square, the Bolshoi Theatre, Dzerzhinsky Square, Kirov Street, the railway stations, Sokolniki – all this is very familiar and yet somehow new … The crimson sun is rising, fragments of clouds are scudding past and thinning as more and more blue skies are opening up. The day seems to be a good one.

In his fifth-floor apartment in Chkalovsky, Gherman Titov had also finished packing and was saying goodbye to his wife, Tamara. Their marriage had always been close, but in the months since their infant son had died it had become even closer. As always, she knew everything that he knew, just as she had known about those medical tests fifteen months earlier, just as she had also known from the beginning about this strange and dangerous career her husband had embarked upon. 'I guess I was more aware than any of the other wives,' she said many years later. 'He shared everything with me.' But the one thing her husband could not share was what he still did not know himself: 'He told me that nobody yet knows who it will be. Of all the guys, nobody knows. And then he flew away. And I thought to myself, God forbid if it's him.'

In the apartment next door, Gagarin was also saying goodbye to his family, his wife Valentina and their two daughters, Elena, now nearly two, and Galina, not yet a month old. For the rest of her life until she died in 2020, Valentina would give almost no interviews to historians or the press. But after years of building her trust, a Russian journalist, Yaroslav Golovanov, was one of the very few to whom she did agree to speak. She told him the story of her husband's departure early that morning, one she had never revealed to anybody. This is how Golovanov relates it:

> 'Take care of the girls, Valyusha,' Yura said quietly, and suddenly he looked at me very kindly.
>
> That night we had talked about all sorts of things, we couldn't stop talking. In the morning he checked over his things again – had he missed something? – he clicked the lock of his little suitcase. He kissed the girls. He hugged me tightly.
>
> I suddenly felt weak at the knees and quickly said, 'Please be careful, keep your cool, and don't forget about us …' And I said something else, something incoherent that I can't remember now. But Yura reassured me. 'Everything will be fine, don't you worry …'
>
> And then I asked what I probably shouldn't have asked, the question that was burning me up inside:
>
> 'Who will it be?'
>
> 'Maybe me, maybe someone else.'
>
> 'When?'
>
> He paused a second before answering. Just a second.
>
> 'The fourteenth.'
>
> Only later did I realise he'd given me this date so I wouldn't worry on the real date.

The tentative date for the flight was now April 10 or 11. But Kamanin had instructed all six cosmonauts to tell their wives that it would be the fourteenth, and for precisely the reason Valentina gives.

THERE WERE THREE aeroplanes altogether, small, propeller-driven Ilyushin IL-14s, and they were waiting on the snow-decked apron at Chkalovsky. The cosmonauts were too precious to lose in a plane crash and so they were split between two of the aircraft. Gherman Titov would fly in one, along with the stolid, imperturbable Andriyan Nikolayev and Valery Bykovsky, the man who had replaced Varlamov in the Vanguard Six after the latter's diving accident the previous summer. Gagarin, Grigory Nelyubov and Pavel Popovich would travel in the second plane. They would be joined by

Kamanin along with Dr Vladimir Yazdovsky, the autocratic chief at the Institute of Aviation and Space Medicine. Yazdovsky would lead the medical team at the cosmodrome. His voice was decisive. Whoever ended up riding the rocket into space would have to be cleared as 100 per cent fit first. And not just physically fit either: the team would also need to make a detailed psychological assessment of the chosen candidate before he could be allowed to fly.

The third aeroplane was reserved for Yazdovsky's doctors, nurses and additional camera crews to record the events of the next few days and create a carefully edited version for posterity; that is, should everything be successful. Vladimir Suvorov was not among them – he was already at Tyuratam – but his colleagues took a moment to film the six cosmonauts before they left. Six decades later some of the raw colour footage of that moment has survived in a Moscow archive. The cosmonauts pose beneath the wing of one plane, standing on the fresh snow in their thick army greatcoats, grinning for the camera in the morning sunshine like young men taking selfies before a holiday. The moment is brief; the hand-held camera pans to the ground and abruptly cuts. Then they all boarded their separate planes, taking off at fifteen-minute intervals.

The last time the cosmonauts flew to Tyuratam three weeks earlier, they had stopped for the night halfway at Kuybyshev; but this time, with a generous tailwind urging them south, they flew the two thousand five hundred kilometres in one long leap. Footage from the flight has survived too, more fleeting moments: Kamanin hunched over a game of chess – his favourite hobby – his figure trim and fresh despite that early-morning start, his hair neat and receding, impressively bemedalled in his lieutenant general's uniform. He plays with Popovich while Nelyubov checks maps of their route. Something in Nelyubov's expression chimes with Kamanin's judgement of his character, one that many shared: a pout perhaps, a subtle curl of the lip suggesting that arrogance Kamanin noted in his diary. And then there is Gagarin, at one point sitting in the co-pilot's seat and having a go with the controls, something of a treat for him or any of the cosmonauts since they rarely, if ever, got the chance to pilot planes, unlike their American rivals with their NASA supersonic jets. Gagarin turns round suddenly to smile at the cameraman; and his smile, as ever, is dazzling. The camera seems to linger on him more than on the others, as it also does on Titov in the other aircraft, who examines his map, or chats to Bykovsky or, most often, looks out of the window, his handsome face noticeably pensive.

'All the time now, as I am writing these lines,' Kamanin confides in his diary on that same April 5, 'I have been haunted by the thought: Who should be sent on the first flight, Gagarin or Titov?'

> Both are excellent candidates, but in recent days I have heard more and more statements in favour of Titov, and I myself have increased my faith in him. Titov performs all his training exercises more precisely, more sharply, and he never uses more words than necessary … Titov has the stronger character. The only thing that keeps me from deciding in his favour is the need to keep the stronger cosmonaut for the second, much longer, flight that will last a whole day. This second flight of sixteen orbits will undoubtedly be more difficult than the first flight of a single orbit. But mankind will never forget the first flight and the name of the first cosmonaut. The second and subsequent ones will be forgotten as easily as the subsequent records are also forgotten … I have just a few more days finally to resolve this issue.

The decision would not be exclusively Kamanin's to make, even as the head of cosmonaut training. Others would be involved, notably Korolev now waiting in Tyuratam. Above all there was the Soviet political hierarchy whose stake in the matter was paramount. Before he also left for Tyuratam, Yevgeny Karpov, the chief of the Cosmonaut Training Centre, took a portfolio of photographs to the Defence Department of the Communist Party Central Committee. The photos were of Gagarin and Titov – notably none of Nelyubov – dressed in both uniform and civilian clothes. It was the first step in a labyrinthine and gloriously Soviet trail to the top, as the journalist Golovanov describes:

> Yevgeny Anatolyevich Karpov took the photos of Yuri and Gherman along to the Defence Department of the Communist Party Central Committee to show them to the comrades there … Each of them silently examined the photos with a serious expression, moving their eyes from the uniformed to the civilian shots and back again: both candidates were photographed twice. Then Ivan Dmitrievich Serbin took the photos to show Frol Romanovich Kozlov, Secretary of the Central Committee and in those days the second-in-command of the country. Frol Romanovich later showed them to Nikita Sergeyevich [Khrushchev]. He took one look and said: 'Both youngsters are great! Let them pick one themselves!'

Which, if true, was not much help to Kamanin. Of course it may also not be true: Khrushchev's son, Sergei, later claimed that Khrushchev never saw any photos, dismissing the whole story as 'one of those fairy tales that usually spring up around especially remarkable events'.

And yet, whatever really happened, it is almost incredible to reflect that this final, crucial, choice had still not been made, especially when compared to the Mercury astronauts; not even now, just a week or so before a cosmonaut was expected to fly in space. Perhaps the sheer magnitude of the decision was enough to keep putting it off until the very last moment. But the waiting was especially hard for Gagarin and Titov. They were close friends who had shared intimacies and tragedy. They were also the most competitive of rivals, both reaching for the same ultimate prize. But only one of them could be first.

Now, as the three planes landed at Tyuratam that same afternoon, everybody found themselves in a world far removed from Moscow's snow. The cosmodrome was enjoying its fleeting moment of spring. Everywhere the steppe was ablaze with tulips. It was a beautiful sight;

but Kamanin was still oppressed by the question that kept circling in his brain. He returned to it in almost the last words of his diary entry that night, weighing the enormous fame against the enormous risk:

> So which one will it be, Gagarin or Titov? ... It is difficult to decide which one of them to send to their certain death, and it is just as difficult to decide which of these worthy men should be made world famous, whose name will be forever preserved in the history of mankind.

NINETEEN

JUST IN CASE

April 5–7, 1961
Tyuratam Cosmodrome
Soviet Republic of Kazakhstan

SERGEI KOROLEV WAS there to greet the cosmonauts as soon as they landed. He was smiling and made a joke. But behind the smiles and the jokes lay all the worries Kamanin had also been worrying about day and night, and the two began discussing one of the biggest of them before they had even left the airport. The Vostok's dehumidifying, or air-regeneration, system was *still* not working properly. Every test so far had totally failed to absorb the water content in the cabin's air over any length of time. They always left those puddles of brine in the cockpit, puddles that in weightless conditions would not be puddles at all, but free-floating drops of chemicals that could easily short-circuit critical electrical systems as well as enter the cosmonaut's respiratory system and possibly poison him. And even if that did not happen there was the outstanding matter of the cosmonaut's sweat, the perspiration from his own body that the system would progressively be less capable of absorbing the longer he remained in space, gradually raising the cabin's humidity to dangerous and ultimately fatal levels. This was not an air-regeneration system Korolev was dealing with; given sufficient time, it was quite possibly an air-killing system.

There was no question that OKB-124, the research institute responsible for this life-support system, had failed dismally in its task. Its chief designer, Georgy Voronin, had more than once felt the lash of Korolev's tongue. But there was no more time to fiddle with tests. Voronin's system would have to do. At a key meeting of the Vostok state commission on April 6, the day after the cosmonauts arrived, Korolev insisted on retaining it for the first human flight, flawed though it was. As always his powerful personality dominated the room. Nobody wanted to quarrel with their bull-necked chief or face one of his lacerating tempers, of which there had been rather too many recently. The deadline Korolev had promised the Presidium would be met, even if it put the cosmonaut's life at even further risk.

And it did put him at further risk. By agreeing to retain the defective system, the state commission was effectively staking everything on the Vostok remaining in orbit only for the hundred-odd minutes it was supposed to – which meant that the Vostok's single braking engine *had* to work. If it failed there was no guarantee the cosmonaut would remain alive for the ten days it would take for gravity to get him down. And yet this same braking engine had at least partially failed once before, on a Vostok dog flight the previous December. On that occasion the two dogs inside, Pchelka and Mushka, had been destroyed by the on-board bomb. A human cosmonaut faced with a similar failure would at least escape that fate, despite the best efforts of the KGB to have it otherwise. But just how reliable was this engine? It was hardly encouraging that even its own designer, Aleksey Isayev, had fretted about it from the start and had even had to be corralled by Korolev into taking on the job: 'If a man fails to return to earth because of me? All I can do is blow my brains out!' Such sentiments were not guaranteed to inspire confidence.

But it was all they had. Or almost all. If Isayev's worst nightmare ever came true and his engine simply refused to start, the options were limited. But other failures, potentially also fatal, might even yet be turned around if the design philosophy that had shaped and sustained Korolev's entire thinking about the human role in space flight over the past decade were changed, and changed now at the eleventh hour: if, that is, the cosmonaut himself were allowed to switch off the automatic systems in an emergency and attempt to *fly* his spacecraft home.

There were two critical operations that had to work perfectly before the Vostok could return safely to earth. One was starting that braking engine. The other was earlier, when the capsule needed to orient itself into the correct angle for re-entry. That angle had to be *precise*: too steep and the capsule would burn up on its plunge back into the atmosphere; too shallow and it would skip higher into space. For the cosmonaut inside, the outcome in either case was the same. The only difference between them was how long it would take for him to die.

Like everything else with this flight, both these systems – orienting the capsule correctly and firing the braking engine at the right moment – were fully automatic. Almost every system on board the Vostok was automated: 'Even the seatbelts were tightened automatically,' one engineer remembers. On his left the cosmonaut had a black panel with a few switches he could be trusted with: he could do things like turn on the tape recorder, for instance, or his air circulation fan; he could adjust the cabin lighting or turn the volume of his radios up or down – not much more than a modern airline passenger. When others had dared to challenge Korolev over this degree of automation, he had rammed through their concerns. But one concession had been made – a '*just in case*' scenario, as the engineer Boris Chertok put it – in which the cosmonaut could, in an emergency, access a *manual* control system that was normally kept locked with a three-digit code. Once the correct digits had been entered, the relevant controls would unlock. The control system was limited, far more so than in the Mercury capsule where there was also no lock; but it would allow the cosmonaut to perform those two crucial actions himself if the automatic system ever failed: to manoeuvre his spacecraft into the correct re-entry angle and then to press the button that started the braking engine. In other words, it would allow him an outside chance to get home.

Now, with so much at stake, Korolev broke with his own philosophy – *just in case*. And he lost no time doing it. Hardly had the cosmonauts set foot on the ground at the cosmodrome airport before he told Kamanin to get them to practise emergency manual descents and to make sure they did it 'thoroughly'. They had already had some training on the crude simulator back in Moscow, but clearly not enough. The thirteen-page Vostok manual authorised by the Communist Party's Central Committee – itself shortened on Korolev's insistence – contained just two paragraphs on hand-flown descents. To perform a successful descent from orbit to earth was a delicate and complex process. The Mercury astronauts, all highly experienced test pilots, had been practising the procedures in their more sophisticated simulators for almost a year. The cosmonauts had less than a week.

WHY DID KOROLEV change his mind? Why make a decision that flew in the face of his core principles, and make that decision so late?

Korolev had always pushed himself to his limits; but in those final days before the first human space flight he worked, if that were possible, even harder. And where he went others followed. Almost every witness from that time recalls the relentless pace demanded by their chief. 'We worked fifteen days a week,' said Oleg Ivanovsky, who had first seen Korolev tearing past in his German Opel years before. Ivanovsky and his men grabbed whatever odd hours of sleep they could in corners of the workshop. Others did the same. 'Negligence was a crime,' said Vladimir Yaropolov, another engineer. 'Were a specialist to show any negligence, he would be immediately fired.' Everybody felt the pressure from above. And just like their chief they felt that pressure from the Americans too, 'in much the same way as an opponent's breath', as Mark Gallai, the cosmonaut instructor put it, 'on a runner's back'.

But however tough or backbreaking the work, those same witnesses also recalled something else from that unforgettable period in their lives: a dramatic sense of purpose, of exhilaration even, as they prepared to put the first human being in space. 'Of course everyone understood what this meant,' wrote Boris Raushenbakh, a department chief at Korolev's headquarters. 'The first manned flight in space! Everyone was fully aware of just how extraordinary a moment this would be.' Again Gallai vividly captured the mood: 'There was something special that could not be accurately described, but it was clearly felt by everyone and anyone. This time it wasn't dead dummies or even experimental animals that were going to be flying in space. This time a man would fly!'

And that, of course, was the point. Korolev's responsibilities were not only to his teams or his ambitions or even his lifelong vision. *This time a man would fly*. Any mistake, any piece of negligence, any hour of the day or night not worked by him and his engineers, might cost that man his life. And these were men he knew and liked, perhaps even loved, men who also trusted him. They were his 'Little Eagles', brave young warriors who would soon go where he had dreamed of going almost his whole life, but now never would because he was too old.

Two of those little eagles, the two finalists, he had grown especially close to. There was Titov, clever, cultured, questioning, never afraid of saying what he believed even to his own disadvantage – 'I like that brat,' Korolev once told a colleague. And there was Gagarin, who had caught his attention in that first meeting over a year before, and who caught it again three months later when the young man removed his shoes before stepping into one of Korolev's pristine Vostoks. Later Soviet accounts almost invariably mythologise the relationship between the two men. But something in Gagarin unquestionably struck a chord with the older man, some echo perhaps of his younger self. 'Although they belonged to different generations,' says Rostislav Bogdashevsky, a psychiatrist who came to know both men well, 'Korolev was like a father to Gagarin and Gagarin was like a son to Korolev.' That may be a professional's interpretation; but Korolev's own daughter, Natalya, also said it: 'Sergei

Pavlovich loved Gagarin as a son. Literally as a son.' In almost every photograph of the two from this period up until Korolev's death, the warmth and tenderness between them is unmistakable. They really do look like father and son.

And so in those final days before the flight, Korolev worked himself to the limits of the tolerable – and beyond – in order to bring his young warrior home alive; even if that meant in the end uprooting long-held principles about the level of cosmonaut control in one of his Vostoks. He was gambling with a man's life; at the very least he would give that man his best chance of surviving the ordeal. But the burden of that gamble cost Korolev his peace of mind, his sleep, his temper, some of his friendships, and above all his health. He was now fifty-four. Despite his square, imposing figure and his show of energy, the old pains from his time in the Gulag had never healed and they were getting worse. He needed medication for his heart, and by the time the cosmonauts arrived at the cosmodrome he was also ill. But he kept his illness hidden from everyone except his wife, Nina, the one person in the world he could confide in. Her worries spill over in a letter she wrote to him from Moscow on April 6, the day after the cosmonauts had landed in Tyuratam:

> My dearest Kotya
>
> You are most likely sick and hiding it from me as usual until you feel better. And I am completely powerless to help you.
>
> I am sending you cough medicine ... Please take care of your throat, and wear a warm scarf if you're going outside, but better still don't go out, although this is probably impossible right now.
>
> How nervous I am right now, but this is only a feeling inside. I try not to show it. My dearest, I believe very much that all will be well ... All my thoughts are with you right now, even though we are far apart.
>
> I know how difficult it all must feel. But you must be calm, otherwise your nervousness (which your voice will betray) will show itself to others immediately.

I'm sending you a big hug and warm and tender kisses. And I wish you huge success.

Yours always,

Nina

TWENTY

THE EGG TIMER

April 7, 1961
MIK Assembly Building
Tyuratam Cosmodrome

DESPITE HIS CHIEF'S urgency at the airport, Kamanin did not begin the cosmonauts' manual descent training straight away.

Not until April 7, two days later, did they start. Even then only three hours were allocated, shared between Gagarin, Titov and Nelyubov. The training did not take place in the Vostok itself, now being checked out in the MIK assembly hall, nor even in a full-scale simulator – which only existed in Moscow – but in a laboratory where an instrument display and a control board had been hastily fitted up. The three men took it in turns to practise the manoeuvres required to descend from orbit to earth in an emergency. Since only one of them could practise at a time, the others went upstairs to another room in the intervals to rehearse their radio calls. 'The cosmonauts sent off reports to base and got used to their callsigns,' wrote one of the engineers, Vladimir Solodukhin. 'Gagarin was *Cedar*. Titov was *Eagle*.'

The entire scenario has a surreally casual, even play-acted quality, performed on a set that appears to be essentially jury-rigged for the occasion and far removed from the terrifying realities a cosmonaut would have to face up there in a genuine emergency: alone in his tiny sphere above the earth, possibly out of radio contact, racing at five miles per second in the most hostile of environments, and somehow getting back alive.

One by one, sitting at their instrument display in the laboratory, Gagarin, Titov and Nelyubov went through the steps. First, the controls had to be released by punching the three-digit code on a keypad to the left. This unlocked a little lever on the right – like a modern gaming stick – that gave the cosmonaut limited control in two axes, firing jets of nitrogen gas into space that would, *if* he got it right, steer his craft into the correct angle for a safe re-entry. In the automatic system this angle was achieved by sun-seeking sensors and a crude analogue computer. In the manual system, the cosmonaut would have to rely on the Vzor – the word means 'Gaze' – a porthole in the floor that also acted as an optical orientation device. Using tiny, delicate twists and turns of his little gaming stick, he would have to try to line up the porthole with the earth's horizon using only his eyes until all eight lights around it flicked on, meaning the correct re-entry angle had been achieved. And he would have to do it fast before he ran, literally, out of gas. Otherwise he was not coming home.

That was half the battle. The other half came next: firing the braking engine itself. This would normally fire automatically at a pre-set time triggered by an on-board sequencing device called a Granit, a sort of advanced – but perhaps not all that advanced – egg timer, which switched on as soon as the Vostok reached orbit and then mindlessly clicked its way down the minutes of flight, turning on various automatic systems at pre-set times until finally, at the appropriate moment, it turned on the braking engine's ignition. But if the egg timer failed the cosmonaut would have to start the engine himself. That same three-digit code that unlocked his little gaming stick also unlocked the switch that fired the engine: his *get-home* switch. To know just when to fire it he was supposed to press a button that spun the miniature schoolroom globe on his instrument panel

– basically a clockwork device – to the point on earth where he was predicted to land. In this light it is not altogether surprising why the KGB deputy chairman Pyotr Ivashutin was so keen to retain the on-board bomb with the human flight. In an emergency it was theoretically perfectly possible, albeit very unlikely, for the cosmonaut to defect and steer his spacecraft directly towards America – without anybody being able to stop him.

After their three allotted hours of practice – roughly an hour for each of the three cosmonauts – Kamanin was confident the job was essentially done. 'Gagarin, Titov and Nelyubov know manual descents perfectly,' he wrote in his diary. There would be further training on the following day, this time in the Vostok itself, but Kamanin was clearly keen to get on with the sort of training the cosmonauts were much more familiar with: keeping fit. After they all left the laboratory there were two packed hours of sports, including badminton, one of Kamanin's favourite games and which, in his own words, he 'began to train the cosmonauts in', rather as they had just been trained in emergency descents from space.

Some of that afternoon's activities were filmed and, like the material from the plane journey, the original footage – unseen for almost sixty years – has survived. Its very ordinariness is a revelation, plunging the viewer back in time and bringing these men, all of whom are now dead, back to vivid life. Shot from various angles, the six cosmonauts jog beside the Syr Daria river, past their dormitory building, past a rather fine wooden gazebo on the riverbank providing shelter from the glaring sun, a nostalgic piece of old Russia planted incongruously in this alien landscape. In the silent footage – the sound has long since vanished, if it ever existed – the men run and lark about, as if this were a Soviet holiday camp on the Black Sea. A cleaner rests on her broom and watches them curiously. It is warm; Titov plays badminton bare-chested and pairs with Gagarin against Kamanin and Popovich; the cameraman makes sure to get several takes of the two front-runners leaping for the shuttlecock, heroic low-angle shots against a wide sky. All the men look incredibly fit, 'joyful and healthy', as Suvorov wrote in the diary he was obliged to deposit with the KGB for safekeeping. They also look very young.

By this time the cosmonauts knew some – but not necessarily all – of the dangers one of them would shortly face, because both Kamanin and Yevgeny Karpov, the head of the Cosmonaut Training Centre, had told them. On their last visit to the cosmodrome Korolev himself had spoken to them about the Vostok dog flight failures. He spoke to them this time too. 'They had a real men's talk,' Karpov told the engineer Oleg Ivanovsky. Korolev 'did not deny that he could not give a 100 per cent guarantee of safety'. And of course the men had now just spent three hours practising emergency descents. But in those unedited clips from the archive's vaults, that carefree mood and sense of fun only days before the actual flight feels authentic. Perhaps it helped that Kamanin kept them busy, working their young bodies hard, allowing them space to let off steam rather than brood on the realities that lay ahead. Later that same evening he took them all off to watch the movie *Ostorozhno, Babushka! – Be Careful, Grandma!* – a hit new comedy about an old lady who takes enormous risks to help build a collective workers' club. The ironies are almost too perfect to believe they were lost on the six men.

And yet in the miasma of Soviet accounts it can sometimes be hard to know what their private thoughts and feelings were in those final few days. Gagarin's *Pravda*-ghosted 'autobiography' *Road to the Stars* contains stacks of sentences about duty and his devotion to the Communist Party, yet frustratingly little else. Reared – one might say immersed – like every Soviet child of their generation on heroic tales of self-sacrifice and even 'kamikaze'-style attacks by Soviet pilots on the enemy during the Great Patriotic War, the cosmonauts were no doubt ready to die for their country should that prove to be necessary. They were soldiers after all. But that did not prevent them from feeling fear. Possibly a more revealing witness is Pavel Popovich, the group's sunniest member, who in later, post-Soviet, years was able to be honest: 'You have to push that fear away, force it down and crush it with all your willpower,' he said in a television interview in 2009, the last year of his life. 'I can't say we were not afraid – no, of course we were – I'm a human not a piece of wood. But using our willpower we could crush our fear. And that's what we did.'

TWENTY-ONE
CHIMP BARBECUE

April 6–9, 1961
Naval Air Development Center
Bucks County, Pennsylvania

WHEN JOHN GLENN was flying sleek F-86 Sabre jets in the Korean War back in 1953 he often flew north all the way up to the Yalu river along the Chinese border, to the place the pilots called 'MiG Alley' because that was where the Soviet-built enemy jets often flew too. The dogfights were fast and furious up there, killing fields in the skies with planes twisting and screaming at speeds of over 600 mph as they tried to shoot each other down. But sometimes the MiGs failed to appear and there were no dogfights, and so Glenn and his squadron would go searching for 'targets of opportunity' instead, hunting down North Korean troops or trucks or tanks on the roads and shooting them up.

One time he was there with his squadron commander, John Giraudo. Giraudo had just completed his strafing run when an anti-aircraft shell knocked out his controls and his plane tumbled helplessly towards a mountain. He ejected, parachuting safely to the ground; but he was two hundred miles inside enemy territory. Glenn decided to stay with him. He flew down, circling Giraudo's position to keep the enemy troops away until a rescue helicopter could come. And even though he was beginning to run out of fuel he kept circling, waiting for the helicopter that had still not arrived, protecting his friend for as long as he could. He stayed so long that when he was finally forced to head back to base before his fuel ran out, his fuel *did* run out and he had to 'dead-stick' home, gliding with his engine stopped and useless. And when, incredibly, he managed to land safely without his engine he hung around just long enough to report where his downed commander was, and then jumped right into another plane and flew north to get back to him – and protect him again. But by then the North Koreans had captured Giraudo.

Glenn did not see his commander again until after the war. But the story reveals something fundamental in the man. Just as he was once there for Giraudo, against the odds and against even his own interest, so he was there now for Shepard, the man who had jumped ahead of him in the line to be America's first astronaut. Back in January he had tried, and failed, to get the decision to place Shepard first overturned. Then in February, after he, Shepard and Gus Grissom had been announced jointly in their press conference as the 'First Team', he had had to suffer the private agonies of everyone thinking he *was* first when he was not – when, in fact, he was third after Gus. And then of course Shepard was almost everything Glenn was not, a man whose values, personality, taste in cars and indeed whole way of life was nearly exactly opposite to his own. Nobody had forgotten the fraught tempers between them over the nameless astronaut who had got caught with a prostitute. There was no reason for these two men to like each other very much, nor even to function well together, despite their common profession.

And yet that is exactly what they now did. In the weeks since that February press conference, and still more since Shepard lost his space flight to Wernher von Braun's unmanned Redstone test in March, the two men became unexpectedly close. Glenn was there for Shepard, just as he had once been there for his squadron commander in Korea. He was there to shadow him, to advise him, to help him, to take the meetings Shepard could not take, to make the phone calls he had no time to make. With a packed schedule of training, the pair were now spending more time with each other than they were with anyone else, including their own wives. 'I don't think two people could have worked more closely together than we did,' Glenn wrote later. As Shepard's back-up, he was 'Al's alter ego, his virtual twin.' Like everything else he did, Glenn put all he had into the job – all 100 per cent.

And in his own way Shepard returned the compliment. From time to time, always in private since the news was not yet public, he could not resist teasing Glenn as 'my back-up'; but he also, explicitly, paid tribute to Glenn's kindness, not the sort of epithet that Shepard, cool, wary, perhaps even icily reserved as he could be, often used – but when he did he really meant it. At the Cape, Shepard and Glenn had taken to running together before breakfast, hammering the miles of Atlantic surf just as Glenn had been doing on his own for the past two years. Sometimes they even chased crabs on Cocoa Beach like a couple of kids; but most of the time, and by now almost all of the time, they were working very hard.

On April 6, the day after the six Soviet cosmonauts had landed in Kazakhstan, Shepard, Glenn and Grissom arrived at the Naval Air Development Center in Bucks County, Pennsylvania, to begin four concentrated days of final 'refamiliarization' acceleration trials on the facility's centrifuge. To the astronauts the centrifuge was a familiar sight – not to mention a familiarly unpleasant experience – but the purpose of their visit this time was for all three, and for Shepard in particular, to get used to the varying G-forces they could expect to encounter in different phases of the actual flight by replicating its profile. Everything was set up to condition the astronauts to the real thing. They wore the silver aluminised-nylon spacesuits that had been specially made for each of them by the tyre company B.F. Goodrich in Akron, Ohio. Each suit had been tailored exactly to each individual astronaut's body by plastering every inch of that body, while he was naked, in strips of wet paper like papier mâché to ensure the perfect fit. Unlike the cosmonauts, there were enough suits for everybody.

In 1961 the US Navy's centrifuge in Bucks County was the biggest in the world. At the end of its fifty-foot arm was a gondola, inside of which was a replica of the Mercury's cockpit. The gondola was pressurised just as the capsule would be, and filled with pure oxygen for the astronaut to breathe just as he would do in the real thing. The centrifuge was capable of spinning the gondola and its occupant up to speeds of 175 mph – it was so powerful that its base had to be anchored into bedrock in case it tore itself loose – but for these refamiliarisation trials the accelerations were limited to approximately 11 Gs. This would still make the 165-pound John Glenn temporarily weigh eleven times his weight or 1,815 pounds, as much as three grand pianos: more than enough to pummel the breath out of him, pool his blood in his abdomen and screw his eyeballs into his skull.

At the same time as Gagarin, Titov and Nelyubov were learning how to make manual emergency descents in Tyuratam, Shepard, Grissom and Glenn were taking turns to ride the centrifuge in Bucks County. But they did it over four days, not three hours. They practised all the things they would have to do on the real flight, which included the actions of flying the capsule manually, using a little hand controller on their right side not dissimilar in size and shape to the hand controller on the right side of the Vostok: two almost parallel actions on opposite sides of the globe, except of course that the Americans had been practising for nearly a year. These were final *refamiliarisation* trials. They were simply polishing off their skills.

But then the Soviets were committed to putting a human in space within days. The Americans, however polished and well-honed their astronauts' skills, were not. In Glenn and Shepard's growing friendship, in their mutual burial of old hatchets, lay the plain fact that they now had a mutual enemy: not the Russians, but people inside their own government. By the time they were training in Bucks County, Professor Donald Hornig's Mercury investigation panel had been up and running for two months, and its ten members, two technical assistants and two special consultants were still active – very active. They had already got Shepard to show them how he coped with the

Multiple Axis Space Test Inertia Facility, the 'vomit machine' that spun crazily in all three axes at once; they had got him into NASA's water tank in Langley, fully dressed in his spacesuit, to show them how he managed to exit from a mock-up of the Mercury capsule; they had watched him ride the centrifuge all the way up to a brutal 15 Gs to show them how he bore that too. And on and on it went. The members of Hornig's panel in their suits and ties – they were all men – never seemed to stop taking notes and asking more questions. 'What the hell can we tell these ... people that we haven't told them ten times?' Shepard complained to a NASA official. The answer was clearly not enough.

Meanwhile those rumours about the panel intending to recommend more chimpanzee flights refused to die. Sometimes it was another thirty chimps, sometimes it was even fifty, but whatever it was it made Shepard mad. Since Ham's flight he had become hair-trigger sensitive on the subject of chimpanzees. Nor did it help that a British cartoon was doing the rounds in which a chimpanzee was depicted before a blackboard lecturing a very attentive Shepard, Glenn and Grissom, and saying: '... then, at 900,000 feet, you'll get the feeling that you *must* have a banana!' Shepard was so sick of the whole thing that he told Glenn he was ready to have 'a chimp barbecue'. And when on one training session in the simulator somebody joked, 'Maybe we should get somebody who works for bananas', Shepard picked up an ash tray and threw it at the man's head. He only just missed.

NASA was still publicly saying that the flight was 'tentatively scheduled for this spring' – an opaque pronouncement at best. The Redstone rocket that *might* carry Shepard into space had already arrived at the Cape at the end of March; within that first week of April it had been erected on the pad, shortly to be mated to the Mercury capsule that Shepard might get to fly. A special 'white room', the access chamber to the capsule, was now being hastily rigged on the third level of the gantry. The capsule itself was undergoing final checks and tests in Hangar S. Just as in the Soviet cosmodrome, the pace of activity was quickening – it was even frantic. The engineers were very nearly ready. The mission controllers were ready. The astronaut was certainly ready. But still there was no definitive date or decision. As those early April days slipped past, Shepard could only pray that the Soviets were experiencing delays of their own. Things had been very quiet since their last dog flight in March. Perhaps there was even yet a chance? The problem was it was impossible to tell. The Soviets kept so many secrets. 'We don't even know,' said John Glenn, 'what we're racing against.'

TWENTY-TWO

CHOSEN

April 8, 1961
Site 2
Tyuratam Cosmodrome

ON THE EVENING Shepard, Grissom and Glenn completed their centrifuge training in Bucks County, a meeting of the Vostok state commission in Tyuratam finally fixed the launch window for the world's first human space flight: between Tuesday, April 11 and Wednesday, April 12. This was halfway within Korolev's proposed timeline given at the end of March to the Presidium of the Communist Party Central Committee. It was also, at most, less than a hundred hours from the time of the meeting.

The seventeen members of the commission had been swollen by the recent arrival of Kirill Moskalenko, a fifty-eight-year-old veteran of the great wartime battles of Kursk and Stalingrad, a Marshal of the Soviet Union and, since 1960, the commander-in-chief of the USSR's Strategic Rocket Forces. His very presence highlighted the military's grip over the entire undertaking; not that anybody needed reminding. Along with everyone else, including Korolev and, as ever, the chronicler Kamanin, Marshal Moskalenko signed his name at the bottom of a document approving the decision to make the flight of a single orbit of the earth. The expected parameters of that orbit were 182 kilometres at its lowest point, 217 kilometres at its highest. The landing place would be safely inside the USSR.

By now it will hardly come as a surprise that this meeting was held in secret. As always, the cosmonauts themselves were not invited. They were anyway occupied practising manual procedures once again, and for the first – and only – time in the real Vostok itself inside the MIK. Meanwhile the state commission members hammered out a number of key issues that had still, even at this very late stage, not been resolved and were clearly causing something of a collective headache. One of those headaches was the landing site. The question was simply this: should it be revealed? And for that matter, since the issue was obviously linked, should the launch site, the secret missile complex where they were now having their meeting – should *that* be revealed too?

Two men argued vigorously that they should not: one was Moskalenko, unsurprisingly given his rank and military background. The other was Mstislav Keldysh, a mathematician and leading member of the Soviet Academy of Sciences. In many of the film clips shot by Suvorov and his colleagues over the days before the flight, Keldysh, fifty, elegant, slim and generously silver-haired, is a standout figure, largely because he is one of the very few (along with Korolev) who wears civilian clothes. Among all those uniforms he looks bizarrely out of place; and in many ways he was, hailing from aristocratic stock, and having had four members of his immediate family at various times arrested by Stalin's secret police, including a brother who was shot. Given that this executed brother had been suspected of being a German spy and a second brother had once been arrested on suspicion of being a French one, perhaps Keldysh had learned from experience to toe the line. At any rate he now did and sided with Marshal Moskalenko.

But Korolev disagreed with them both. It had already been decided the previous month to register the flight after it took place for a number of world records, including the world altitude record, with the international body responsible for such matters, the Swiss-based Fédération Aéronautique Internationale or FAI. This would officially formalise the USSR's trumping the US in the ranks of aeronautical achievement and nullify any capitalist accusations of fake news. But when it came to altitude records the FAI had a number of awkward rules. One of them was that both the launch and landing sites had to be revealed. Another was that the 'pilot' had to land in the same 'vehicle' he also took off in. In the case of the cosmonaut in his Vostok that would not happen; he would certainly be going *up* in the vehicle but he would not be coming *down* in it, not at the very end, because he would have to eject and land separately beneath his own parachute, in the process destroying any chance of the Soviets claiming the world record.

Unless of course they just lied about it. Which is what the state commission now decided to do. It was agreed at the meeting that the official announcement would state that the cosmonaut *had* landed in his capsule. The whole ejection business would be covered up. To add weight to this 'fact', the USSR's sports commissar, Ivan Borisenko, a man whose official role was Executive Secretary at the Commission for Sporting Technical Matters – in itself a helpfully impressive title – would sign a formal report at the landing site confirming that the cosmonaut had indeed landed safely inside his capsule.

The location of that landing site would be given to the FAI, but not the launch site. As the Soviet Union's latest and largest missile complex, Tyuratam's secrets had to be protected. And so another lie would be concocted, placing the launch point approximately 280 kilometres north-east of the real thing at a tiny Kazakh settlement called Baikonur. In order to put American spy planes off the scent a fake launch pad, made out of wood, had been constructed there. When photographed from high altitude it was set up to look like the real thing. The fact that the CIA already *had* excellent surveillance photographs of the real thing was of course not known to the members of the state commission. In a truly Soviet zeal to cover everything up, the fake wooden launch pad was their best bet. It even had its own armed guards, although in truth they were not there to add to the illusion of reality. They were there to stop the locals stealing bits of the 'launch pad' for firewood.

Having disposed of that headache the commission turned to another: the three-digit code that the cosmonaut would need to unlock his manual controls in an emergency. The difficulty here concerned the cosmonaut's state of mind. The worry that he might go insane in space had never disappeared. There were those isolation chamber experiments, of course; but being alone in a chamber anchored to the ground floor of a medical institute in Moscow was not the same as being alone in orbit above the earth. There was even a respectable theory for going mad in space: it was called *space horror*, and its aetiology had been set out in a number of reputable American psychological studies that were available to trusted Russian psychologists, like this one from 1957:

> Man's mental state is dependent on adequate perceptual contact with the outside world … Isolation produces an intense desire for extrinsic sensory stimuli and bodily motion, increased suggestibility, impairment of organized thinking, oppression and depression, and in extreme cases, hallucinations, delusions and confusion.

Hallucinations, delusions and confusion – and in case all that was not enough to alarm bone-headed strategic missile chiefs like Marshal Moskalenko, there was also at least a decade's worth of American science fiction movies to paint those words in garish Technicolor. Perhaps the most famous was the 1955 Hollywood blockbuster *Conquest of Space*, in which the ship's commander, General Samuel T. Merritt, goes completely insane on a trip to Mars, endangering the mission and his fellow crew members in a fit of religious paranoia. Soviet space movies tended to dwell less on the madness and more on Soviet technological triumphs – which included any opportunity to triumph over the Americans – but, as Mark Gallai later wrote in his memoirs, the 'invisible shadow' of 'space horror' nevertheless 'hovered over some of us!'

Here was the problem's kernel: by giving the three-digit code in advance to the cosmonaut he might, in a fit of insanity not unlike General Samuel T. Merritt's, actually *use* it, take over control of the spacecraft and do something terrible both to it and himself. But if he did not have the code in advance, how was he going to be given it if there were a *genuine* emergency? This particular conundrum had taxed the state commission's members for some weeks by now, if not months, and various options had been proposed, all of which were aired in this meeting. One was to send the three numbers by radio. But what if radio communications failed, a highly probable scenario given the enormous distances and frankly patchy record so far? An alternative proposal, supported by Korolev, was to place the numbers in a sealed special envelope and then put this envelope in the capsule before launch. But that raised the question that the cosmonaut might, in a mad seizure, open the envelope in space and unlock the controls, which brought the issue right back to where it started. Indeed, it quite possibly brought it to an even worse place, as Sergei Khrushchev, the premier's son and himself a rocket engineer at the time, remembers:

> The thing they most feared was that something might happen in orbit, that he'd press the wrong buttons. Or the worse fear was that he would go mad and off the rails and press all the buttons.

There was also, as Gallai reminded the commissioners, the danger that in the weightless conditions of orbit the envelope 'might still sail away into some nook and cranny of the cabin', beyond the cosmonaut's reach. Back and forth throughout the evening the question was argued, challenged, examined and turned upside down by some of the nation's most brilliant scientists and senior military leaders until finally Korolev, in a wholly understandable fit of exasperation, called a halt. 'That's it, the case is settled!' The sealed envelope idea would stay.

But it was not quite it. A final refinement would be to fix the envelope somewhere in the cockpit that would be tricky for the cosmonaut to access, the bizarre logic being that if he were somehow able to get to it, open it, remove its contents and read the three numbers that – surely – would be proof that he was not mad. And thus it was agreed, and duly noted, that 'the sealed envelope would be glued to the inner lining of the cockpit next to the cosmonaut's seat'. This being the Soviet Union, Korolev next appointed a special commission to oversee the operation of the special envelope. The commissioners would need to check that the correct code – it was 1-2-5 – was correctly placed inside the envelope, that it was then correctly sealed and correctly fixed inside the Vostok. Gallai himself was deputed to glue it in the day before the launch. And that was the end of the matter. Needless to say the cosmonaut himself would not be told the code. In an emergency he would just have to open the envelope and find out. And if he died trying to rescue his ship at least he would die sane.

There was one other major question that the commission needed to resolve and it was settled right at the start of the meeting, when Kamanin stood up to make a statement.

> On behalf of the Air Force, I suggested that the first candidate for the flight be Yuri Alekseyevich Gagarin, and Gherman Stepanovich Titov as his backup. The Commission unanimously agreed with my proposal.

Three days after flying to Tyuratam, Kamanin had finally made up his mind on the issue that had gnawed at his mind – and perhaps at his conscience – for three months. It took the commission only a few seconds to settle it.

IT WAS INEVITABLE of course. In abilities, intelligence, practical skills and in physical fitness, the two men, Gagarin and Titov, were evenly matched; or not quite evenly, as Kamanin himself noted, since Titov was if anything the stronger character but for that very reason needed to be reserved for the second, more challenging, manned flight later. With Nelyubov becoming something of an also-ran, the two leading cosmonauts had been observed closely at the cosmodrome. Psychologists had watched them for any signs of depression or anxiety. Kamanin had watched them too. And if Gagarin's state of mind had concerned him the first time they travelled to Tyuratam to witness Ivan Ivanovich's launch, he was able to put that down to natural human causes, with Gagarin's wife just out of hospital and his daughter barely a week old. Gagarin had anyway quickly bounced back, a central characteristic of his, enthusiastically joining in the team's activities as Titov had sometimes not, on occasion preferring his own company to, say, a group trip to the movies. A small point perhaps – but one that Kamanin did not miss.

These occasional jarring notes in Titov's personality were matters of concern and sat poorly next to Gagarin's sunnier disposition, his willingness to please, his cheerfulness, optimism and huge charm. Almost everybody seemed to like Gagarin. And if there was no one outstanding quality in which he shined, no single great talent, that too spoke to his benefit. The journalist Yaroslav Golovanov, who got to know him later, once wrote that while Gagarin was not the best at anything he was good at everything. And his worst offences were venial: getting caught playing Battleships during a lesson, as he once was, was unlikely to ruin his chances for immortality. His record as a loyal Communist, both as a Komsomol leader when a teenager and as a faithful party member stood for itself. Later Soviet hagiographies, and especially his 'autobiography' make a point of this – 'I wanted to make my flight into space as a member of the Communist Party', 'he' writes, and 'his' description of the day in June 1960 when he was finally admitted to the party boils with excitement. Yet there is more than a kernel of truth in this conviction. It is hardly surprising either, given that after the horrors of his childhood under Nazi occupation the Soviet state had rescued him, educated him and allowed him to pursue the career of his dreams, which was to fly.

But beyond all this, beyond too his almost-filial relationship with Korolev, there was something essential Gagarin possessed that Titov did not, and that was his biography. He was *the* perfect representative of the Soviet state. His peasant background, his vocational training in industrial foundry work, even his place of birth in Smolensk, the region west of Moscow that was more essentially 'Russian' than Titov's birthplace in remote Siberia – all of these factors were paramount in his ultimate selection. Gagarin's father was a carpenter who worked with his bare hands. Titov's father was a schoolmaster. Gagarin knew how to cast iron. Titov knew how to recite Pushkin. Titov's erudition, with all his clever literary display, could be suspiciously bourgeois. And then there was the question of family. Gagarin had two healthy daughters. Titov's only infant child had recently died and now he had no children, a bleak narrative for a would-be supreme representative of the Soviet state. And Titov's very name was a problem: Gherman sounded far too much like *German*, and it was only sixteen years since some twenty-seven million Soviet soldiers and civilians had died in the war against the Germans. Was this really the right name for the glorious title of first Soviet cosmonaut? Even Khrushchev worried about that.

This flight was intended to change history; but it was also intended to demonstrate that the Soviet state, and everything it stood for, was the greatest social, political and human experiment in history. Gagarin, more than any other of the nineteen cosmonauts, personified, or could be made to personify, the hammer and sickle to their core. If his flight succeeded he, more than any of them, would show the world what the USSR was and, even more, what it could yet be. He would, almost literally with this stunning technological triumph, show that world the future – *its* future. In the Soviets' battle with the Americans to conquer the heavens, he was their ideal gladiator. His open, handsome face with its wonderful smile would touch millions, if not billions, everywhere. 'I have never seen such a smile anywhere,' remembers Popovich's daughter, Natalya. 'Direct, like the sun. Like a light bulb. Everything around it lit up.' And perhaps, in the end, that smile was everything.

BUT BOTH MEN still had to be told.

The day after the state commission meeting, Kamanin summoned Gagarin and Titov to his office. Nelyubov was not invited. As soon as they arrived, Kamanin got straight to the point. He told the two men that the state commission had agreed with his personal opinion that Gagarin should fly the first flight and Titov should be his back-up. And that, in a word, was that.

For Gagarin this was the greatest of accolades. Among this elite group of men it placed him at the very top of the top. His name, as Kamanin well understood, would very shortly be written into history – if he succeeded. But Gagarin was not the man to brag nor to wear his emotions on his sleeve, especially before his fellow cosmonauts. Repeatedly his contemporaries stress his modesty and self-restraint, although quite what lay behind this would often perplex even his closest friends. 'He was like a sphinx,' says Aleksey Leonov, who probably knew him as well as anybody. 'And as impenetrable.'

For Titov the decision fell like a terrible blow. Even though both men had long known the likely decision, there was always the chance it might go another way. In later years Titov revealed that 'up to that last minute I thought my chances were high enough that I could have been the commander of the Vostok capsule. We were all young and we all wanted to be first.' But now that last minute had arrived and he had not been chosen. Kamanin recorded both men's reactions: 'The joy of Gagarin and Titov's slight annoyance,' he writes briefly, 'were noticeable.' By the time *Pravda*'s journalists described Titov's reaction in his own sanitised late-1961 'autobiography', Kamanin's verdict of 'slight annoyance' had translated into this:

> We all warmly supported the candidature of Yuri Gagarin, who had been nominated by our commanders. This excellent man, an absolutely sincere and honest Communist, enjoyed well-deserved respect among us cosmonauts ... When I was appointed reserve space pilot, I was inexpressibly glad.

The reality was different. There was no inexpressible gladness and there was no slight annoyance. Titov was devastated. 'All the journalists,' he said years afterwards, 'wrote that I was so happy for my friend that I hugged him. But that wasn't true at all.' To a British historian in the 1990s he said the same thing, even more emphatically. There was no hug. It was all 'nonsense! There was none of that.' These two men, Titov and Gagarin, had been friends almost from the start of their training. They lived next door to each other, they shared confidences across their joint balconies, they were constantly in and out of each other's homes. When Igor had died at the age of eight months, Gagarin and Valentina had been rocks of support to Titov and his wife Tamara. But still, despite all the bonds and history that linked the two men, the prize of being *first* – the first human ever to break free from the earth, the first to fly in space – was too tremendous for Titov to give up without real anguish, especially when that prize was so tantalisingly within his grasp. To Igor Kurinnoy, a general of the Soviet Space Forces who later knew Titov well, the decision announced in Kamanin's office that day would scar Titov for the rest of his life. 'I'd say,' he declared in an interview on Russian television, 'that he departed his life with that memory. It never left him.' Pavel Popovich, who saw Titov after he had been told the news, felt this too:

> That wound was there for the rest of his life. And then he said – who discovered America? I said Columbus. And who was the second? And I didn't know. And he said – now you understand.

But if Titov was devastated he could also be generous. 'Yura turned out to be the man everyone loved,' he told his British biographer in one of the last interviews he gave before his death. 'Me, they couldn't love. I'm not lovable.' And then he added, 'I'm telling you, they were right to choose Yura.'

THAT SAME DAY, within hours, and possibly only minutes, of Gagarin being told that he would be the USSR's first cosmonaut, an R-9 intercontinental ballistic missile with a dummy warhead roared into the sky from Site 51 in Tyuratam on a proving test flight. Exactly 153 seconds after launch the missile's second-stage engine malfunctioned; less than two minutes later, and much too early, it shut down completely. Approximately 3,850 kilometres east of the cosmodrome the failed booster impacted the ground.

The R-9 was a different rocket from the R-7 that Gagarin would very shortly ride into space and on a quite different mission. Nor can we be sure if he even knew about the failure, although one engineer distinctly remembers seeing the six cosmonauts huddled together in a corner of the assembly building, 'hotly discussing' the accident. Yet even if the cosmonauts were not aware of what had just happened, Korolev was. He ordered an immediate investigation. There were practical issues that needed to be resolved. But coming when it did, so close to the first human flight, the accident unnerved him and everybody involved.

And there was something else that was unnerving: Site 51, the pad from where the R-9 had launched, was just four hundred metres from Site 1, the pad from where Gagarin would launch. It was almost next door. Few of Korolev's colleagues would ever have suspected their hard-boiled chief of being superstitious, but they were wrong. His own daughter never forgot how, in private, he sometimes nailed horseshoes to trees, and he always kept a couple of one-kopeck coins in his pocket for luck. He would need those coins very soon.

TWENTY-THREE
OPENING PITCH

April 10, 1961
Griffith Stadium
Washington, DC

THE DAY AFTER Gagarin learned of his selection President Kennedy stood up in the stands of Griffith Stadium to pitch the opening ball of the 1961 major league baseball season. At least sixty photographers and almost 27,000 fans were there to watch him do it. With the new Washington Senators team ready to take on the Chicago White Sox, the buzz and excitement in the stadium appeared to have gripped the president himself. Despite the cold and the damp he took off his coat, handed it to an aide, and with an expression that was somewhere between a grin and a gritting of teeth, drew back his right arm as far as it would go and hurled the ball into the air. Up it went, sailing over the field as the crowd roared their approval and the players all leaped and clutched for it, but it was Chicago's speedy outfielder Jim Rivera who finally got it. There were those who said that Kennedy had just thrown the longest and hardest ball ever tossed by a president in an opening season game, a tradition that went all the way back to President Taft in 1910. There were others who said he had put on weight – maybe as much as fifteen pounds since his inauguration in January.

Vince Lloyd of Chicago's WGN-TV managed to grab a moment in the game to ask Kennedy about his new national physical education plan and the president, speaking over the noise of the crowd, was happy to respond. He had always been interested in sports, he said, but it was vitally important that the US be 'a nation not only of spectators but also a nation of participants'. That was why he was promoting this plan. But the president said nothing about another plan, this one to send CIA-backed fighters of the Cuban brigade to invade Cuba and get rid of its leader in a week's time. And yet, even as he was chatting amicably to Vince Lloyd about the nation's fitness, events were rapidly gathering momentum.

That same day the brigade's troops had begun to move by truck from their base in Guatemala to the embarkation point at Puerto Cabezas in Nicaragua. The aircraft carrier USS *Essex* and seven navy destroyers were already in the Straits of Florida and steaming south: this was the fleet designated Special Task Group 81.8 that would secretly rendezvous with the brigade and escort it towards the Bay of Pigs; despite the president's anxieties about direct US involvement in the operation, this task group would provide indirect support. So far the sailors had been told nothing about the mission. Instead they were told they were there to study Atlantic tides and currents. To lend weight to the story there were even marine scientists on board. But few of those sailors were fooled, especially when a squadron of Skyhawk jets screamed from the skies to land on the deck of the *Essex* – and stayed there. For some weeks now newspapers in America and around the globe had been thick with rumours about a possible US-backed invasion. On the Soviet side even *Pravda* had signalled alarm, warning darkly four days earlier on April 6 of American 'anti-Cuban plotting'. Everybody knew that something was up, despite the president keeping his mouth shut – or opening it only to lie. And far from fading, the rumours only got thicker.

But now, on this Monday afternoon, there were rumours about something else. At the end of the second inning, as Kennedy sat enjoying the game in the presidential box, his associate press secretary Andrew Hatcher came up to him, leaned over and whispered something in his ear. The president straightened up and appeared to squint in concentration; he nodded without saying a word. Hatcher departed immediately, leaving Kennedy to eat a hot dog and watch the White Sox go on to beat the home team 4–3. How far his mind was still on the game we will never know, but possibly less than before. Hatcher had just told him that United Press International was about to report a rumour that the Soviets had sent a man into space and returned him back to earth alive.

The story appeared to have sprung from a CBS report earlier that day by its Moscow correspondent Marvin Kalb. Kalb had sent his report to New York in which he wrote that at 3 p.m. Moscow time an announcement would be made that the Soviets had successfully placed a man in space. But there was no announcement at 3 p.m. or at any other time that day. The story nevertheless quickly gained traction, reaching Hatcher, and then Kennedy, by mid-afternoon. Adding fuel to the frenzy was the fact that parts of Kalb's report had been heavily censored. The Soviets had only recently lifted their censorship restrictions on foreign correspondents' stories. Now they had suddenly re-imposed them. Clearly something major had either just happened – or perhaps was soon to happen.

IF KOROLEV'S NERVES were on edge after the R-9 accident on April 9, they were even more so on April 10. And if he managed for the most part to keep his hair-trigger temper under control, the slightest provocation could set it off.

As Kennedy watched his baseball match in Washington, the R-7 was finally 'mated' to the Vostok in the cosmodrome's assembly building, a prolonged and delicate operation in which every bolt, clamp, plate and cable joining the capsule and rocket had to be perfect. There was more than enough here to worry about, but when Korolev turned up early that morning to watch the start of the procedure he was horrified to find the Vostok surrounded by a pile of connectors and cables that appeared to have been ripped haphazardly from its guts. Not surprisingly, he exploded. And as always his temper was terrifying.

While workers cowered under his onslaught, he yelled most of all at the supervisor Oleg Ivanovsky. Ivanovsky tried to explain. The capsule had been weighed the night before, he said. To his considerable alarm it turned out that it was too heavy, possibly by as much as fourteen kilos, or thirty pounds. *Fourteen kilos* – it could make all the difference between success and failure. So Ivanovsky had decided to improvise. There was only one way to get the weight down at this very late stage, and that was to hack out any surplus cables and remove any odd bits of inessential equipment. So out went the electronics for the deactivated on-board bomb. Out went a gas analyser. Out went a food warmer. Ivanovsky had spent the whole night on the job. The only problem was that he had done it without telling Korolev.

Korolev raged in disbelief. There were just two days to go. A part of the Vostok's intestines now lay spewed across the assembly hall's floor. He threatened to fire Ivanovsky on the spot and send him on the first train back to Moscow. But he did not. Ivanovsky was one of his best engineers – as Ivanovsky himself was well aware. More to the point the capsule was now at the right weight. But every clipped cable had to be checked and re-tested before the Vostok could finally be mated to the R-7, and all within the fast diminishing timescale. In their haste to get it done nobody noticed that two sensors, one measuring temperature and the other pressure, remained disconnected. Fortunately there were back-up sensors on board. But the oversight was telling. At this speed mistakes happened.

FROM THE ASSEMBLY building, Korolev drove to the cosmodrome's town of Leninsky. His destination was the wooden gazebo on the banks of the Syr Daria river, opposite the cosmonauts' dormitory. A select group of some twenty-five individuals had been invited to gather there before lunchtime to celebrate Gagarin's upcoming flight. Most of the state commission members joined the party, including Marshal Moskalenko, the USSR's strategic missile commander. All six cosmonauts were also invited, along with several photographers and cameramen, among them Vladimir Suvorov. The whole event was arranged as a photo opportunity to be used in future Soviet propaganda films, assuming Gagarin was not killed. Korolev, fresh from his bruising encounter with Ivanovsky, also took his place under the gazebo's sheltering roof while the sun beat down on the river. In the footage he is all smiles.

The table was heaped with luxurious bowls of fresh oranges and bottles of champagne. Long, solemn and wildly optimistic speeches were made by the participants about the glories of Soviet space flights to come. Korolev, wearing a hat to protect himself against the sun, proclaimed that as many as ten Vostoks would be ready within the year and further two-seater ships by next year. Toasts were given to Gagarin, sitting in the place of honour to the marshal's right. 'Fly, dear Yuri Alekseyevich,' Marshal Moskalenko grandly declared, 'and return to Soviet soil and the arms of all our people.' Everybody raised their glasses; somebody popped a cork and managed to squirt champagne all over his uniform; and somewhere in all these speeches and celebrations Suvorov's camera caught Titov, framed against the brown river and the hot sky. The moment is brief but illuminating. He gazes down at the tablecloth lost in thought, oblivious to all the chatter and laughter and congratulations around him.

For Titov it was doubly difficult. Not only was he now the number two; but as the back-up he had to follow Gagarin's every step, shadow his every move, be his 'twin' in every action right up to the moment of launch – do everything, that is, except actually fly. As with John Glenn and Alan Shepard, nothing could change his fate unless Gagarin were suddenly to fall ill, develop a cold, fall over and break his leg or succumb to a last-minute fit of nerves or panic. But Gagarin was always, at least outwardly, calm. And he never fell over.

At five o'clock that afternoon Titov had to suffer again – this time more publicly. A meeting of the state commission was convened in a large hall to confirm Gagarin's selection as the first cosmonaut. Of course Gagarin's selection had already been confirmed two days earlier and by exactly the same body of people. But that had been behind closed doors. Now the whole thing was to be 'reconstructed' on camera, again for future use in films glorifying the world's first space flight. Once again Suvorov set up his cameras. This time there were microphones to record every word. 'There will be many speakers,' Korolev had told Suvorov. 'And we will not limit the time. It's history!'

Korolev was not joking. Speaker after speaker stood up at the table to add their interminable contributions to history. Seventy people filled the room to hear and applaud them, a 'concert' of bigwigs, as Suvorov wrote in his diary, including chief designers, ministers, air force officers, strategic missile forces officers, representatives of the Soviet Academy of Sciences and the Communist Party, a roll-call of the great and powerful, and all of them there to hear Gagarin announced as the world's first cosmonaut-in-waiting. 'The ship is ready, all the equipment has been tested and works perfectly!' pronounced Korolev, not exactly ingenuously, as the cameras rolled for posterity. The date of the flight was finally confirmed – for the morning of April 12, less than forty-eight hours from this meeting. Once again Kamanin stood up to declare that all six cosmonauts were fully prepared and trained for the flight, but since only one of them could be selected, 'the Air Force command recommends First Lieutenant Yuri Alekseyevich Gagarin for the first space flight', with Titov 'as the back-up pilot'. Once again the state commission unanimously agreed. And once again Titov is caught in Suvorov's lens, sitting motionless, with that same miles-away look in his eyes.

He and Gagarin both had to stand up and make speeches too. But of course it was Gagarin who riveted the room's attention. 'How young he looks!' Suvorov noted later in his diary. 'Like a boy with this fascinating smile of his and very kind eyes … Today he speaks slower than usual and this betrays his deep emotions.' A hush filled the hall as Gagarin got to his feet. 'Permit me, comrades,' he said,

> to assure the Soviet government, the Communist Party and all the Soviet people that I will accomplish with honour the mission entrusted to me and will pave the first road to space. And if I encounter difficulties in doing so, I will overcome those difficulties in the manner that Communists do.

And then, because the film ran out in the middle, Titov had to watch him say it all over again.

TWENTY-FOUR

ROLL OUT

April 11, 1961
New House Office Building
Washington, DC

LESS THAN TWENTY-FOUR hours after Marvin Kalb had filed his story for CBS, the rumours surrounding the Soviet space flight had thickened, multiplied and spread around the world. Newspapers everywhere picked up the story, from Stockholm ('A Man is Already in Space') to Iran ('the whole world is engulfed in rumours about a successful flight of a Soviet man into space'), to London ('Man in Space Alert') and to New York, whose *Daily News* cited 'unofficial sources' claiming that 'Russia may have put the first man in space and brought him back alive'. Now a French reporter, Eduard Bobrovsky, revealed the man's name: apparently he was Lieutenant Colonel Vladimir Ilyushin, the son of the famous Soviet aeroplane designer, and according to Bobrovsky he had flown not the day before but back in March and had also been injured in his flight, possibly suffering some sort of mental breakdown as a result of his time in space. Meanwhile Dennis Ogden, the Moscow correspondent of the British Communist newspaper the *Daily Worker*, was busy preparing his own story for the next day's edition, asserting that the flight had in fact been made three days previously on April 7 and that the cosmonaut had gone on to orbit the earth three times in a spaceship named, appropriately, 'Rossiya'. And so it went on, each report adding unverified tantalising details and imaginative splashes of colour until the tale of the Rossiya and Ilyushin's somewhat mysterious condition raced from country to country like wildfire, becoming, as we would say, viral. Nor did it help that the Soviets neither denied nor confirmed these reports, sending foreign correspondents in Moscow, and especially regime-sympathetic ones like Dennis Ogden, into fevers of fantasy.

Kennedy himself knew the Soviets were close to putting a man in space – extremely close. His own intelligence people were telling him so all the time. But they also told him Vladimir Ilyushin was not that man. Ilyushin did exist but he was not and never had been a cosmonaut and he had certainly never flown in space, whether in March or on April 7 or at any other time. He was subsequently traced to a Chinese health resort in Hangchow where he had already been resident for some time nursing a leg injury from a road accident. But behind those swelling rumours of his 'flight' lay blunt realities about a system that systematically and deliberately withheld or distorted truths concerning practically everything to do with its space programme. There *was* a dead cosmonaut, and his name was Valentin Bondarenko, and he had been incinerated on the ground in the isolation chamber; there was a cosmonaut who *looked* like he was dead, and his name was Ivan Ivanovich, the dummy that had fooled many of the villagers who saw him; and there *was* a human space programme, one that was poised to launch its first human into space, not on April 7 but on Wednesday, April 12. Almost certainly the story of Ilyushin's flight and subsequent injury was born in some witches' brew of all these things, and then caught fire.

On the day after the baseball game at Griffith Stadium that fire reached the august setting of Room 214B of the New House Office Building in Washington, DC, where the deputy administrator of NASA, Hugh Dryden, was facing questions from the House Science and Astronautics Committee. In the middle of a discussion about NASA's unmanned space probes, he was suddenly asked this:

> *The Chairman* (Rep. Overton Brooks): Talking about a man up there, have we got anything this morning about a Russian man in space?
>
> *Dr Dryden:* We saw the press reports which I am sure you have also seen. According to these reports, as near as we can tell it is purely rumor based on conversations with Russian taxicab drivers who had conversations with somebody else.
>
> *The Chairman:* The usual Russian first?
>
> *Dr Dryden:* To the best of our knowledge this is purely rumor.
>
> *Rep. Walter Moeller:* It made a good story.
>
> *Dr Dryden:* Yes.

But even as Hugh Dryden dismissed both that story and the Russian cab drivers helping to spread it, a real rocket was already standing on the launch pad at Site 1 in Tyuratam, lit by powerful arc lights and pointing at the night sky; and on the following morning, at just after nine o'clock Moscow time, a real cosmonaut, Yuri Gagarin, was hoping to ride that rocket in humanity's first attempt to leave the earth.

THE FIRST STEP of that journey had begun earlier on that day.

Shortly before sunrise, the great doors of the MIK assembly building were rolled open and the R-7, now mated to its Vostok capsule, began the two-kilometre trip to the pad in advance of the next morning's launch. As with all the unmanned missions, the rocket lay horizontally on a trailer, pushed from behind by a diesel locomotive. At barely more than walking pace the whole assembly eased forward into the cold pre-dawn air: first the grey nose cone within which the Vostok was protectively housed, then the long grey body of the rocket with its four colossal strap-on booster engines, and finally, at the very rear, the grinding, grunting, squealing locomotive itself, absurdly small against the huge vehicle it was transporting. The drama of this opening act was not lost on any of those who witnessed it. One was the engineer Vladimir Solodukhin. 'Those walking alongside it,' he remembered, 'could not tear their eyes away from the magnificent sight.'

Another witness was Korolev. He watched the procession pass without speaking a word. As the eastern skies began to lighten over the steppe, he got into his car and drove along a road paralleling the rail track, stopping a little way ahead of the rocket. Mark Gallai observed him:

> He pulled over at the side of road while the rocket passed by. He looked at it for a minute or two, deep in thought, then he got back into the car and drove to the next point further up the road. Perhaps, once upon a time, generals did the same when they stood on a hill and watched their troops leaving for battle.

Gallai could see how emotional this moment was for his chief – the adjective is his own. Korolev was closer to his dream than he had ever been. And yet over and again the single thought drilled through his brain: in the rush to get everything done was there anything he and his teams had missed, anything they had overlooked? His nerves were frayed, but he had to make every effort to keep his fears to himself as Nina had urged in her letters. Above all he had to keep the fear from his voice. As he watched the rocket – *his* rocket – slowly cross the steppe towards Site 1, there was no doubting, as Solodukhin said, that it was a magnificent sight. But one account claims that even while it was happening Korolev suddenly announced that perhaps the launch ought to be postponed altogether.

His colleagues looked at him in astonishment, wondering if he were joking. As quickly as it was raised, the idea was dropped. But Korolev then insisted that two of his senior engineers, Boris Raushenbakh and Konstantin Feoktistov, go over every detail of the flight plan again with Gagarin and Titov. Both engineers were startled at the request. As far as they were concerned the two cosmonauts knew the plan perfectly by now, but here was their chief telling them to go over it all over again. Having known Korolev for decades, Raushenbakh knew better than to argue. So at ten o'clock that morning the two engineers sat down with the cosmonauts to do as their chief had ordered.

But Raushenbakh could not keep his mind on the job. Even now, with so little time left, the reality of what they were about to attempt was almost too much for this level-headed engineer to grasp:

> I looked at [Gagarin] and in my mind I understood that tomorrow this kid was going to awaken the whole world. But at the same time I just could not make myself believe that … this First Lieutenant sitting in front of us would tomorrow become the symbol of a new epoch. I would start giving him instructions, such as, 'Turn this on, don't forget to switch this on' – all these normal, pedestrian, even boring remarks, and then I would become silent and some sort of devil would begin whispering to me, 'This is all a bunch of crap. None of this is going to happen.'

ONCE THE DIESEL and its outsize trailer reached Site 1, a huge hydraulic jack slowly hoisted the R-7 with its Vostok to the vertical. Now it stood in all its grandeur, more than thirty-eight metres high and gleaming in the spring sunshine. When fuelled, it was almost nine times heavier than the Mercury-Redstone, far too heavy to sit on the pad without collapsing. But an ingenious feat of engineering meant that the R-7 never actually *sat* on the pad, but was instead *suspended* over the flame pit by four steel arms. These would spring back like petals once enough thrust had built up at launch. The giant rocket literally hung there in space, adding to the impression that somehow, even though it was not yet moving, it was already on its way.

At one o'clock that afternoon, after finishing with Raushenbakh and Feoktistov, Gagarin, Titov and the other cosmonauts were driven out to Site 1 to meet the pad workers. Having watched the rocket's procession at dawn, Vladimir Solodukhin was there to see them arrive. They were met 'with thunderous applause. Everybody's spirits were high. People were proud of the fact that they had contributed their part to this common goal.' With Suvorov filming, Gagarin made a speech of thanks. He was greeted with rapturous cheers. A crowd of excited, grinning pad workers in blue overalls jostled around him, all of them eager to be in his presence. In the unedited footage his charisma burns through the frames. He grins too. People slap him on the back, others shake his hands, everybody wants to touch him. He looks every inch the superstar, the very quality that Titov, for all his cleverness and chiselled looks, lacked – and knew that he lacked. Standing by the launch pad in the centre of this thrilled, adoring crowd, Gagarin already appears to inhabit his new role: the man of destiny, chosen for a unique task that will change history. And all this before he had even flown.

A bouquet of wild tulips picked from the steppe was pressed into his hands. Then, together with Korolev, Gagarin walked up to the rocket and took the lift to the top. Like a racehorse sniffing the first fence, he had come to take a closer look at the Vostok he would ride to space. He was still smiling. Korolev, with a touch of irritation, asked him why he was smiling so much. 'I don't know,' Gagarin answered. 'I must be rather frivolous', a reply that was unlikely to soothe Korolev's jangled nerves.

Standing beside the Vostok, Korolev insisted on going over procedures yet again. Far below they could see the pad workers they had left moving like ants over the launch site. Stretched across part of the flame pit they could also see a steel net. If anything were to go wrong with the rocket in the first seconds after ignition that net would be

Gagarin's only hope. His hatch would blow open and he would be ejected backwards out of the capsule, too low for his parachute to open. The net was supposed to catch his freefall. It was a last resort, another 'just in case'. In the race to get a man into space there had been no time to develop a more advanced pad abort system. The net was the best they had.

They could only pray it would never be needed. Even if Gagarin managed to survive the twelve-storey drop to the net, there was still no obvious means of getting him down from there. In the end it had been decided that the only way was to use a rope attached to an ordinary domestic bathtub. Once he had catapulted onto the net, the cosmonaut was somehow supposed to crawl across to the tub and get inside it. Then a rescue team would pull him down to safety.

There was one other decision made concerning this procedure, according to Gennady Ponomarev, an engineer assigned to work on it, and that was not to tell the cosmonaut it even existed:

> Many years later at an informal dinner at a restaurant, when they told cosmonaut Gherman Titov about this method of evacuation, he laughed to the point of tears. But then he said, in all seriousness, 'It's good that we did not know this.'

KOROLEV AND GAGARIN had already spent an hour at the top of the rocket when Korolev became weak with exhaustion. Perhaps the mounting stack of 'just in cases' was suddenly too overwhelming. He had to be helped down from the gantry. He went back to his cottage to rest briefly and take those pills for his heart. Gagarin also left. He would be spending the night at a cottage next door. We have no record of Gagarin's response to Korolev's moment of frailty but it must have been alarming. Few got to see it. Only later would Gagarin learn just how ill his friend and mentor really was.

At some point after the two men had gone, Mark Gallai also rode the lift up to the Vostok. With him was the envelope containing the secret code to unlock the manual controls. He was joined by the other members of the special commission Korolev had appointed to oversee this operation. One can safely assume that the commission had carefully inspected the envelope and its contents, ensuring that Gallai had written down the code correctly and then sealed the envelope in the officially approved manner. While the commission members huddled by the hatch, Gallai climbed part-way in and reached across to the numerical keypad on the panel to his left – like the buttons on a hotel safe. He first checked that the three-digit code was working by pressing 1-2-5. Then he tried various alternative three-digit combinations to check that they did not inadvertently unlock the controls. They did not. The safe was locked. Next came the envelope. Gallai slipped it into the cabin lining next to the seat. We do not know if he also glued it down. But the envelope was there for Gagarin to open if he ever needed it in an emergency, and thereby prove he was not insane.

That done, Gallai pulled himself out of the hatch. But for a moment longer he and the others lingered by the Vostok at the top of the rocket, unable quite to leave:

> We have nothing more to do here. It's time to go down. And then I see that all the members of the commission have the same reluctance as I do. I really do not want to tear myself away from the ship – the first manned spacecraft in the history of mankind!

TWENTY-FIVE

NIGHT

April 11, 1961
Nedelin Cottage
Site 2, Tyuratam Cosmodrome

IT WAS AN odd place for a man to spend his last night before leaving the planet, in part because of its incongruity, its piece of old Russian rusticity, with that rocket just two kilometres down the road. The cottage was almost an exact replica of Korolev's own cottage, just twelve metres away: the same simple peasant-style bungalow, the same steeply pitched wooden porch, the same four basic rooms inside. There was one room for the doctors, and another for Kamanin and Yevgeny Karpov, the head of the Cosmonaut Training Centre. There was a third for Gagarin and Titov. The fourth was barely a room at all but it did contain the luxury of luxuries in the cosmodrome, an actual bath with hot running water fed by a primitive gas cylinder. The plainness of decoration matched that of Korolev's cottage, with the same striped wallpaper and cheap curtains, except that here it was thick with the scent of fresh flowers. Adilya Kotovskaya, the doctor who had supervised the cosmonauts' brutal centrifuge training in Moscow, had shown a softer side to her character by spending most of the afternoon picking tulips. She had placed vases of the flowers in the cosmonauts' bedroom. They were all of the purest white.

After eating a late lunch of 'space food' – two types of pureed meat in toothpaste-type tubes and another toothpaste tube of chocolate sauce, which Kamanin describes as 'filling' but 'not particularly tasty' – Gagarin and Titov were each thoroughly examined for well over an hour by the doctors. As always, Titov was subjected to exactly the same procedures as Gagarin. But Gagarin's medical was near-perfect. His blood pressure was perfect, his pulse was a respectable sixty-four beats per minute, his temperature was normal, his ECG – checked by Dr Kotovskaya after her afternoon's flower-picking – was excellent. *Everything* was normal. Even his psychological state was impressively normal. His medical report from that afternoon has survived. The doctors are quick to note his 'liveliness and self-confidence' and his 'cheerful mood', with 'no visible signs of despondency, depression or irritability'. And they add with a flourish in the last line: 'Ate with great appetite.' He even seems to have loved that meat puree.

The medicals concluded with the fitting of five electrodes to both men's bodies in advance of the flight, an operation intended to speed up preparations in the morning. Then they changed into blue tracksuits, 'like gymnasts,' wrote Gallai, 'waiting to perform'. Whether either man took a luxurious hot bath we do not know. Later in the evening the other four cosmonauts joined them at the cottage. A curious companionable ease, not entirely manufactured, seems to have descended on the group, almost as if these old friends really *were* away for a weekend in the country. But there were anomalies that punctured that myth: those electrodes attached to the two men's bodies, Kamanin checking his watch to ensure that nobody stayed too long, or the sudden appearance of Suvorov and his camera crew turning up to film them all. 'They came in like mice,' recalls Dr Kotovskaya, 'like invisible beings.' But they also flooded the bedroom with lights while they filmed Gagarin and Titov playing chess or quietly listening to music from a reel-to-reel tape recorder. The music was mostly Russian folk songs, Gagarin's choice. 'He loves

Russian songs very much,' wrote Kamanin. 'The tape recorder runs continuously. Yura sits across from me and he says, "To fly tomorrow ... I still cannot believe I will fly. And I'm surprised at how calm I am."'

While Suvorov was filming, Korolev also showed up briefly at the cottage. His earlier turn on the gantry appeared to have been forgotten. Now he made jokes about all the fuss over this first flight – in five years' time, he said, the unions would be subsidising trips to space for their workers. Again the atmosphere was kept artificially casual. There were no more lectures about flight plans. Before he disappeared, Korolev took Gagarin and Titov outside for a last breath of evening air. Suvorov grabbed his camera. He was determined to miss nothing, to capture every moment even though there was hardly any light. His diary is injected with a sense of immediacy that leaps across the years. 'The yard is hopelessly dark,' he writes:

> There is almost complete silence: all the traffic has been detoured in order not to disturb the cosmonauts ... The light meter is surely dead, I do not even glance at it. I insert a new roll of film and follow the cosmonauts and Korolev: they have gone to the main road. I cannot afford not to shoot these three silhouettes against the grey sky dissolving into darkness.

By 9.30 p.m. Gagarin and Titov were in bed. Both men had been offered sleeping pills and both had declined. What neither of them knew was that strain gauges had been discreetly fixed underneath each of their mattresses. This was the result of a secret order by Dr Vladimir Yazdovsky. Wires from the gauges were fed through tiny hidden holes in the wall, then out into the yard and across to another building where a technician and a psychologist were sitting at a bank of instruments. They would monitor every turn and toss the two men made during the night. Even now Gagarin might lose his chance to fly simply by moving in bed too much. But when Karpov checked on them half an hour later, both men appeared to be sound asleep.

Gagarin might have made jokes about his own calmness, but he was far from nerveless. Lying in his suitcase was a private letter he had written to his wife Valentina and their two infant daughters the night before. It was to be given to her only if he failed to come back.

> My sweet and much loved Valechka, Lenochka and Galochka,
>
> I've decided to write you a few lines to share the joy and happiness I felt today … A simple man has been entrusted with a great national task – to blaze the trail into space! Is there anything greater to wish for? This is history, a new age! … I wish I had a chance to be with you for a little before I go, to talk to you. Sadly you are far away. But I always feel you by my side.
>
> I trust the machinery completely. It will not fail … But sometimes things go wrong. If they do, I ask you not to waste yourself in grief. Life is life … Take care of our girls. Love them as I do … As for your personal life, you must settle it the way your heart tells you, the way you feel is right. I do not think it is fair to hold you under any obligation.
>
> I hope you'll never see this letter and I'll never have to be ashamed of this moment of weakness. But if something goes wrong, you have to know it all …
>
> Goodbye my dearest ones. I hold you all tightly and kiss you.
>
> Your Dad and Yura

UNLIKE GAGARIN AND Titov, Korolev did not even try to sleep. He lay on his cot alone in his own cottage, listlessly flipping over the pages of *Moskva* magazine as the hours slipped by. At two in the morning he finally got up and went back to the cosmonauts' cottage. Kamanin and Karpov were also awake. They told him that Gagarin and Titov were still sleeping soundly. Korolev did not stay long. 'Sergei Pavlovich went off silently,' said Karpov later. 'He went straight from us to the launch pad, where the final tests began at three o'clock in the morning.'

Suvorov was already there, filming as always. The sight through his viewfinder left an impression he would never forget.

> Night. The thick violet velvet of the sky is bestrewed with diamonds of stars. There is now a wind, and the air is cold. At the launch pad the gigantic rocket is lit by the powerful search-lights and towers over the deserted silent steppe and into the dark violet sky. The sight is unreal and even scary, as if all this is taking place on an alien planet.

'So tomorrow will be the greatest of feats – the world's first manned flight into space,' wrote Kamanin in the quiet hours during that night, while the two cosmonauts slept next door. 'And this humble Soviet man, First Lieutenant Yuri Alekseyevich Gagarin, will accomplish this feat. Now his name means nothing to anyone. But tomorrow it will fly around the whole world and humanity will never forget it.'

ACT IV

LAUNCH

APRIL 12–14, 1961

Earth is the cradle of humanity, but one cannot remain in the cradle forever.

Konstantin Tsiolkovsky, Russian scientist and visionary, 1911

Can a man dream for more?

Yuri Gagarin, nine days before his launch, April 3, 1961

TWENTY-SIX

A MOTHER'S LOVE

April 12, 1961
Nedelin Cottage
Site 2, Tyuratam Cosmodrome

AT 05.30 LOCAL time Yevgeny Karpov entered the bedroom where Gagarin and Titov were sleeping. He touched Gagarin lightly on the shoulder. Gagarin's eyes opened at once. 'It's time,' said Karpov.

In the next bed Titov was also quickly awake. Both men were asked how they had slept. Both said they had slept excellently. Only later would Gagarin admit that he had not slept at all. Nor had Titov. Neither of them had known about the hidden strain gauges under their mattresses, but every time someone had come into the room to check on them during the night – Karpov, Kamanin or one of the doctors – Gagarin had concentrated hard on lying absolutely still. A bad night's sleep was one thing; but admitting to a bad night's sleep could cost him his place, even now.

As the two men were dressing Karpov gave them each a bouquet of wild steppe tulips – yet more to add to their collection. The flowers had been brought earlier that morning by Klavdia Akimova, an elderly woman who looked after the cottage. She wanted the cosmonauts to have them. Perhaps the young men reminded her of her own son who had also been a pilot but had been killed during the war. He would have been a similar age to Gagarin and Titov. At any rate, in all the grey masculinity of the morning they added a welcome touch of freshness and colour. Then Vladimir Suvorov and his film crew turned up. The cosmonauts pretended to be astonished: 'What, you here already?' they said, and everybody laughed.

By now none of Suvorov's crew had slept for well over twenty-four hours. Yet here they were again to record every last detail of history. 'We are shooting everything,' wrote Suvorov in his diary. Film cans piled up in the cottage, ready to be flown later that day to Moscow. They filmed Gagarin and Titov doing their exercises – even on *this* morning neither would be spared that routine – washing, shaving, and sitting down to a light breakfast of 'space' food. This time the menu was chopped meat, mashed potatoes, blackcurrant jam and coffee, all of it packed into those toothpaste-sized tubes. Similar tubes had been placed inside the spacecraft earlier that morning. Everybody maintained the carefully light-hearted mood. 'Instead of advice and farewells, all I could do was joke and tell funny stories,' said Karpov. Gagarin made fun of his breakfast, saying it was the most delicious meal he had ever eaten and that he would have to take some of it home for Valentina to try. But Valentina was two and a half thousand kilometres away in Moscow, unaware that her husband had been chosen or that this was the day.

ALMOST AS SOON as he had driven to the launch pad in the very early hours of the morning, Korolev had gathered his entire team together and stood them in a line. Then he made a speech. It was brief and unforgettable. Vladimir Yaropolov was one of the engineers who heard it:

> Standing before the line-up he reminded each crew member of their special responsibility: to follow each step of the launch timetable efficiently as they prepared the rocket's engines and the spacecraft. 'Honesty is above everything,' he said. 'If you

make a mistake – report it to your superior. There will be no punishment. A dishonest engineer is a recipe for failure.'

It was Korolev to the core. Simple, direct, unembroidered and fundamental to the mission's chances of success. In a society that tended to punish error and condemn initiative, it was also distinctly un-Soviet. But Korolev swept such niceties away. Here was the man who had once given one of his workers an engraved watch for admitting to an error. It was an essential part of his leadership, a keynote of his whole character; and right now, as he stood before his men in the pre-dawn chill at the foot of the rocket, his words hit home. It was already past 03.00 local time – 01.00 in Moscow. All timings would now be referenced to Moscow time, a practice we shall also follow in the events of this day. Lift-off was set for 09.07. A loudspeaker blared suddenly: 'Readiness for launch in eight hours!' The line-up dissolved and everybody went to work.

That launch was time-critical. Once in orbit, the Vostok's automatic orientation system would use the sun as a point of reference to align itself correctly before re-entry. As long as the rocket left on time the sun would be where it needed to be to ensure a landing inside the borders of the USSR. Too late and the Vostok could end up in the ocean or the wrong country. Little wonder Korolev stressed the timetable in his speech. From this point on it drove everything.

At 05.00, just as the first rays of the sun were peeping over the horizon, the rocket's fuel tanks were filled with kerosene and cryogenic liquid oxygen, the two propellants which, when mixed together and ignited in its five initial-stage engines, would blast it off the pad and on its way to space. Instantly the pale-blue liquid oxygen began evaporating in the relative warmth of the air, venting through open valves in the tanks and enveloping the launch pad in those familiar clouds of hissing steam. It was an especially beautiful sight in the early morning light when the low sun tinged the swirling clouds gold, but its beauty masked a cost because the liquid oxygen would need to keep being topped up until almost the point of launch. Fully fuelled, 90 per cent of the rocket's 291-tonne mass was essentially just highly flammable kerosene and liquid oxygen. And all of that 90 per cent would be consumed in the first eleven minutes of flight.

At 06.00 the state commission met for the last time, this time not in a committee room but in a dugout near the pad. One by one Korolev asked for reports from his heads of department. In turn they each gave a green light. Even the weather was co-operating. The winds were light and there were, as Yaropolov noted, 'beautiful clear skies'. So far everything looked good. But Soviet protocol had to be observed and Korolev duly observed it, formally asking the commission for permission to launch. The request was 'accepted without further discussion'. If everything went to plan there were just over three hours to go.

TITOV AND GAGARIN were not long at breakfast. Within half an hour they left the cottage to go to the MIK, the spacecraft assembly building a few hundred metres down the road. In one of its rooms they sat down for their final medicals. Each of the biomedical sensors that had been attached the night before needed to be re-checked and both men thoroughly examined one last time. Here was another of Titov's dwindling opportunities to replace his friend, but once again Gagarin proved almost indestructible. His blood pressure was a perfect 120/70, his temperature, lung capacity, weight – all perfect. 'Decision: Healthy' was the unambiguous conclusion of his medical report. But even though Gagarin smiled dutifully for Suvorov's camera during his medical, he was noticeably less ebullient than usual. One of the doctors was Adilya Kotovskaya:

> Usually he was a smiley and cheerful person but on that morning he was very self-controlled. I would say that he was shut in on himself. While we were dressing him, we asked questions and he either nodded or just said yes. And I can definitely say that he was greatly changed that morning. Actually I didn't see anybody who wasn't worried. But he was certainly worried.

At one point during his medical Gagarin began singing softly to himself, a popular Russian song; perhaps one of the same love songs that had been playing on the tape recorder in the cottage the night before. Dr Kotovskaya found herself deeply moved by this young man about to face an unknown fate within the next few hours. 'I did my job steadily,' she told the author in an interview over half a century later, 'without fuss but with love. I understood everything. I was like a mother.'

The medicals did not take long. Next the two cosmonauts were led to another room to begin suiting up. As the back-up, Titov went first to prevent Gagarin from overheating in his own suit before he was plugged into a ventilation unit for the short bus journey to the launch pad; yet another reminder of Gagarin's primacy, like the prized animal in the herd. While Titov was being dressed 'he kept silent,' remembers Dr Kotovskaya. 'He was clearly disappointed. But we understood.' Once the four technicians were finished with Titov, Gagarin was next. First came the sky-blue thermal undergarment, then a rubberised pressure suit to seal in the air, next a steel-roped restraint garment, and finally the bright orange coverall, its colour chosen to make its wearer easier to spot by search and rescue teams. All together the suit weighed more than twenty-two kilos – a hefty fifty pounds. In orbit it would weigh nothing. Next came the boots, high-laced with thick soles designed to absorb the shock of landing. Gagarin's parachute instructor, Nikolai Nikitin who would later die in a parachute accident, was with him now, quietly giving him last-minute tips and advice while the technicians continued dressing him. Gallai was there too, watching Nikitin:

> Why did he do this? … It was a precise psychological calculation. To concentrate the cosmonaut's attention not on the immense Unknown, but on something specific, and most importantly, something already tried and tested and therefore doable. An excellent teacher was Nikolai!

Briefly Korolev came in from the launch site to see them. Gagarin was struck by how exhausted he looked. In his 'autobiography' he describes the moment, without of course revealing Korolev's name. Despite the presence of his ghostwriters, in this instance the description seems to ring true:

> Clearly he'd had a sleepless night. I wanted to give him a hug, just as if he were my father. He gave us some useful advice about the coming flight, and it seemed to me that talking to us cosmonauts cheered him up a bit.

In a touching reversal Gagarin and Titov found themselves being the ones to console Korolev. 'Everything will be all right,' they told him before he returned to the launch pad. 'Everything will be normal.'

Now suited up, Gagarin was handed his identification card, a revolver and a hunting knife. He had already been given instructions on what to do if he landed in the sea and was attacked by sharks: he was supposed to slap 'flat objects' on top of the water to scare them off – if, that is, his unreliable dinghy did not sink first. He was told not to shoot polar bears if he landed in the Arctic, as that would only make them angry. And he was given final instructions on what to do – or not to do – if he landed outside the Soviet Union: he was not to reveal any details of the launch site or the R-7 rocket or the Vostok or the 'military or civilian leaders' involved in the space programme or any of its institutions; and if asked, he was to give his address only as 'Moscow. Cosmos.' Then something unexpected happened:

> The ground staff who helped me into my spacesuit began holding out sheets of paper to me. One of them offered me his service certificate – I gathered they were asking for my autograph. I could not refuse and signed the papers.

Unlike the Mercury Seven astronauts, for long famous in America if not throughout the world, nobody had asked Gagarin for his autograph before. He was an unknown first lieutenant in the air force. It was unsettling, as Yevgeny Karpov could see. 'For the first time since his arrival ... he was at a loss, unable to give his usual instant replies,' said Karpov. 'He asked, "Is this really necessary?" I said, "You'd better get used to it, Yura. After your flight, you'll be signing millions of these things."'

Only the helmet was left. Carefully the two technicians lowered it over Gagarin's head and sealed it in place. For now the visor was left open. But Karpov was suddenly concerned. Dressed in his spacesuit and white helmet, Gagarin bore an alarming resemblance to the U-2 spy pilot Gary Powers. What if Gagarin were mistaken for a spy himself when he landed – and in his own country? It had almost happened to Ivan Ivanovich in March, even with the 'dummy' sign in his helmet. Who could vouch that it might not happen again? Gagarin might be attacked, with disastrous consequences.

But what was to be done? With only minutes left before they had to leave for the launch pad there was no time for standard Soviet procedures, no time for a special commission to be appointed, no time to ponder the problem, arrive at a decision and then request approval from some committee of the Communist Party for doing it. The bus was waiting outside. The rocket was waiting. *History* was waiting. So Karpov did the only thing possible. He ordered brushes and a can of red paint, pointed at Gagarin's white helmet and told Viktor Davidyants, a technician who happened to have a 'good hand', to paint the letters CCCP – for USSR – on the helmet. 'But it won't have time to dry,' someone said.

'Don't worry, it will dry on the way,' said Karpov. 'Let's go.'

IN WASHINGTON IT was still the evening of April 11. Earlier that afternoon President Kennedy's burly press secretary Pierre Salinger swung his feet onto his desk in the White House, lit an Upmann cigar and addressed his fellow press aides from both the State and Defense departments. The latest intelligence reports, he said, suggested that the Soviets might attempt to put a man in space within the next few hours. Of course it could be yet another rumour – but it was better to be prepared with a press statement just in case. The task now was to set the right tone.

But the right tone proved difficult to find. To help set it Kennedy had recently asked the advice of CBS's veteran reporter Ed Murrow, whom he had appointed to run the US Information Agency, effectively a government mouthpiece. On April 3 Murrow had suggested a strategy in a memorandum whose cynicism is flagrant even today. 'In the event of a Soviet manned shot failure,' he wrote, 'we should express, with all the sincerity we can muster, the deep regret and distress of the President and the people of the United States.' But he also suggested that 'covertly, the US might encourage commentators in other countries to deplore the low regard for human life which prompted the Soviets to attempt a manned shot "prematurely"'.

Should, however, the shot be successful Kennedy was keen to give the Soviets credit, just as he had with the recent Venus probe. So Salinger and his team put together a draft and brought it to the president at the Oval Office. Jerome Wiesner, Kennedy's Mercury-sceptic science advisor, was there too. Wiesner also had a strong hunch that the Soviets would attempt a launch very soon, and possibly later that night. He then disappeared to have dinner with Homi Bhabha, the secretary of India's Atomic Energy Commission. Salinger went home to an early supper in Lake Bancroft, Virginia. The president dined quietly with his family in the White House.

At eight o'clock that evening, just as Gagarin and Titov were sitting down to their unappetising breakfast in Kazakhstan, Major General Clifton, the official responsible for ensuring that intelligence reports reached the president, approached his chief and asked if he wanted to be woken up if the Soviets launched a man into space during the night. 'No,' replied Kennedy. 'Give me the news in the morning.' The president retired to his living quarters on the second floor of the White House. As always the day's newspapers were spread across his four-poster bed and around his room. Kennedy was a famous devourer of newspapers. The day's news was dominated by the first day of the trial of Adolf Eichmann in Jerusalem, the SS officer accused, in the words of the New York *Daily News*, 'of masterminding the Nazi massacre of millions of Jews'. Eichmann had 'shuffled into court' to hear the six-thousand-word indictment, blinking under the courtroom's bright lights inside his bulletproof glass cage.

At ten o'clock that night, NBC-TV broadcast an eye-opening hour-long special report about the Kennedys' life inside the White House. Millions watched Jackie worry about the day when her three-year-old daughter Caroline was old enough to go to kindergarten. 'I rather hold my breath,' she said. 'If she's in the papers all the time that will affect her classmates and they'll treat her differently.' Jackie also said she wanted to bring a new warmth to the White House with fresh flowers and burning logs in the fireplaces in winter. Whether or not she or her husband actually watched the show is unknown. Ten minutes before it was over Yuri Gagarin had just stepped off the bus carrying him to the launch pad. But perhaps by then the president was already in bed.

TWENTY-SEVEN

KEY TO START

April 12, 1961: 06.30–09.07
Launch Site 1
Tyuratam Cosmodrome

THE BUS RIDE only took a few minutes. Gagarin sat at the front, plugged into his ventilator to keep him cool. Behind him, also with his ventilator, sat Titov. A knot of excited people huddled around them, jamming the narrow aisle. Three were fellow cosmonauts, all wearing classic pilots' leather jackets: Nikolayev, the calmest, most imperturbable of the group but now unusually animated; Bykovsky, thin and whippet-like as always; and Nelyubov, the once-front-runner relegated to third place. Along with their chief Karpov, there were several engineers and two doctors on board including Adilya Kotovskaya, still keeping her mothering eye on Gagarin. Even in these last minutes he was being monitored for any untoward changes in behaviour. There were moments on the short journey when he remained distant and shut in but then, abruptly, he would banter and joke with the other cosmonauts. Like actors they responded in kind, sometimes playing up for Suvorov's camera, who was there filming it all. At one point a grinning Nelyubov leaned over and popped a sweet into Gagarin's mouth. Gagarin grinned back. And so the little blue and white bus with its complement of excited passengers rumbled across the steppe until its destination appeared through the windscreen, a moment Gagarin himself would never forget:

> The closer we got to the launching-pad, the larger the rocket grew, just as if it were changing size. It looked like a giant beacon and the … rising sun shone on its pointed peak.

From the top of the launch pad's service tower, like lookouts on a ship, Vladimir Solodukhin and his engineering crew could see the bus crawling down the road towards them: '"They're coming! They're coming!" we shouted at the top of our lungs from the very highest point of the tower', and they watched as the bus entered the launch complex to pull up less than a hundred metres from the rocket itself. The door swung open and after a brief pause Gagarin stepped out, followed shortly afterwards by Titov, both men moving slightly awkwardly in their bulky spacesuits.

A group of senior figures was there to greet him: among them Kamanin and Marshal Moskalenko, along with other members of the state commission, and of course Korolev himself, wearing a hat and his 'happy overcoat', as he called it. No doubt he was also carrying those two one-kopeck coins in his pocket. They all sported the red armbands designating their elevated status, permitting them to remain on the pad until the final few minutes. But before all else, Soviet protocol once again had to be observed. Gagarin's first action was to walk up to Konstantin Rudnev, the chairman of the state commission, and formally report for duty. He also declared his readiness to undertake the mission ahead of him.

Almost the instant that he finished speaking the group dissolved into a scrum of hugs, handshakes, kisses and embraces. Titov tried to give Gagarin the traditional Russian farewell of three kisses on alternate cheeks, an impossible feat in their spacesuits. Suvorov filmed them: 'So they simply clang each other with their helmets, and it looks very funny.' The normally reserved Nikolayev now lost himself completely and flung his arms around Gagarin. 'I was so nervous,' he said later, 'that I forgot he was wearing a helmet and tried to kiss him. I knocked my forehead against it so hard I even had a bump there.' The grizzled Marshal Moskalenko, one of the most powerful military figures in the USSR, seized Gagarin's hand and shook it with great warmth. With so much at stake and so many dangers facing the smiling young man, everybody felt the emotion of the moment. 'I still feel his hands on my shoulders,' wrote Gallai in his memoirs, a quarter of a century later.

The hugging and kissing went on so long it threatened to delay the launch. Nobody wanted to let him go – they might never see him again. 'It was difficult keeping any kind of planned order,' Kamanin recorded drily in his diary. 'Instead of just wishing him a good trip some were saying their farewells and even crying. I had to force the cosmonaut out of their embraces.' But before he did Korolev himself came up to Gagarin. He too shook the cosmonaut's hand and then, throwing all formality to the winds, hugged him tightly.

Together with Korolev, Kamanin and Oleg Ivanovsky – the engineer who had ripped out parts of the Vostok's guts two days before – Gagarin walked up to the foot of the giant rocket. A set of steel steps led to the gantry lift. Ivanovsky would accompany him to the top and help him into the Vostok's cabin. Now, finally, Kamanin grasped Gagarin's hand. 'See you in a few hours,' he said.

Gagarin climbed the steps, using both hands on the rails to help himself up, gloves dangling from each arm. The R-7 towered over him, still wreathed in white steam, pointing at the powder-blue sky. At the top of the steps, by the entrance to the lift, he grinned and waved at the small crowd of well-wishers below. Korolev took off his hat and waved it back. For a fleeting moment Gagarin appeared to hesitate. He raised his hand a second time in farewell. Then he disappeared into the lift.

Titov returned to the bus. There was nothing more for him to do. His role was played out. 'What could happen at this late stage?' he said many years later. 'Was Yuri going to catch flu between the bus and the launch gantry? Break his leg?' But even now, even this late, a glimmer of hope still remained, teasing his brain while he sat on the bus and Gagarin rode the lift to the rocket's summit: 'Probably nothing will happen, but what if? No, nothing can happen now, but *what if …?*' Even the normally stern Dr Yazdovsky was moved by Titov:

> Of course he was hoping that when Gagarin went up to the capsule, a small tear would appear in his spacesuit or something, and immediately the Number Two would be in command of the flight. But Gagarin went into the launch-tower elevator cage very carefully, ascended to the capsule and sat down in the cabin, and when he reported to me that he was safely strapped in place, I gave the order to Titov to remove his spacesuit.

Dr Kotovskaya was also moved, but hers was a softer voice than Yazdovsky's. Just as she had mothered Gagarin during his medical, so she mothered Titov now. 'I watched Gherman,' she told the author more than half a century later. 'He realised that the possibility of his taking part in the flight was approaching zero. We helped him to take off his spacesuit. He was very, very disappointed. We tried to comfort him.' Once his spacesuit was off, Titov settled down at the back of the bus where, for the next half hour, he had a nap. After all, he had barely slept all night.

LOOKING LIKE AN oversized cupboard, the lift carrying Gagarin and Ivanovsky rose slowly to the top of the rocket's service tower. The doors snapped open and the two men stepped onto the platform and straight into a battery of film lights. The ubiquitous Suvorov had got there ahead of them:

> Gagarin looks in my direction, spots me and stops for a moment, then shakes his head as if saying, 'Again the movies!', waves his hand to me and comes to the [spacecraft] hatch. My hands are busy with the camera and I cannot shake his hand in farewell. He grabs the upper side of the hatch and after a short

> lag pulls himself easily inside the capsule and into the seat … I put my camera on the platform floor. 'Good luck! See you in Moscow! I will meet you there,' I yell to him.

Ivanovsky helped strap him in. Gagarin lay on his back in his ejection seat with the open hatch behind him. A small mirror on his sleeve allowed him to see what was going on behind. He now found himself inside a partly padded ball, approximately two metres, or seven feet, in diameter. The interior padding itself was an orange-yellow foam-rubber, perhaps not the most restful of hues but certainly cheerier than the gunmetal grey of the Mercury capsule – which was also much smaller. Directly before him was the sparse instrument panel with its little schoolroom globe, its warning lights and its four dials. To his right was the gaming stick, the manual controller he had been practising with earlier in the week, now locked by the special code. Here too was the radio which, to modern eyes, looks alarmingly like a car radio from the same era: a couple of knobs, three buttons and a simple tuning display. Below it was an old-fashioned telegraph key to be used if – or rather when – voice communications broke down, in which case Gagarin would be expected to tap out messages in Morse code and hope that somebody on the ground received them. His food and waste containers were also on this right side. There was enough food and water to last for ten days should his braking engine fail to fire but, thanks to the defective dehumidifying system, he would probably be dead before he managed to eat everything on board. The faecal waste unit was also there in case he ended up in orbit longer than the hundred-odd minutes expected. His own spacesuit contained a urination device.

To his left was the black panel with its switches controlling such things as cabin temperature – a simple rotary knob offered options from 'warmer' to 'colder' – as well as cabin and TV camera lighting. As with the dog flights, a TV camera pointing at Gagarin's face would transmit live pictures to the ground. In addition there was a tape recorder into which he was supposed to record his experiences during the flight. The same black panel also contained the numeric keypad required to unlock the spacecraft's manual controls. Somewhere near his ejection seat was the sealed envelope containing the three numbers that Gallai had placed in the cabin the day before. There were three portholes: a small one above his head, another to his right and a third one, the Vzor, above his feet, the device that also acted as an orientation aide for the critical manoeuvring phase before re-entry. All three portholes were currently blocked off by the nose cone shielding the Vostok. If everything went to plan that cone would fall away two and a half minutes after launch, revealing a view of the earth no human eye had ever seen.

The time was 07.10 – less than two hours until launch. Gagarin switched on his radio to test communications with the men in the control bunker a hundred metres away. His call sign was *Kedr* – Cedar. The bunker's was *Zarya-1* – Dawn-1. 'Do you hear me?' he asked.

Kamanin answered. 'I can hear you well. Start checking your spacesuit.'

'Checking.'

Ivanovsky was still standing behind Gagarin. He suddenly leaned into the open hatch and made a signal. Clearly he wanted to say something and not be overheard on the radio link. He leaned in closer, speaking almost in a whisper. 'Yura,' he said, 'the numbers are 1-2-5.' Even though the sealed envelope with its code was somewhere in the capsule, Ivanovsky wanted Gagarin to have the numbers immediately at hand in case of an emergency. His action was criminal and might have landed him in prison had it been discovered. But Gagarin's response was unexpected. 'You're too late,' he grinned. 'Gallai's already told me.' So, according to some accounts, had Kamanin. So had Korolev. They had all broken their own rules.

Ivanovsky squeezed Gagarin's arm one last time and stepped back onto the platform. Assisted by two fitters, he began to secure the hatch. First they checked the electrical contacts around its circular rim; these would signal that the hatch was correctly sealed and airtight when locked. They also primed the three small explosives that would blow the hatch before landing or in an abort – allowing

Gagarin to eject in his seat moments later. Then came the hatch itself with its thirty bolts. One by one the fitters began the laborious job of screwing them down, locking Gagarin inside his padded sphere.

ONE HUNDRED METRES from the rocket was the entrance to the launch bunker. Down a steep flight of concrete steps a door led to a complex of three underground rooms buried deep enough, in theory at least, to protect the men inside from an explosion on the pad. One of those rooms was now filled with a row of operators whose task would be to monitor the rocket's telemetry data during the launch, observing every millisecond of its performance as it tore into the skies. A second room was reserved for VIP observers. This was where members of the state commission and other grandees like Marshal Moskalenko now gathered. Only one person in here would actually be able to see the launch because there was only one periscope. Manning it would be Valentin Glushko, the chief designer who had built the R-7's powerful mainstage and booster rocket engines and whose accusations had once sent Korolev to the Gulag.

The third room was the most important. This was the control bunker, the nerve centre of the launch operation. Its director was Anatoly Kirillov, a former artillery officer who had worked with Korolev for years and worshipped him. Korolev surrounded himself with people he trusted from long experience, and for this launch in particular he had made sure he had the very best men available. Rockets were in Kirillov's blood: he had spent the war commanding a Katyusha battalion, firing devastating rocket salvoes on German troops from trucks. Like Glushko in the other room, Kirillov would also be watching this launch from a periscope. His right-hand man was Boris Chekunov, just twenty-five years old and another of Korolev's star picks, who would start the launch sequence using a special key. Korolev had chosen him expressly because of his 'light touch'.

In this room too was Gagarin's fellow cosmonaut Pavel Popovich, just as he had been in the mannequin flight. The other four would be watching the launch from a safe observation point but Popovich was in the bunker for a specific purpose: to communicate with Gagarin, a role that NASA also employed and called 'CapCom' – the capsule communicator. One can assume that Popovich had been chosen because of those attractive characteristics that had always marked him out, his warmth and humour, even his Ukrainian folk-singing voice, qualities that might help to ease nerves and calm fears as the minutes ticked towards zero – and beyond. 'Was Gagarin frightened?' Popovich said in later years. 'Well, imagine. There's a huge colossus of a rocket, twenty million horsepower, and you're sitting somewhere on the top of it. And you know that if she explodes nothing will be left of you.'

At 07.34, Popovich flicked on his microphone and greeted his friend.

'Hey, Yura, how's it going?'

And back came Gagarin's amused, if careful, reply.

'Just as they taught us.'

To one side of the control room and sitting at a green cloth-covered table, as if he were about to begin a game of bridge, was Korolev himself. On this table were just a microphone and a single telephone. The microphone allowed Korolev to speak directly to Gagarin. The telephone was there for Korolev to give the password for an abort in the first forty seconds after launch. If that happened an operator waiting at a guarded command post would first have to identify and confirm the correct password before then executing the abort by remote control, flinging Gagarin out of the Vostok in his ejection seat. The system was hopelessly cumbersome, and like the steel net draped across the flame pit that was supposed to catch Gagarin on the way down, hopelessly ill-thought out. By the time Korolev picked up that telephone Gagarin would almost certainly have been killed. Nor could Gagarin command an abort from the capsule himself. Unlike the Mercury astronauts he had no chicken switch; no say over his own safety.

But now, ten minutes after Popovich, Korolev also tested his link to the spacecraft.

'How are you feeling?'

'All's good here, am feeling great. They're about to lock the hatch.'

Compared to Mercury's state-of-the-art mission control everything about the launch bunker reeked of old technology. With its banks of cramped equipment, giant dials, heavy-duty switches and periscopes, it strongly resembled the cramped interior of a Second World War submarine. Many of the operators wore leather helmets like old-fashioned pilots. Korolev's little bridge table with its telephone and microphone added a further home-grown touch to the room. Not only did this not look like NASA's mission control, it was not mission control. There was no worldwide tracking network to follow Gagarin's voyage around the globe. There was no electronic wall map signalling his real-time position. There would not even be guaranteed two-way voice communications much beyond the Soviet Union. On a good day the longest-range short-wave ground transmitters could only reach five thousand kilometres. That left a lot of world where Gagarin would have nobody to talk to.

At best, and if everything worked, Korolev would still be able to follow the Vostok's *path* part of the way by utilising thirteen missile tracking stations strung across the country to the Kamchatka Peninsula in the far east. Additional ships in the Pacific and Atlantic disguised as commercial vessels would, it was hoped, provide further tracking information. But first the raw data from all these stations had to be transmitted to a secret missile computation centre at Bolshevo outside Moscow codenamed NII-4 for number-crunching on computers far more primitive than NASA's IBMs. Flight data had to be recorded on special paper treated with chemicals so toxic that operators had to wear masks. Only then could the results be passed on to the cosmodrome itself – a delay of several minutes at best. And there was just a single telephone line between NII-4 and the cosmodrome down which to send all this mass of information. Theoretically Korolev may have had the final decision-making authority over the mission. In practice for much of the time neither he nor anybody else would have very much clue what was going on.

At 07.54, while the hatch was still being screwed down, Popovich came back on the line.

'Yura, the lads here are sending you a big collective hello … Are you receiving me?'

'I receive you. A big thank you. Please send them my warmest greetings.'

There were seventy-three minutes to go. The critical launch window, so essential if Gagarin were to land back on Soviet soil, was still on target. The last bolt of the hatch was torqued tight, cutting Gagarin off from the outside world. But then, abruptly, everything changed. One of the technicians in the bunker shouted a warning. The engineer Vladimir Solodukhin heard it too:

> Suddenly we hear, 'No KP-3! no KP-3!' … One of the three displays on the central control board, signalling that the hatch was properly closed, was not lighting up. We would not be ready for launch in time! The entire team froze.

'I went cold,' said Ivanovsky, who had just supervised the hatch's closure. 'KP-3 is an electrical contact-sensor, signalling that the cover is pressed against the hatch frame.' Its failure to light up *could* mean that Gagarin's hatch was not airtight, with consequences too horrifying to contemplate. His cabin would instantly depressurise in flight; the oxygen would be sucked out, leaving his spacesuit his only protection against the vacuum of space. But the clock was running down. Perhaps this was just an instrumentation problem? Should they risk leaving the hatch and hope that it was? As Ivanovsky stood there at the top of the rocket not knowing what to do, Korolev's voice burst over the radio.

> 'Will you have time to remove and reinstall the cover?' he said. I looked at the guys … We needed to do this calmly. Not to rush. But there was almost no time. The three of us merged into a single six-armed creature. Not only now, but even then it was impossible to understand who was doing what. We unscrewed

all thirty bolts from the locks and removed the hatch cover. I only had time to notice that Yuri, slightly raising his left hand, was looking at me in a small mirror sewn on the sleeve. He was quietly whistling a tune.

He was whistling 'The Motherland Hears', as casually as if he were on a road trip:

The Motherland hears, the Motherland knows
Where her son is flying through the clouds.
With friendly embrace
With tender love
She watches over you.

'I took one last look at Yuri,' said Ivanovsky. 'There was no time to say goodbye again. I just managed to catch a knowing look in his mirror.' Gagarin went on whistling. The man's nerve was extraordinary. The six-armed creature cracked on with the work. One of the men, Ilya Kholstov, was famous throughout the cosmodrome for his strength. That strength was needed now. The hatch was heavy – sixty kilos, a hundred and thirty-two pounds, like a massive manhole cover. Back on it went, shutting Gagarin in again. Next came the thirty bolts. It was already 08.14. Fifty-three minutes to launch and Ivanovsky's team was still not finished.

Suddenly Gagarin's voice came over the loudspeaker in the bunker.

'If you've got a bit of music that wouldn't hurt.'

It was, at the very least, unexpected. From his card table Korolev shot straight back.

'Give him some music, give him some music!'

'Do we have any music?' asked Popovich, a not unreasonable question given they were all sitting in a missile launch bunker.

On the other side of the Vostok's hatch Ivanovsky's team raced on with the bolts.

Somebody managed to find 'Lilies of the Valley', a love song. They played it down the radio.

'Is that OK?' said Popovich. 'You're not bored in there?'

'No, it's good. They're singing about love.'

The closure of the hatch was almost complete. Ivanovsky kept going: 'Hands like automatons quickly screwed on the lock nuts … The seconds pounded at the temples … the fifteenth nut … the twenty-third … There is the last – the thirtieth!' In the bunker all the contact lights suddenly flicked on. Gagarin whistled his love song. There was one final check: a hermetic seal test. A suction cup was clamped over the hatch. No leaks. The hatch was airtight. Relief swept across Ivanovsky's team and all three rooms of the bunker. Korolev glanced at the closed-circuit TV screen relaying live pictures from inside the Vostok. He called Gagarin.

'Hermetic seal is checked. Everything is normal.'

'Got you.'

'I've just seen you on the TV monitor. Pleased to see how you're looking. You look perky!'

The time was 08.27. Up on his platform Ivanovsky patted the Vostok one last time for luck. 'We loved our Vostok,' he wrote later. 'We loved it with that love that comes with an inevitable parting.' Then he descended to the ground. Hydraulic motors began to pull the service tower and its platform slowly away from the steaming rocket. Gagarin could feel it in his seat, a slight swaying beneath him. A loudspeaker blared: 'Readiness in thirty minutes!'

OUT ON THE launch pad with his film crew Vladimir Suvorov glanced up at the R-7, now free of its ugly service tower, 'shining in all its futuristic beauty in the yellow-brown landscape'. It hung over the enormous flame pit, held in place only by its four suspension arms. Suvorov was working fast too. A final cable was plugged into his console confirming all his cameras were ready. They were everywhere: two at the very top of the searchlight tower, strategically positioned to capture the rocket as it blasted off the pad; four more fixed beneath the rocket itself; several others inside the flame pit, protected – he hoped – from the coming fireball under heavy metal covers with their lenses poking through tiny holes. The one place he would not be able to film was inside the bunker itself. There were too many people down there and his lights would get in the way. In the best – or worst – of documentary traditions he would have to reconstruct the events immediately afterwards, making everybody say all the same words they had just said in the hope that nobody watching the finished movie would ever notice.

A jeep was racing across the sand towards him and his crew, throwing up dust from its wheels. It pulled up sharply and an officer leaped out. 'Who are you?' he yelled. 'Quickly, get to the shelter!' Suvorov flashed him a piece of paper: written permission from Korolev himself that they could remain outside. The officer calmed down. 'Good luck!' he said. Then he roared off as hurriedly as he had come. Suvorov turned back to his men setting up their last camera. They swivelled the tripod until it was pointing directly at the R-7 spitting and hissing on the pad. Clearly that officer had thought they were mad. He was probably right. To get the money shot Suvorov and his team were now closer to the rocket than anybody else. Apart, that is, from the man at its top.

TWENTY-TWO MINUTES BEFORE launch, Gagarin put on his gloves. Ten minutes later he closed his faceplate and checked it was secure. Then he turned up the volume on his radio to maximum. This would prevent the rocket's noise from drowning out the sound of his voice. In the bunker Dr Yazdovsky was watching his pulse on a monitor. It was still low – sixty-four beats per minute. 'So my heart is still beating?' Gagarin joked. At 09.00, seven minutes before launch, a warning siren wailed across the launch pad. Korolev spoke into his hand-held microphone.

'Everything is normal. Another couple of minutes before operations start.'

'Message received. Feeling well, mood is cheerful, am ready for launch.'

For the next two minutes the radio was silent while Gagarin waited quietly inside his sphere.

The bunker was packed, three times as full as in any other launch. Everybody felt the tension; it was almost unbearable, a physical presence in all three rooms but especially in the control room. 'There wasn't a single person,' remembered Vladimir Yaropolov, 'who was not on edge. All the operators at their panels were feeling the stress.' 'Our nerves were strained to breaking point,' said one of those operators, Vladimir Khilchenko. 'Korolev looked as tense as a coiled spring.' Standing at his periscope, Anatoly Kirillov also glanced at Korolev. 'I remember his unblinking eyes, ever so slightly fading, as if his face were turning to stone.' The red abort telephone sat on his table directly in front of him. Earlier he had discreetly taken a tranquilliser pill to calm the pounding of his weak heart.

'We cannot afford even the tiniest mistake!' he had told Kirillov. 'I repeat – not one mistake!' But as always the list of possible mistakes piled up in his brain. The psychiatrist on the programme, Rostislav Bogdashevsky, stated the chances of Gagarin's success as just fifty-fifty. A later analysis would put it at 46 per cent, meaning that the mission about to start within the next five minutes was more likely to fail than not.

Korolev believed in his rocket, the magnificent machine his engineers also called 'our dear one'. But he had less faith in its third-stage engine, whose failure had hurled Zhulka and Alfa into Siberia in December, leaving the dogs stranded along with their on-board bomb in the snow. This was the engine that was needed to get Gagarin into orbit. For weeks Korolev had been worrying about it. He worried particularly that its failure might land Gagarin in the Atlantic near Cape Horn on the tip of South America, a place notorious for its terrible storms; and he also knew that Gagarin would most likely sink in his unstable dinghy and drown long before anybody could get to him there. Korolev's worries had nagged and gnawed at him, so much so that he had insisted on installing a special telemetry system in the bunker exclusively to monitor the performance of that engine during each one of its planned 353-seconds' burn. If everything was good it would print out a series of 'fives' on tape. If everything was not good, it would print 'twos'. Anything in between was potentially a disaster waiting to happen.

Korolev picked up his microphone again.

'One minute readiness,' he said. 'His voice,' remembered Khilchenko, 'gave away his nervousness' – just as his wife Nina had feared it would.

'Roger one minute,' replied Gagarin.

On Boris Chekunov's panel was a simple keyhole, like the kind used to start a car. This one would start the most powerful rocket on earth. Chekunov picked up the key, ready to insert on Kirillov's command. Turning it would initiate a four-minute automatic chain of events inside the R-7's guts culminating in launch. If, that is, nothing went wrong somewhere in the chain. Kirillov found himself gripping the worn handles of his periscope more tightly than ever. 'My palms were sweating. But it wasn't possible to loosen the white-knuckle grip of my fingers.' He kept his eyes riveted on the viewfinder, scrutinising the rocket standing a hundred metres away on the pad, watching for anything that did not look right. Then he gave the command. 'Key to start.'

It was 09.03.

Chekunov inserted his key. 'Everything inside me was trembling, although I tried not to show my fear.' But his touch was light, just as Korolev knew it would be. He turned the key to the right. The four-minute sequence had begun. The second hand of the clock began its slow sweep around the dial. 'We could hear our own heartbeats,' said Vladimir Solodukhin. The telemetry specialist Vladimir Kraskin felt suddenly sick. He had not slept all night. He tried to concentrate on his two oscilloscopes but all he could think about was how his wife had cleaned the blood off his clothes after the R-16 missile disaster six months earlier.

'Key to drainage.'

Chekunov turned his key further to the right. A light in his panel blinked on.

'Drainage valves shut.'

Two minutes to go. The vents allowing the rocket's liquid oxygen to boil off into the outside air had just been closed. An umbilical cable providing external power fell away. The rocket's own batteries were now in charge. Valves inside the tanks began opening, feeding the first tonnes of flammable fuel and liquid oxygen into each engine's combustion chambers, ready to be ignited. The rocket was coming alive. Kirillov could see it through his periscope. Up in his shrouded metal sphere, Gagarin could see nothing outside but he could feel it – and he could hear it too, a grinding, whining, rumbling symphony of sounds from somewhere deep under his feet. Now he braced himself in his seat, tensing his muscles just as he had been trained. Directly in front of him was the red ejection warning light. If it flashed he had less than two seconds before his hatch blew and he was punched out of the capsule backwards into whatever inferno awaited him. But two seconds might be too late.

Korolev called: 'Ignition.'

'Understood,' replied Gagarin.

The clock crept past 09.06. The four minutes were almost up. Inside each of the rocket's thirty-two combustion chambers, the most complex and advanced power units yet invented, were crude wooden sticks shaped like a 'T'. They were made of birch, the symbolic Russian tree with whose wood Gagarin's father had once built his house. Now they would start his son's journey into space. In an instant, pyrotechnical charges attached to each stick automatically fired, igniting the mixing fuel and liquid oxygen like giant matches. This was a very dangerous moment in the four-minute chain – every chamber *had* to ignite simultaneously; just a single unlit chamber could easily explode in the cascading waterfall of flames from the others, blowing up the rocket before it ever got off the pad. Kirillov screwed his eyes even more tightly to his periscope. 'At that moment nothing else exists in the world. My eyes were watering from stress, my blood pulsed violently, my heart beat so quickly and loudly I thought it might rip itself from my chest.' Gallai glanced at Korolev. He was breathing heavily. A blue vein was throbbing on his neck. His abort telephone waited in front of him.

'Preliminary interim combustion.' All five engines had ignited perfectly. Flames began spewing from the rocket's base. Chekunov waited for the next signal on his panel. It lit up.

'Main combustion.'

Screaming jets of hot gases roared from all twenty nozzles, a volcano of fire pouring from the R-7, filling the enormous flame pit with thick clouds of black smoke. The sight was awe-inspiring, as was the sound. In the bunker the men could hear it through the several metres of soil and concrete above their heads. The three rooms began to shake – 'like a fever', as Kirillov remembered. Up above, Suvorov's crew kept their camera trained on the rocket even as the noise assaulted their ears and the ground trembled like an earthquake. At a wooden observation platform 1.5 kilometres away Titov and the other cosmonauts were also watching. They saw the brilliant flash of fire – and then felt the wall of sound hurtling towards them like a physical assault. 'It hammers your ribcage, shaking the breath out of you,' Titov said. Everything was 'shaking with the noise ... there were stones and pebbles flying through the air.' It was almost impossible to imagine that a human being could be sitting on top of that monstrous fireball, the first ever to do so – or that he could survive it.

But still the rocket had not quite left the pad. The four suspension arms kept holding it in place, balancing it over the flame pit until its five engines had built up to their maximum thrust, nearly a million pounds of raw unsurpassed power. All thirty-eight metres of the R-7 strained and shuddered in their grasp in these final few seconds. Gagarin could feel the vibrations hammering into his seat. The noise inside his cabin was rapidly increasing. The red ejection light remained off, but his heart was suddenly racing. From its earlier 64 beats per minute it now shot up to 157 beats. The sensor fitted on his chest flashed the numbers back to the bunker. Dr Yazdovsky saw the sharp spike on his monitor. For several minutes the doctor had been biting his lips so badly that Korolev had asked him why they were not bleeding. But Korolev was not looking at Yazdovsky now. He was looking at Chekunov by his panel. At exactly 09.06 and 59.7 seconds the final indicator in the launch sequence blinked on. Chekunov shouted out: '*Lift-off!*'

The four suspension arms fell away at last, releasing the rocket from their embrace. For the briefest moment its enormous mass seemed to hang in the air. Then, in a visceral demonstration of Newton's Third Law, the downwards thrust from the R-7's five engines exerted its tremendous force on the rocket in the opposite direction, forcing it slowly upwards into the cloudless sky. In a moment of pure exhilaration Gagarin yelled the word over the radio that would shortly sweep around the world and into history:

'*Poyekhali!*' he cried. *Let's go!*

TWENTY-EIGHT

THE FIRST SEVENTEEN MINUTES

April 12, 1961: 09.07–09.24
Tyuratam Cosmodrome

IN THE BUNKER they heard him over the loudspeakers, the loud whoop of joy that for one brief moment smashed through the tension and electrified everybody in there. Even Korolev could not contain himself. 'That's my boy!' Gallai remembers him shouting. 'A real Russian hero!' But there were still six hundred and seventy-seven seconds – more than eleven minutes – before Gagarin made it to orbit and his rocket was hardly off the pad. Kirillov did not take his eyes off the periscope, straining to pick up the slightest anomaly in the rocket's trajectory, any faint deviation from its proper path that could indicate imminent disaster. After six seconds it cleared the searchlight tower, grasping for height in a tornado of smoke and flame while outside a battery of exotic tracking devices – interferometers, cinetelescopes, cinetheodolites – followed its every move. Then it began to pick up speed. The deafening thunder of its engines tore across the steppe as it accelerated away from the pad.

'Goodbye until we meet again dear friends!' radioed Gagarin over the noise.

'Goodbye,' said Korolev. 'See you soon!'

From their observation platform the cosmonauts watched Gagarin go in stunned silence. Only a couple of hours earlier they had all been joking with him on the bus. Now the only thing Bykovsky could say to himself was '*Don't let anything happen to him, don't let anything happen.*' And he kept on saying it to himself, over and over, as his friend's rocket ripped on upwards into the blue.

Then Korolev switched his microphone on again.

'Seventy seconds.'

Gagarin came back, his voice almost drowned by the static and engines. 'Understood, seventy. I'm feeling well, continuing the flight.'

The rocket was now streaking into the sky, haloed by a brilliant ring of flame, its twelve steering engines gradually flattening its trajectory and pointing it east to take advantage of the earth's rotation for an extra kick of speed. Already it was visible well beyond the limits of the cosmodrome. From her home in Leninsky thirty kilometres to the south, Khionia Kraskina also watched it go, a sight she would remember for the rest of her life. She was staggered by its beauty even as she was terrified for the young man on board. She found herself clenching her fists and praying – 'Please God nothing goes wrong'. In the bunker's telemetry room her husband Vladimir and his colleagues sat hunched over their oscilloscopes, each electronic waveform on the screens exposing what was happening inside the guts of the machine; or what was not happening. From seven hundred sensors on the rocket a torrent of data poured into the room: fifty thousand measurements per second recording vibration levels, six thousand measurements per second recording engine performance, impersonal numbers masking the furnace of chemistry in each engine combustion chamber with temperatures half as hot as the surface of the sun. Vladimir realised that his whole back was soaking in sweat.

'One hundred seconds,' said Korolev. 'How are you feeling?' Popovich was supposed to be the person communicating with Gagarin but in his excitement Korolev had hogged the microphone.

'I'm feeling good,' replied Gagarin. And then, with a flash of humour: 'How are you feeling?'

'Attaboy!' cried Korolev.

But in fact Gagarin was struggling to speak. The G-forces were increasing fast, constricting his breath and tightening the muscles in his face. The vibrations were building even further, the noise was even louder. He tensed himself for the moment of first-stage separation, squeezing his abdomen as hard as he could just as he had been trained to do on Dr Kotovskaya's centrifuge. It came on cue at 119 seconds. A sudden drop in the G-forces as the four strap-on booster engines shut down, then a feeling 'like that of being ripped away from the rocket', as he would later describe it in a secret briefing, declassified only after thirty years; then came a violent jolt as all four spent booster engines tumbled from the rocket to fall back down to the steppe. Gagarin had been flying for just two minutes. Nine minutes left until orbit.

He called over the radio: 'The first stage has separated!'

The drop in G-forces was brief. They intensified again as the mainstage core engine kept thrusting the now much lighter rocket higher and faster, accelerating it to over 12,000 mph. And this time the Gs were worse – Gagarin felt himself pressed down so hard in his seat he could hardly speak or even breathe; the weight was so heavy he could barely lift his arms – but at least the noise level was lower now with only one engine firing beneath his feet instead of five. He could still see nothing outside. His portholes remained shrouded by the rocket's nose cone. Then, at 154 seconds, the cone jettisoned in two sections, two petals dropping away, graceful in the symmetry of their fall. But for Gagarin sitting inside his little aluminium sphere, there was nothing graceful about it. The moment was abrupt, striking, astonishing. Light suddenly poured into the cabin despite a filter on the portholes. And something extraordinary filled the window.

'I can see the earth!' he yelled into his microphone. 'I can see rivers, folds in the earth. It's all easy to make out. Visibility is good. Everything is wonderfully visible!'

He was already leaving behind the Kazakh steppe for the greener wildernesses of Siberia, fast approaching over the horizon. And he was already in space. He had just raced past the border between the earth's atmosphere and space that the Fédération Aéronautique Internationale had only recently, and somewhat arbitrarily, fixed at an altitude of a hundred kilometres, or sixty-two miles. But he was not yet in orbit. His trajectory was still ballistic, like a missile. He was on the upwards slope of a path that would, under the laws of physics, necessarily slope back down, like Ham in his Mercury capsule. To reach orbit Gagarin was going to need more speed – much more speed – and for that he would need the R-7's third-stage engine, the engine Korolev was so worried about.

But that view – it was *mesmerising*. He was higher than any human had ever flown, and he was climbing still higher with every second. The river that snaked across the landscape far below – was it the Ob or the Irtysh? – he could not tell. And there were clouds down there too, tiny cumulus ones with their dotted shadows on the ground beneath. 'How beautiful it is, what beauty!' he exclaimed into his microphone. It is not clear if they heard him on the ground, but his on-board tape recorder picked up the elation in his voice. 'I feel great!' he said. Within seconds the clouds beneath him were thickening as he hurtled north-eastwards across Siberia. There was a little vibration, but not as much as before. He reported the fact – and almost at once the mainstage engine shut down on schedule. He felt the same sharp drop in G-force, the same bang, the same sensation of the rocket running away from him; the huge spent barrel, now useless without its fuel, fell away towards Siberia; there were a few seconds of silence – and then the final third-stage engine kicked in. Just over five minutes had passed since launch. In six minutes Gagarin should be in orbit. Everything now hung on that last engine.

THE MOMENT THE third-stage fired Vladimir Kraskin felt his knees trembling. 'The strain was indescribable.' Kraskin knew all about that engine. He had witnessed its failures before, just as he had witnessed too many failures with too many other missiles. But this one had a man on top, which made everything different, and especially when the prize of orbit was so tantalisingly close.

In the control room next door all of Boris Chekunov's attention was on the telemetry machine that had been specially installed to monitor this engine throughout its almost-six-minute burn. Right now it was going well. 'The numbers were being repeated out loud. '*Five ... five ... five ...*' Everybody watched the machine spewing out its tape of numbers and the man reading them out; Korolev most of all. The seconds passed. The tape kept spewing. The fives kept coming. And then, to Chekunov's horror, everything went wrong. The numbers dropped to threes.

> That meant we had an emergency situation in the third-stage engine. I saw Korolev's face and did not recognise it. I had never seen him like that. Grey and severe, lips pressed tightly together. The error lasted around two seconds. But such seconds shorten your life.

The numbers abruptly shot back up to fives. Nobody could explain what had just happened. Had they dropped to twos that would have been an unequivocal sign of an engine malfunction. But *threes*? And for just two seconds? Had they got away with it? Over the loudspeaker Gagarin's voice crackled, halfway into the engine's six-minute burn. 'Everything is working well, on we go!' There was no hint of a problem up there. So far it all sounded OK. But now Gagarin reported that he could not hear the ground very well. At his incredible speed he was moving quickly out of range from Zarya-1, the bunker's radio designation. Like a relay, it would soon be time to hand over communications to Zarya-2, the next VHF station along the line of his flight, based at Kolpashevo, a small town in Siberia. From Kolpashevo the transmissions would still be relayed back to the loudspeaker in the cosmodrome bunker. But Korolev would no longer be able to speak to Gagarin directly. He would be too far away.

The handover was supposed to happen at 09.16 – nine minutes after launch, two minutes before orbital insertion. Just before it did Zarya-1 called Gagarin for the last time.

'How are you doing?'

'I can't hear you very well. My mood is cheerful, I'm feeling good, continuing the flight. Everything is going OK.'

And then, seconds later, all radio communication was lost.

This time it was Kamanin who was watching Korolev:

> I don't know how I looked at that moment, but Korolev, who was standing next to me, was very agitated … His hands trembled, his voice broke, his face was twisted and changed beyond recognition.

For Korolev it was almost the last straw. There was less than a minute to go before the critical third-stage engine was supposed to shut down. But Gagarin could no longer be heard.

IN THE FINAL seconds before his engine shut down Gagarin was not only the highest man in history but also the fastest. He was accelerating to an astounding eight kilometres per second or 17,895 mph, more than three times as fast as Ham at his fastest. At such a speed it would take less than twelve minutes to travel from New York to London. But Gagarin needed every last mile per hour in order to enter orbit and not drop back down to somewhere on earth in one long ballistic curve – perhaps to some Arctic polar bear-inhabited waste or those terrible Pacific storms off Cape Horn in Korolev's worst nightmares.

He called the ground.

'The third-stage is working. The light of the television is working. I feel wonderful, I am in great spirits. Everything is going well.'

But there was no reply.

At just after 09.18, eleven minutes after launch, his engine shut down, apparently on cue. As expected he heard a loud bang. Ten seconds later came a jolt, as the third-stage automatically jettisoned from the Vostok. All of Korolev's R-7, all six of its engines and its fuel, had gone in a matter of minutes. Its job was done. The spacecraft was now on its own. It began to spin slowly. Swiftly Gagarin made notes of the temperatures and pressures on the gauges in front of him as his checklist demanded. He opened the visor of his helmet. He loosened his straps. As he did so he felt himself floating gently off his seat; only the straps held him down. Then he turned his head to look through his porthole:

> The Earth was moving to the left, then upwards, then to the right, and downwards … I could see the horizon, the stars, the sky. The sky was utterly black, black. There were huge numbers of stars. Their brightness was made sharper against the blackness and the speed of their movement across the 'Vzor' and the right-hand portholes. I could see the very beautiful horizon, I could see the curvature of the Earth. The horizon had a beautiful deep blue colour. The very edge of the Earth had a soft blue colour, which gradually darkened further, turning to a shade of violet, and then turning black.

He was in orbit.

No human had ever seen what he now saw. As the Vostok spun slowly on its axis the earth disappeared from his porthole to be followed by the sun gliding slowly past, almost blinding his eyes in the pure unfiltered light. Then came the sky again, blacker than any sky he had ever seen. He stared in wonder through the window. The TV lamp was getting in his way; he lowered its brightness to get an even better view. He felt utterly euphoric.

As in a slow dance, the earth drifted once again past his window. Its beauty was overwhelming. It is perhaps one of the greatest paradoxes in this story that Gagarin was able to witness all this beauty of the planet only by riding a missile designed specifically to destroy it. Now transfixed, he watched the world slide past. Somewhere down there was the eastern landmass of the USSR. Then it was gone and he saw the stars again, thousands of brilliant, glowing points of light that did not flicker like they did on earth. There was no sensation of weight, no pressure on his body. He had experienced a few seconds of weightlessness before in those aerobatic training flights, seconds he had found exhilarating; but this was something different, and now it did not stop. 'Everything is floating,' he recorded shortly afterwards on his tape machine, 'absolutely everything is floating!' And everything was floating – because it was *falling*. It was not that there was no gravity in space. Gravity was still pulling his Vostok back to earth; but at his astonishing speed the earth was curving away beneath him before he ever got there. This was what induced the sensation of weightlessness, not the absence of gravity. He was, quite literally, falling around the world.

IN THE BUNKER the suspense was almost unbearable. Over the relay they could hear the radio operator in Kolpashevo calling Gagarin.

'How are you feeling?'

And suddenly Gagarin's voice burst through loud and clear over the loudspeaker.

'I can hear you well! I'm feeling great! I can see the earth through the Vzor. Everything is OK. Hello everybody! Are you hearing me?'

'We're hearing you!'

The entire bunker broke into applause. 'There were hugs, kisses, congratulations,' remembered Kirillov. 'And there were tears. Tears of joy.' Korolev stood up at his table. Ivanovsky remembers people wanting to lift him up in their arms 'as the old tradition goes' but 'the ceiling was too low'. Instead Korolev went round the room 'and shook everyone's hand,' said Chekunov. 'Then he said simply, "Thank you."'

From the tracking station where they had watched the launch, Titov and the others also heard Gagarin:

> We could hear Yuri's voice – he was telling us that all was well and that he was going into orbit. Everyone was silent … and then the place erupted with congratulations. I stood up. I was trying to make sense of it all for myself. But I simply couldn't get my head around it. He was here not very long ago – and now he was up in orbit? That quickly? I felt bewildered … How could it be, that our Yuri was up in orbit? He was only just here – and now he's in orbit? I simply couldn't make sense of it!

Korolev immediately telephoned Khrushchev, on holiday at his dacha in the Black Sea resort of Pitsunda. He told the premier that Gagarin had just reached orbit. So far everything was going well. He would call again after Gagarin had landed. Khrushchev had been spending the morning working on an important party speech but his mind was not on the job. According to his son Sergei, he would sit out the next ninety minutes 'in nervous expectation while Gagarin circled the earth'. And he would spend most of it looking at the telephone.

Korolev then left the bunker. He drove straight to the command centre a few minutes away, where operations would be monitored for the rest of the flight. This was the room with the single telephone line connecting the cosmodrome to NII-4, the secret missile computation centre two thousand five hundred kilometres away near Moscow. Inside was something resembling a school's world atlas – possibly an *actual* school atlas if Mark Gallai, who was present, is to be believed. On this atlas 'a small bright red flag on a pin was stuck into an ordinary school eraser', which was supposed to indicate Gagarin's estimated position over the earth. Currently he was somewhere over the eastern USSR. Now everybody waited for NII-4's primitive computer to calculate the predicted parameters of Gagarin's orbit. It took seven tense minutes. But when the figures did finally arrive down that single phone line something was wrong. The Vostok was supposed to fly no higher than 217 kilometres above the earth at its apogee or highest point. But NII-4's initial calculation predicted that it would be 302 kilometres, and a later calculation would put it at 327 kilometres. At the very least, Gagarin was some eighty-five kilometres – fifty-two miles – too high.

The culprit, it would later transpire, was not the third-stage engine – despite that heart-stopping series of threes – but the R-7's second, mainstage engine. A radio command from the ground to shut it down at the correct time had failed. A back-up on-board system had kicked in; but it had shut down the engine a fraction of a second too late, over-accelerating the rocket and flinging it too high – rather like Ham had been flung too high in his Mercury capsule. And even though nobody in the command bunker knew any of those causes yet, what they did know – what Korolev certainly knew as he stood by that atlas with its pin and its red flag – was that Gagarin's higher orbit had two crucial consequences. First, he would land in the wrong place, and not where any of the rescue teams were expecting to find him. And second, if his braking engine failed and he had to rely on gravity to get him home, he would unquestionably die.

It was no longer a question of his defective dehumidifier unit quite possibly killing him in the ten days it would take for gravity to pull him back to earth. With this much higher orbit that ten days had just become as much as *thirty* days – by which time all his food, water and oxygen would have long since run out too. His braking engine was scheduled to fire in less than sixty minutes, when he reached the west coast of Africa on the other side of the world. If it refused to start he would die slowly, stranded alone for days, if not weeks, in space. His incorrect orbit had suddenly exposed him to the most terrible risk.

But as he sped in all his euphoria thousands of kilometres eastwards across the USSR, his route marked in the command centre by the changing position of the little red flag on the school atlas, Gagarin himself had no idea he was too high. He had no equipment on board his capsule to show him. And nobody was about to tell him.

TWENTY-NINE

THE NEXT THIRTY-TWO MINUTES

April 12, 1961: Morning
Gzhatsk
180 kilometres west of Moscow

YURI GAGARIN'S FATHER Aleksey was also travelling that morning, except in his case on foot. It would take him longer to walk the twelve kilometres across the fields from Gzhatsk to Klushino than it would for his son to fly round the world. Not that Aleksey knew a thing about that. But there was work for him in Klushino, the village where the family had once lived. Aleksey was helping to build a club for the collective farmworkers. Before he left, his wife Anna had prepared a simple lunch for him, boiled eggs, bread, a few potatoes. She watched him go, carrying his carpenter's toolbox, with his axe stuck in his belt. His job would take him almost the whole day and he would not get home until evening.

But the walk was hard going, especially for a man of nearly sixty. The spring floods from melting ice had come early this year. The main street in Gzhatsk was almost a river of oozing black mud that clogged Aleksey's boots as he set off. At least it promised to be a glorious morning. There was not a cloud in the sky. It was still cold but the scent of spring was everywhere. Anna could smell it in her garden and it made her happy. She loved this time of year when everything felt fresh and full of promise. Soon she would feed the piglets, the chickens and rabbits that were kept behind the house. But first she made breakfast for her daughter Zoya and her grandchildren, eleven-year-old Yura and fourteen-year-old Tamara. Tamara would be off to school shortly. Zoya would leave later for her job at the local hospital. Young Yura would spend the day doing his schoolwork quietly at home.

Anna was always happiest when she had her large family around her. Two of her three sons also lived in Gzhatsk; her eldest Valentin occupied the house next door with his wife Maria and their own child. Boris also lived in the neighbourhood. Only Yuri, her fighter pilot son, was not nearby. And right now he was not even at his home in Chkalovsky near Moscow. She knew that because the last time they had spoken he had told her he would be leaving soon on a work trip. 'How far away?' she had asked him. 'Very far,' he had replied. But of course she had not asked him where he was going.

AT 09.25, EIGHTEEN minutes after launch, Gagarin raced out of range of Zarya-2, the ground station in Siberia. A further thousand kilometres east was Zarya-3, the third and final link in the VHF radio chain, located in the small town of Yelizovo on the Kamchatka Peninsula. The Vostok covered the distance, a two-hour plane flight, in just over two minutes. This was the far eastern edge of the USSR, a wild and remote region of forests, lakes, mountains and bears. The Soviet border ended here. Beyond was the Pacific Ocean.

In Zarya-3's tiny monitoring station at Yelizovo the cosmonaut Aleksey Leonov stared in amazement at the flickering black and white pictures on his TV screen. Two days earlier he had been sent out to Yelizovo to help man the radio station there. Nobody had told him who would be on board the first manned space flight. Gagarin's selection had been kept hidden from every one of his fellow cosmonauts who was not part of the Vanguard Six. But the grainy images Leonov was seeing on his TV revealed the truth.

> The quality was poor but I could just make out the shape of a man strapped inside the Vostok. I could not make out his facial features but I could tell by the way he moved it was Yuri. I felt enormously proud. We had become very close friends.

Leonov was not surprised. Ever since he had watched Gagarin take off his shoes before stepping inside that Vostok in front of Korolev he was sure his friend would be the chosen one. Now he watched Gagarin inside his sphere as it approached the Yelizovo outpost. And then he heard him. Gagarin had just one question.

'Zarya-3, Zarya-3, what can you tell me about my flight path?'

Leonov did not know his friend's orbit was too high. But he was also forbidden to reveal any details of the spaceship's trajectory over the open airwaves. This was yet another secret to be kept from eavesdropping foreigners, as well as from Gagarin himself. But Gagarin was waiting for an answer.

> The radio operator by my side did not realise his finger was depressing the button that opened the radio link with the module when he turned to speak to me, 'What shall I tell him, Commander?'
>
> 'Tell him everything is going fine,' I replied.
>
> Yuri overheard me speaking. 'OK message understood. Give my regards to Blondie,' he said, using the nickname I was given because of my fair hair – although I had lost much of it.

Across hundreds of kilometres from earth to space, it was a touching bond between these two friends who had first met in their pyjamas on the hospital ward before their medical selection tests. But despite the answer he had just been given, Gagarin still did not seem fully satisfied. He was clearly 'anxious' about his orbit, Leonov recalled in an interview half a century later. And he kept on asking about his flight path. In the short five minutes that he was within range of Yelizovo he asked about it six times, two of them in the final minute before he disappeared over the horizon. His last two calls to Yelizovo reveal a man who was both exhilarated – and worried.

'How are you feeling?'

'I'm feeling superb, great, great, great. Please give me details about the flight.'

'Repeat, I can't hear you very well.'

'Feeling very good, very good, good. Please give me your data on the flight!'

But nobody told him. 'Even if I'd known his orbit were bad,' Leonov admitted in his interview, 'I would still have told him it was good.' What would have been the point of telling the truth? There was nothing Gagarin could have done. Even if he had used his code and accessed his limited manual controls, he could not change his orbit. The Vostok did not have that capability. So instead he was left in a state of ignorance. If ever proof were needed that Gagarin was not the pilot of his vessel but a mere passenger, then this, surely, was it.

By the time of his last transmission at 09.30 his voice was becoming faint. Leonov's radio operator tried one last time to raise him. 'Are you receiving me?' he called. But this time the only response was static. Gagarin was gone. He had crossed the far eastern border of the USSR. He was now out over the Pacific, coursing south-east, a quarter of the way into his flight. Ahead of him were seventeen thousand kilometres – over ten thousand miles – of open ocean before he skirted the northern tip of Antarctica on the bottom of the world. It would take him thirty-five minutes to get there, just a few minutes longer than the bus ride to the pad earlier that morning. He was no longer in contact with his friends. But now other eyes were watching him.

THE FIRST SIGNALS reached the US ELINT – electronic intelligence gathering – station on the windswept island of Shemya in the Pacific midway between the USSR and Alaska at 09.26 Moscow time, one minute after Gagarin had first radioed Zarya-3. The American monitors instantly recognised the signature on their oscilloscopes. It was the same eighty-three-megacycle TV transmission as with the previous Soviet dog flights, which meant only one thing: another orbiting Soviet spacecraft was coming their way. This one was already approaching over the western horizon. Even with their new sets it would still take approximately twenty to thirty minutes to demodulate the signal and get to watch the actual moving images themselves, but everything pointed to the likelihood that this time they would show, not a dog, but a human. With every day the news had suggested that possibility; and these waveforms rippling across their oscilloscope screens were now surely proof. They would know very soon.

But while they waited for the signals to be converted into TV images it was necessary to alert the relevant authorities in Washington. Nine minutes after the Soviet spacecraft's TV transmission had been intercepted, the phone rang at the home of President Kennedy's press secretary, Pierre Salinger. In the nation's capital it was 1.35 in the morning. The jangling noise woke up Salinger and his whole household. At the end of the line was Jerome Wiesner, the president's science advisor. He told Salinger he had just been informed that within the past half hour the Soviets had launched another of their giant rockets, and that it was in orbit. He believed this time there was a man on board. So far the Soviets had not confirmed the shot. But that confirmation might come any minute.

Salinger grunted his acknowledgement and put down the phone. Then he went back to bed and waited for the flood of phone calls to begin.

IT WAS HIS first time outside the USSR. In his isolation chamber experiment, Gagarin had imagined this moment, crossing continents, crossing countries and cities, crossing borders. Now it was real: in his tiny sphere he was sailing away from his own country over the ocean towards places he could never have expected to see as an ordinary Soviet citizen. There were no physical borders down there, at any rate none that he would see from his height. There was just the earth. And now there was just sea, endless sea, and its colour was strangely not blue but grey. His spacecraft was still rotating slowly, each revolution taking between two and three minutes, changing the view from his portholes. After losing contact with Yelizovo he tried to reach the long-range high-frequency station at Khabarovsk in south-east Russia on his car-style radio. To attract Gagarin's attention it was continually broadcasting 'Amur Waves' on a loop, the most quintessentially Russian of Russian waltzes by the composer Max Kyuss. Several times Gagarin tried to contact Khabarovsk using its call sign Vesna, or Spring – but without success. For twenty-three minutes after he had last spoken to Yelizovo, he heard nobody from the ground.

So he lay in his sphere listening to music as if it really *was* a car radio while the cabin's noisy cooling fans whirred away; and he had lunch. He practised drinking from a tube and squeezing food out of those toothpaste tubes into his mouth, as his checklist required. 'Swallowing,' he noted, 'is possible', thus ending once and for all one of the big debates about flying in space. He watched with fascination as bubbles of water from his drinking tube floated lazily away from him, drifting around the capsule. 'You watch them,' he wrote, 'as in a dream.' He noted too the effects of prolonged weightlessness: it had not stopped his blood from circulating nor his heart from beating – another debate settled – but it did present a different kind of problem when he briefly let go of his pencil while he was eating and the pencil floated away, ending up maddeningly out of reach near a porthole. That was the end of his written notes. Still he had his tape recorder; but whoever was responsible for it had forgotten to put in enough tape for the world's first human space flight. On its noise-activated automatic setting it was picking up all those noisy cabin fans and kept recording non-stop. Halfway across the Pacific the tape ran out. So Gagarin made one of his very few autonomous decisions, rewinding the tape part of the way back and starting it again. A substantial chunk of what he had just recorded was thus erased for ever.

The tape recorder ran out moments before he reached the night side of the earth – although for him the clock had moved on only thirty minutes and it was still morning. At his speed the transition was startling. On one side was the brilliance of the sun, far brighter than from the ground. On the other side was the earth and its thin band of atmosphere, 'a beautiful blue halo' darkening to turquoise, to violet and finally to 'coal-black' as it met the edge of space. And then, abruptly, instantly, everything went 'pitch dark' in all his portholes, as if all the lights had just been switched off. For an instant he wondered what had just happened. Then he realised he was in the night.

At 09.49, twelve minutes after entering night, he tried the Vesna station yet again.

'I cannot hear the earth. I am on the dark side.'

This time the Vesna operator thousands of kilometres away in Khabarovsk heard his voice breaking through the haze of atmospheric static.

'Roger,' he called.

But Gagarin could still not hear him.

Two minutes later Gagarin reported that the automatic solar orientation system had switched itself on. As soon as he re-entered daylight in another nineteen minutes the device would begin the critical process of manoeuvring the Vostok into the correct attitude for re-entry, using the sun as a point of reference. If it failed, Gagarin would have to attempt to do it manually, unlocking the spacecraft's controls with the three-digit code and then using his gaming stick, just as he had been practising, all too briefly, earlier in the week. Vesna heard his report that the orientation system was on and acknowledged it – but again Gagarin did not hear the acknowledgement.

He was now halfway across the Pacific, black and invisible far below. At 09.53, sixteen minutes after entering the night, Gagarin heard a crackle in his headphones. For the first time the voice of the Vesna operator in Khabarovsk came through:

'We are transmitting this: the flight is proceeding as planned and the orbit as calculated.'

'Roger,' replied a relieved Gagarin. And then: 'The flight is going smoothly.'

The Vesna operator was wrong of course, whether or not he was aware of it himself. The flight was far from being in the correct orbit. But Gagarin did not know that. He did not even know where he was. His last call to Khabarovsk was made four minutes later at 09.57. 'I am in good spirits,' he reported. 'I am continuing the flight. I am currently above America.'

But he was nowhere near America. He was somewhere over the southern Pacific, as far from land as it was almost possible to get. He kept dutifully transmitting his reports but all further two-way voice communications ceased as he sped on through the night. He switched off the TV camera light because its brightness was bothering him. Then he switched off the cabin lights to see better outside. His sphere was now in darkness. On he flew high above the planet towards the northern tip of Antarctica. Through his right-hand porthole he could see the stars. They drifted slowly from right to left. 'A little star is going past, it's going past,' he said into his tape recorder. 'What a beautiful sight.' He was now absolutely, utterly alone, as no human being had been alone before.

THIRTY

MOSCOW CALLING

FIFTY-FIVE MINUTES AFTER LAUNCH

April 12, 1961: Late morning
House of Radio
Moscow

YURI LEVITAN KNEW at once that the envelope in front of him contained something important, perhaps the most important piece of news he had broadcast since the death of Stalin in 1953. Not only was the envelope sealed but the studio in Moscow's 'House of Radio' from which he would shortly make his announcement was also sealed, a locked and guarded room on the fourth floor that was only ever used for announcing major events. The fact too that *he* had been chosen for this occasion was telling too. Only the day before, he had been recalled very suddenly from his vacation in the Caucasus. One moment he was lazily doing crossword puzzles and admiring the mountains, the next his chief had telephoned and ordered him back. Within forty minutes a car had turned up to take him to the airport. A special plane had then flown him directly to Moscow. Clearly somebody very senior had decided that his voice was needed.

It was probably the most famous voice in the USSR. Everybody knew it. Many had grown up with it. It was a rich, deep, authoritative but also a trusted voice, one that Stalin himself had liked so much that Levitan became his favourite radio announcer. That voice had broken the news of the German invasion of Russia in June 1941 and the German surrender in May 1945, momentous events in the nation's history. Now aged forty-six, a trim, neat, bespectacled figure, Levitan had been chosen to deliver news of another momentous event to the more than two hundred million citizens of his country and to the world beyond.

He opened the envelope. Inside was one of the three TASS statements pre-prepared and authorised by the Communist Party's Central Committee two weeks earlier announcing the news that a Soviet cosmonaut was at this moment orbiting above the earth in space. This was the statement that the government had agreed would be released while the flight was still happening to prevent that cosmonaut from being arrested as a spy should his spacecraft land in error on foreign soil. Two blanks on the statement had just been filled. One was the spacecraft's orbital parameters. The second was the name of the cosmonaut on board.

The time was just after 10 a.m. Gagarin had been in orbit for fifty-three minutes but only now was Levitan about to break the news. Some of that delay was due to the several minutes it had taken for the slow computer at NII-4 to calculate those orbital parameters. But most of it was the result of something quite different – Gagarin's military rank. Gagarin was a first lieutenant. There had already been some discussion about promoting him immediately to captain, the next rung up the military ladder, if his flight were successful. But somewhere between Gagarin's launch and Levitan's announcement, a high-level decision was made to promote him even further, to the rank of major, bypassing captain altogether – a double-jump. Nobody has definitively verified whose decision that was, or how high that level, but Sergei Khrushchev, the Soviet premier's son, is certain that it was his own father:

> Nikita Khrushchev called Malinovsky, the Minister of Defence, and congratulated him. He said to Malinovsky we should at least raise his [Gagarin's] rank a bit. Malinovsky said, let's make

> him a captain. And my father said, why are you so stingy? Make him a major instead!

Making him a major instead may have rolled easily off Khrushchev's tongue, if indeed he ever actually said it, but in the elephantine bureaucracy of the Soviet military this was no easy task, especially when there were only minutes in which to do it. While Gagarin was sailing across the Pacific a complex chain of permissions, ratifications and formal approvals needed to be followed before his double-jump in rank, one of the fastest promotions in the history of the Soviet Air Force, could be confirmed and finally entered into that second blank space in the pre-prepared TASS statement. At last it was done. Levitan dashed upstairs 'like a whirlwind' to the special guarded studio on the fourth floor. Another reporter, Irana Kazakova, watched him go:

> Suddenly I saw Levitan without his jacket and tie and looking a bit dishevelled running up to the fourth floor with a white envelope in his hand – carrying it like a banner – then running down the corridor towards us and shouting, 'A man is in space!' We knew it before anyone else – the people on the fourth floor.

The guard unlocked the door and Levitan slipped into the studio. He checked the clock. It was 10.02 – five minutes after Gagarin had radioed wrongly that he was over America. The microphone sat on the table, the portal to every loudspeaker in every village, town and city in the country, to every radio in all those apartments, houses, factories, offices, palaces of culture and collective farms across ten time zones of the USSR. The station was currently broadcasting a Soviet patriotic song, '*Shiroka Strana Moia Rodnaja*' – 'How Spacious Is My Beloved Country'. Abruptly the music stopped. Levitan's famous voice came on the air.

> *Govorit Moskva! Govorit Moskva!* This is Moscow calling. This is Moscow calling. We are broadcasting a TASS statement announcing the world's first manned space flight.

Levitan tried to speak slowly and keep his voice as calm as possible but his mind was a whirl of excitement. He kept on reading. First came the once-secret name of the spaceship, its occupant – and his brand-new rank.

> The pilot-cosmonaut of the spaceship 'Vostok' is a citizen of the Union of Soviet Socialist Republics, Major Yuri Alekseyevich Gagarin.
>
> The launch of the multi-stage rocket has gone well ... The spaceship has begun its flight around the orbit of the earth. According to preliminary data, its maximum height is 302 kilometres. Two-way radio contact has been established and is continuing with cosmonaut Comrade Gagarin.

302 kilometres. The orbital parameters were exactly as NII-4's computer had recently calculated them, as accurate as they could be given the time constraints. Millions would therefore now know what only Gagarin did not know – how high he was.

> Comrade Gagarin ... feels well at present. The systems providing all necessary life functions inside the cabin are functioning normally. The flight of the spaceship 'Vostok' with pilot-cosmonaut Comrade Gagarin in orbit continues.

It took Levitan fifty-eight seconds to read the bulletin. Within the next few minutes the story would flash round the world even faster than Major Gagarin was orbiting it. In Moscow crowds began gathering in the streets, electrified by the news. The city had been swamped with rumours about a manned space flight for days. Now it was really happening. *Time* magazine's Moscow correspondent described the moment when not just the capital but the entire USSR abruptly 'came to a halt':

> Streetcars and buses stopped so that passengers could listen to loudspeakers in public squares. Factory workers shut off their machines; shopgirls quit their counters. Schoolkids turned easily from the day's lessons. Somewhere above them, a Soviet citizen was arcing past the stars, whirling about the Earth at 18,000 mph, soaring into history as the first man in space.

But there was one inescapable consequence of the decision to release the news while Gagarin remained in orbit. It may have protected him from being arrested as a spy if he landed in the wrong country, or helped in any international rescue on the high seas. But it left the world, and above all his own family, in an agony of suspense, because he still had to come down.

TAMARA TITOVA WAS at home in her Chkalovsky apartment when she heard Levitan's voice on the radio. 'I froze,' she said. 'My heart disappeared into my toes.' Then she heard the name – Gagarin. Her husband Gherman was not the one. It was Yuri. For a few seconds relief surged through her. 'Lord, thank goodness it's not my husband!' she remembers feeling. And then she remembered Valentina next door.

She raced onto their adjoining balcony and shouted for Valentina to switch on the radio. Valentina came out. She had also just heard the news. One of the other neighbours had told her. 'She was filled with emotions in that moment,' says her daughter Elena. 'She was very worried about my father. It was extremely dangerous.' The news was even more of a shock because Valentina had not been expecting anything to happen until April 14, two days later, the date her husband had given her. But there was no time to think about that now. She had to be practical. Her one-month-old baby Galina was with her but little Elena was at kindergarten and Valentina needed to get her home urgently. She called across the balcony. Would Tamara collect her?

Tamara threw on some clothes and 'shot off like a bullet' to pick up Elena. By the time she got back with the child the stairs up to the Gagarins' apartment on the fifth floor were jammed with people. Neighbours, well-wishers, friends, everybody seemed to have turned up within a matter of minutes. Tamara pushed her way through the crowds – 'Please move aside, I've got a child here!' she kept saying – until she finally made it to the fifth floor. She stepped into the tiny sitting room. 'All the wives were there. All of them. And all of them were crying. We just sat there crying … At that moment I realised that we were all a family, that we were all so close, and that we were all there for each other.' Every one of those wives had long lived with the fear that their husbands could be killed at any moment. Their bonds were as tight, if not tighter, than the men's. The sisterhood had gathered.

But now they could do nothing but wait. There was no phone in the apartment. There was only the radio and the nerve-racking minutes until the next Levitan bulletin, whenever that would be or whatever he would say.

IN GZHATSK, ANNA Gagarina was tidying up the house while her daughter Zoya was getting ready to leave for her shift at the local hospital. Anna's daughter-in-law Maria burst into the room. She had dashed over from her house next door.

> She says, 'Mama, why aren't you listening to the radio?' And she's crying. And I asked her, 'What's the matter?' And she said: 'It's Yura.' And I said: 'What is it? Has his plane crashed?' And she said: 'No – he's in space!' And immediately I started screaming, 'What the hell has he done, he's got two little ones!'

Anna threw on her quilted coat and headscarf. 'I'm going to Valya in Moscow. She'll be alone with the children.' Then she set off as quickly as possible through the mud-choked streets to the railway station. She was going to take the next train to Moscow.

She was so wrapped in her thoughts that she barely remembered buying her ticket. She threw down some money and forgot to take the change. The cashier had to run after her with it. A young woman sitting next to her on the platform bench asked if she was all right. 'Yes,' she said. 'Everything is fine.' But all the time the same questions drilled through her mind: 'How is my Yura? What is Valya doing now? *How ...?What ...?*' Then, at last, the train arrived.

Back in the house, Zoya switched on the radio. She was not going to go to work now. A neighbour came in to sit with her and Maria.

> We listened to the radio. The music with the news reports was cheerful, and we felt a little more at ease. Then the music stopped, and the announcer said that the name of Major Yuri Alekseyevich Gagarin was to be included in the Komsomol [Young Communist] Central Committee Roll of Honour. That's what they do for dead people, I thought.

FOR WELL OVER an hour Aleksey had tramped through the thick mud and slush to his construction job at the workers' club in Klushino. The road wound through gently rolling countryside. The spring floods made the going tough but the weather was still glorious. He passed a village. At the last house a peasant he knew yelled out to him, as he later described it to Anna:

'Aleksey Ivanovich! Isn't your son flying in space right now? I just heard a TASS announcement on the radio! They said the pilot was Major Gagarin!'

'No, that's not my son,' said Aleksey. 'Mine's a first lieutenant.'

'But the name and patronymic are the same – Yuri Alekseyevich.'

'As if the world is short of Gagarins!' said Aleksey, and he picked up his toolbox and continued walking to Klushino.

When he got there everybody seemed to be out in the street. A friend rushed up excitedly. 'Have you heard the news?' he said.

'Sounds as if it's your Yura! They've just been talking about him on the radio.'

'My Yura is a first lieutenant.'

But everyone interrupted him.

'No, it all matches!' they said. 'Yuri Alekseyevich, originally from here in Klushino. It's him! It's our Gagarin!'

Aleksey was bewildered. None of it made any sense. The head of the local collective farm pushed through the crowd towards him. 'They're sending a car,' he said. The chairman of the Gzhatsk Communist Party committee had just telephoned to say they needed Aleksey back as fast as possible. But Aleksey knew the roads were impassable. It had been hard enough to get to Klushino on foot. A car would never make it. What about a tractor? somebody suggested. But Aleksey did not want to wait for a tractor. So they found a horse for him instead. A tractor would catch up with him further on. In a daze, he set off straight away with his horse back the way he had just come.

WITHIN MINUTES OF Levitan's broadcast the telephone next to Pierre Salinger's bed rang for the second time, just as the president's press secretary had been expecting since Dr Wiesner's call half an hour earlier. In Washington it was now shortly after 2 a.m. This time the call was not from Wiesner but from the *New York Times*. Its Moscow desk had just picked up the sensational TASS announcement. Did the White House have a comment? Salinger stalled them and checked back with Wiesner. Wiesner confirmed the report. The Russians *did* appear to have a man up there – and he was still up there. Wiesner said he would call Salinger back as soon as he had confirmation that the man had returned safely to earth. Or not returned safely to earth.

Meanwhile for Salinger any further sleep was over. From then on his phone rang off the hook as every major newspaper, TV network and wire service called for confirmation of the Soviet space flight and for the president's response. Despite the press release that he had drafted the afternoon before, Salinger refused to comment until

Kennedy had approved the final statement. But the president had explicitly instructed his aides not to wake him up. He was still asleep in the White House in his four-poster bed.

LIKE KENNEDY'S PRESS secretary, the NBC Cape reporter Jay Barbree was also woken up in the black hours of the morning. On the end of the line was his editor, Jerry Jacobs. Jacobs told him the news. 'You're kidding,' said Barbree. 'Get on it,' said Jacobs. 'I'm moving, boss,' said Barbree.

He rubbed his eyes and leaped out of bed. 'I had only one thought,' he wrote later. '*NASA could have had Alan Shepard up there three weeks ago*' – a forgivable slip since Shepard's selection was then still secret. He flicked on the radio. 'Excited voices spoke of Yuri Gagarin, of earth orbit.' He splashed water on his face, threw on his clothes, headed out the door. It was a six-minute drive to his office. Cocoa Beach's streets were empty. It was not yet four in the morning. Once at his desk, he phoned Colonel John 'Shorty' Powers, NASA's famously cantankerous press officer, waking him up.

> 'Morning. Shorty,' I said in my most pleasant voice. 'Sorry about the hour.'
>
> He definitely wasn't a morning person.
>
> 'Morning, my ass,' he growled. 'Whatta you want?'
>
> 'The reaction? NASA's reaction to the Russians orbiting a cosmonaut?'
>
> 'Fuck you, Barbree, we're asleep here,' he yelled, slamming the phone in my ear.

It made Barbree's report. And it would shortly make some of the early edition front pages, defining for many Americans the whole sorry story in a single headline:

'SOVIETS PUT MAN IN SPACE.
SPOKESMAN SAYS U.S. ASLEEP'

THIRTY-ONE

A BEAUTIFUL HALO

SIXTY-THREE MINUTES AFTER LAUNCH

April 12, 1961: 10.10
On board Vostok
In earth orbit over the South Atlantic

IF HIS ENTRY into the earth's night had been dramatic, Gagarin's exit back into the day was spectacular. His night had lasted only thirty-two minutes. In that interval he had covered almost sixteen thousand kilometres or ten thousand miles of the planet, from the Pacific Ocean all the way down to the tip of Antarctica before sweeping in a graceful curve past Cape Horn in South America and up into the southern Atlantic. He was now heading towards Africa. Twice he tried to make contact with another long-range Vesna station, this one based in Moscow, but they were unable to hear him. He could hear their music, however: not the 'Amur Waves' this time, but a playlist of popular songs about the Russian capital. Transmitting from almost the other side of the world, the Moscow signal was 'weakly audible', but at least the music kept him company in his darkened sphere as he raced towards the light.

Within minutes of Levitan's broadcast, he was crossing into the new day. It began as a brilliant streak of orange in the distance – an 'iridescent', glowing orange, like a 'beautiful halo' crowning the eastern horizon. 'It looks like a rainbow,' he said into his tape recorder. 'Like a rainbow rising from the edge of the earth. A stunning sight in the right-hand window. And I can see the stars passing too ... A fantastic sight!' The colours were dazzling, from orange to blue to the black of space, colours purer and more intense than anything he or any human had ever seen. 'One doesn't get these colours on earth!' he exclaimed. The sight was 'beautiful' – by now Gagarin was running out of adjectives to describe his incredible ride in the heavens – and he could not take his eyes off it through his portholes. Then, with shocking suddenness, the sun emerged over the horizon, a sunrise in fast motion, flooding his cabin with radiant light. He was on top of the world.

Three minutes into the daylight, at 10.13, he heard a faint voice: the Vesna station in Moscow. A brief, tantalising moment of connection across half the planet.

'How do you hear me?' the voice said.

'I can hear you well. The flight is going –'

But in Moscow they never received the rest of his sentence. Twice more Gagarin transmitted status reports but nobody heard him; nor did he hear them. Again two-way voice contact was lost – and this time for good. But now he had to forget about the radio and the view and concentrate instead on getting home.

As he sped further into the daylight his on-board Granit sequencing device – the clicking egg timer – had already begun activating the spacecraft's automatic sun-seeking orientation system. Gagarin could feel it working, the sensors picking up cues from the sun's position and sending commands to the spacecraft's steering thrusters. Quick jets of cold nitrogen gas squirted into the vacuum outside, gradually manoeuvring the Vostok into the correct re-entry attitude before the all-important braking engine was scheduled to fire, tail pointing towards the sun. Gagarin detected the movements in his cabin, a change of pitch, a slight fishtailing as the machine hunted for the correct angle, like parking a car in the tightest of spaces; except that this was a three-dimensional parking space and the consequences potentially perilous if the machinery got it wrong. Then he would have to use his secret code and try to do it himself.

For sixteen minutes he waited for the steering thrusters to complete their task. At 10.24 the solar orientation system flashed a yellow 'readiness' signal on the panel in front of him. It had performed flawlessly. The spacecraft was now aligned for re-entry. In one minute the Granit timer would command the next step in the sequence: firing the braking engine itself. Gagarin pulled his straps tight and closed the visor of his helmet. He turned the cabin lighting back on. He shut the blinds on his right-side porthole. Then he waited for the engine to fire. If everything went to plan its forty-second burn should slow him down sufficiently to take him out of orbit and begin the long curving descent back towards earth. Less than ten seconds after engine shut-off his sphere would automatically separate from the spent engine compartment and re-enter the atmosphere. He was just thirty minutes away from landing, approaching the west coast of Africa, eight thousand kilometres from home. With his extra height – the height he did not know about – his life now depended on an engine whose own designer had said he would blow his brains out if it failed. There was no other way back. It was, literally, a one-shot deal.

THIRTY KILOMETRES SOUTH of Saratov on the Volga river, where Gagarin had once studied foundry work at technical college, a turning off the main road at Podgornoye on the river's east bank led to a closed military zone. Hidden from prying eyes behind the 'Forbidden Entry' signs and the armed guard posts were six S-75 anti-aircraft missile launchers belonging to Unit 40218, the same type of missile that had shot down Gary Powers the year before, and would later spark the Cuban missile crisis in 1962. Alongside the missiles was a radar guidance and detection unit. Its purpose was to spot enemy planes or missiles and then destroy them.

Since six o'clock that morning the radar operators had been placed on very high alert. The order had come from military district headquarters in Kuybyshev, some four hundred kilometres north. Nobody in the radar unit had been told why. But as the morning wore on they were told to begin searching the skies for an 'object' that might appear at any moment. The unit's commander, Major Akhmed Gassiev, and his team of operators began a circular sweep of their radars over a range of a hundred kilometres. Until an hour earlier they had wondered what kind of object exactly might be heading their way. Perhaps it was another U-2 spy plane? But then, after Levitan's radio announcement at 10.02, Gassiev knew what they were looking for.

They were not the only ones looking. At a military air base near Kuybyshev, helicopters and IL-14 twin-engine search and rescue aircraft stood parked on the apron ready at a moment's notice to recover Gagarin once he had landed. Each aeroplane carried two paratroopers and a doctor specially trained to parachute directly over the target. Equipped with emergency medical kits their task was to handle on the spot any injuries the cosmonaut might have sustained in the landing. Unlike the helicopters, the doctor teams could jump even in the most inhospitable and challenging terrain. Their leader, Vitaly Volovich, had famously parachuted onto the North Pole in 1949, and terrain did not get more challenging than that.

There were seven such teams now waiting, alongside twenty aircraft and ten helicopters, some in Kuybyshev, some elsewhere. Most of the aircraft and helicopters were equipped with homing devices. These were supposed to lock on to both the capsule and a radio beacon installed inside Gagarin's parachute straps as he drifted down. Additional homing capability was provided by anti-aircraft defence radars and direction finders like Major Gassiev's missile battery. The stakes were high – and not just for the man in the spacecraft. One senior commander, General Andrei Stuchenko, had been woken in the early hours of the morning by a phone call from someone in the Kremlin. 'A man is shortly to fly into space,' the voice at the other end of the line had said. 'You are to organise his safe recovery and reception. You answer for this with your head.'

Across the vast Saratov region, an area only slightly smaller than England, all these teams of soldiers and rescue personnel were now preparing for the final act in the drama. Somewhere up there, within the next thirty minutes, a young, newly promoted air force major should be returning home from space. Gagarin's planned arrival spot was at Khvalynsk, a small river port midway between the cities of Saratov and Kuybyshev. But with his higher orbit he might land hundreds of kilometres off target. The dog flights had proved that spacecraft rarely, if ever, ended up where they were supposed to. But they were dogs. This was a human, and already a famous human. They *had* to find him quickly, wherever he landed. Meanwhile Major Gassiev and his operators kept watching their screens while their radars made circular sweeps of the sky.

INSIDE THE VOSTOK a green light blinked on the instrument panel – the descent signal. Gagarin heard a noise, a buzzing. The braking engine behind his back suddenly and 'very abruptly' kicked in. It was *working*!

He zeroed his clock. Approximately forty seconds for the burn. He could feel the G-forces beginning to build up again, pushing him into his seat for the first time since he had first reached orbit. The exhilarating weightlessness disappeared. He was heading back to the world. His engine was slowing him down, its thrust pushing in the opposite direction, *against* his direction of travel. At the point of cut-off it would shave just 304 mph off his speed of nearly 18,000 mph, the tiniest of fractions but enough to take him out of orbit. The seconds ticked down. Everything was looking good.

But inside the guts of the engine everything was not good.

A single valve that was supposed to close at the start of the burn had not closed – not completely. Unknown to Gagarin, fuel was beginning to leak out through the valve, instead of pumping into the engine's combustion chamber to create maximum thrust. With less fuel the engine was developing less power, still slowing the spacecraft down but not by enough. Effectively the brake pedal was not fully on the floor. The difference in speed was minuscule – Gagarin was heading back to earth just *nine miles per hour* too fast. But that nine miles an hour was about to change everything.

For thirty years, until the Soviet Union collapsed in 1991, the truth about this last, and most dangerous, phase of Gagarin's flight would remain a secret. To the world, Gagarin himself would present his entire flight as flawless. But in a confidential report recorded the following day he describes what really happened next.

First there was a jolt as the engine reached the end of its burn. Then a loud thump. And Gagarin's world turned upside down.

> The spacecraft began to spin around its axis at a very high speed. The earth passed by the Vzor porthole above, then to the right, then below and then to the left. The speed of the rotation was no less than around thirty degrees per second. It played out like a 'corps de ballet': head then feet, head then feet, at a very high speed of rotation. Everything was spinning. Now I saw Africa, now the horizon, now the sky. I only just managed to shield myself from the sun, to stop the light getting in my eyes. I stuck my legs over the porthole but did not close the blinds. I was curious as to what was actually happening. I was expecting a separation. But there was none.

Ten seconds passed. Within that ten seconds the braking engine compartment, along with its battery pack, various instruments and additional oxygen supplies, should have separated from Gagarin's sphere. But it did not separate. The two sections of the Vostok remained shackled together, spinning and tumbling in a crazy dizzying dance across Africa. Gagarin kept waiting for the green *Descent* light on his panel to go out, confirmation that the separation had occurred. But the light stayed on. There was still no separation. The seconds became minutes. The minutes became more minutes. Africa, horizon, sky chased each other round and round the porthole – the spinning did not stop.

Gagarin had no clue what was happening, or why. But at its root was that faulty valve. When the engine failed to slow the spacecraft to the precisely prescribed speed, the automatic system had *not* commanded the separation. The Granit egg timer had simply stopped dead in its tracks. Like a stubborn guard it barred the way to the next critical step in the sequence, the separation. Its primitive internal logic dictated that the Vostok had not yet left orbit and that therefore all those high-capacity batteries and oxygen supplies in the engine compartment would still be needed. That logic was trying to protect Gagarin when in fact it might be about to kill him.

He was helpless – a passenger trapped in an out-of-control spacecraft, which nothing in his very basic cosmonaut's manual had told him how to fix. Here was a scenario similar to the one Boris Chertok had spotted on both previous dummy flights, the problem nobody had taken up or bothered to fix in their hurry to press on. The wild *corps de ballet* continued over more than two thousand miles of the African continent, over rainforests, deserts and whole countries, all the way from Angola to northern Egypt; and while it did the spacecraft continued to descend, the downward curve of its trajectory taking it ever closer to the upper boundary of the earth's atmosphere where the temperature would quickly climb to well over a thousand degrees Celsius from the friction of re-entry. And right there was the biggest danger of all. If the two sections plunged into the atmosphere still clamped together anything might happen. The engine compartment was not designed to survive such extreme heat. It would undoubtedly burn up, quite possibly taking Gagarin's sphere with it. Or it might slam into the sphere, smashing through the thin metal wall to Gagarin strapped inside. Even as the Soviet Union was already celebrating his success, as people were joyously filling the streets of its towns and cities, Gagarin was facing the very real possibility of a horrifying death.

And yet he displayed an astonishing cool. He calmly recorded the time. He radioed that the separation had not happened, a message the ground never received. His flawed spacecraft continued to tumble head over heels towards the upper atmosphere and the possibility of imminent destruction, but he never once panicked. He did not even declare an emergency. He sent out another message, this time tapping it out in Morse code as he was spinning downwards. Incredibly, he even found time to make a note of the view.

> With the telegraph key, I broadcast 'VN' – everything is all right. Through the Vzor, I saw the northern coast of Africa, the Mediterranean. Everything was sharply visible. The spacecraft continued to rotate along every axis.

Ten whole *minutes* had gone and still the Vostok's two sections were stuck together. The spacecraft was spinning and dropping at almost five miles per second through an altitude of 100 kilometres, skimming the upper boundary of the atmosphere. At this height the air was extremely thin, almost non-existent, but the few molecules making up that air were already beginning to pile up in the Vostok's path, stealing energy from its tremendous speed and turning it into tremendous heat.

But the heat turned out to be Gagarin's salvation. Sensors on the spacecraft detected the sudden temperature change. They flashed a warning to the automatic system on board: the ship was entering the atmosphere. To this day the exact sequence of events remains unclear but it appears that this warning was able to override the blind logic preventing the separation. Gagarin heard a bang. Then he felt a jerk. The four steel straps binding the two compartments released their grip – although not quite. For a few alarming seconds they stayed loosely connected by a thick electrical cable, just as they had in the two mannequin flights. The tumbling dance continued. But as then, so now the heat seared through the cable, finally breaking the sections apart. The *Descent* light blinked off. Twelve minutes late, the separation was complete.

Gagarin was free. But he still had to face the violence of re-entry:

> Suddenly a bright purple light appeared at the edges of the porthole along the edges of the blinds. I observed the same purple light in the small gaps of the right-hand porthole. I felt the spacecraft rocking and its outer skin burning. I don't know where the crackling sounds came from: either the structure was under strain, or the heat-resistant coating was expanding under the heat, but the crackling was audible.

Down he went in his sphere, hurtling through the atmosphere in a ball of fire, the first human to fall to earth from space. Trailing a brilliant plume, he ripped like a meteor across the Mediterranean, over Turkey, over the Black Sea and into Russia in just six minutes. Outside the temperature soared to approximately 1500°C, hotter than molten lava, hotter than the foundry where he had once cast iron as a student. His only protection from the inferno was the coating of resin-impregnated asbestos surrounding the sphere, eleven centimetres at its thickest point, just four centimetres at its thinnest. As he plunged further into the atmosphere, the asbestos burned off in layers, carrying the heat away from the vulnerable metal shell in a gas trail visible across the sky. That was the crackling sound he could hear. He had been expecting it – but still it felt like he was trapped in a house on fire. He could smell burning too, an alarming odour that came on suddenly and then mercifully disappeared. Crimson flashes glowed at the edges of his portholes. His heart was pounding. The capsule was shaking, juddering, rocking in its earthward-plummet, and throughout it all the G-forces kept building beyond anything he had experienced during the launch, until they reached a massive 10 Gs – making him ten times his normal weight. He could barely move his arms or force out his breath in grunts. The blood drained away from his eyeballs. His vision tunnelled to a point. But the G-forces climbed still higher.

> I could feel that the G-force was past 10. There were two to three seconds when all the readings on the instruments began to drift apart. Things began to lose their colour. I had to strain and exert myself to focus.

He was approaching the dangerous condition called 'G-loc' – the loss of consciousness that comes with extreme acceleration forces as the blood begins emptying from the brain, inducing cerebral hypoxia. But all that tough training on Dr Adilya Kotovskaya's spinning centrifuge was helping to keep him still conscious, and amazingly even now he was able to calculate mentally the approximate arc of the spacecraft's oscillations from the movement of the sun across his porthole. Then the G-forces began to drop from their peak, and they dropped fast. The tremendous drag from the Vostok's dive through the atmosphere had slowed it right down. It was approaching the speed of sound – a mere fraction of the speed it had just been travelling.

For the first time Gagarin could hear the air 'whooshing' outside, 'as audible as in a regular airplane'. He was almost back in the world. In front of him the red *Prepare to Eject* light was illuminated, the next stage of his descent. He checked his leg and shoulder straps were pulled as tightly as possible. Once again he checked his visor was properly closed. He assumed the position for ejection, pressing firmly down into the back of his seat, making sure his pelvis was correctly placed in the seat's base. If he got it wrong, the force of ejection could snap it apart like a match.

At an altitude of seven kilometres, 23,000 feet, the spacecraft's barometric sensors triggered the three pyrotechnic charges in the hatch. There was a bang. The hatch blew off behind Gagarin. Cold air rushed into the cabin. For two endless seconds he waited to be ejected. The thought spun through his mind that the mechanism might have failed. Then, with a roar, his seat rocketed out.

THE TIME WAS 10.42. A hundred and eighty kilometres to the west of the capital, Aleksey Gagarin was riding back to Gzhatsk across the muddy fields on a tractor, his head still full of the astonishing news. Since he had left the village, first on the horse, then on this tractor, he had heard nothing further about Yuri's flight. He knew only that he was in space, and that he had not yet landed. Somewhere ahead, where the roads were better, an official car was supposed to pick him up and take him the rest of the way home. Then he would find out what had happened to his son.

Meanwhile, sitting on her train on the way to Moscow, his wife Anna looked out of the window as they stopped at the passing stations. It seemed to her that the whole world was talking about

Yuri. 'Everybody,' she remembered, 'was smiling and laughing.' But not Anna. She was still afraid, and she would not stop being afraid until her son was safe and home.

THIRTY-TWO

BA-BOOM!

NINETY-FIVE MINUTES AFTER LAUNCH

April 12, 1961: 10.42
Between Smelovka and Uzmorie
40 kilometres south of Saratov, USSR

GAGARIN SHOT THROUGH the sky still strapped to his seat. Within the first three seconds a small stabilising parachute popped out to slow down his ejection speed and stop him from spinning. He was sitting 'very comfortably', but falling fast. The sun was shining. A few clouds drifted below. The landscape unrolled beneath his boots, its details sharpening as he fell. From his seat he had a grand view.

There was a river down there, a big one, and something about its sinuous brown course was familiar. It was the Volga. He could see a substantial town straddling one of its banks, and with a shock he recognised Saratov, the place where he had once learned to cast iron and fly little planes, and where he had also learned to parachute just a year before. He was at least two hundred and fifty kilometres off course – but by an incredible coincidence, an oddly fortuitous combination of his higher orbit and his defective braking engine, he was returning from space to a landscape he knew well and loved. He was coming home.

> The silvery Volga flashed past below me ... Everything here was familiar to me: the spring fields, the thickets, the roads and Saratov, whose houses I could make out in the distance.

His ejection had taken place over the Volga's western bank. But a strong wind was pushing him east, away from the safety of the bank and out across the river. At his location it was at least five kilometres wide. He began preparing himself for the serious challenge of a water landing. He dropped further, through thirteen thousand feet, and suddenly his main parachute jerked open, blossoming beautifully above his head, the size of a substantial living room and conspicuously orange like his spacesuit. At the same time his seat automatically fell away from underneath him, an operation so smooth he barely noticed it. Over to one side, perhaps four kilometres away, he could see the Vostok sphere had already just landed beneath its own parachute. It was lying not far from the eastern bank of the river and even from his still considerable height he could see it was 'black and burned out'. But it was safely down. Now he just had to get safely down too.

He fell another fifteen hundred feet. On cue his survival pack unlatched from his parachute harness to hang from its fifteen-metre lanyard as he descended, along with its emergency radio, inflatable dinghy, knife, first-aid kit and the shark repellent he would no longer need in the middle of Russia. But the dinghy, even with its propensity to leak, might yet be critical. Then he heard a snap and felt 'a powerful wrench'. The heavy survival pack tore from its lanyard and dropped uselessly away, carrying his dinghy with it. A water landing had instantly become much more hazardous.

And there was a problem with his breathing tube too. He struggled to open a special valve in his helmet that let in the outside air but the valve had somehow got itself stuck in the folds of his spacesuit. He tried to free it, using his sleeve mirror to guide him, a demanding operation with his thick clumsy gloves and while still parachuting. It took six minutes. He might have just opened his visor but it was safer to keep it closed in case he did have to land in the river. He was still struggling with the valve when, without any warning, his back-up parachute inexplicably and abruptly yanked open: not fully, but partially. He was now descending under two parachutes, one canopy deployed, the other dangling below him and swelling hesitantly in the breeze, a potentially catastrophic scenario. If the back-up opened much more its suspension lines could easily wrap around the main parachute, the resulting mess of entanglement collapsing both canopies, an unsurvivable condition. Gagarin would plummet to the ground.

The stuck breathing tube, the loss of his dinghy, the two parachutes; after everything he had gone through now, in these last moments, he still had to go through all *this*. But again he did not panic. Panic, he knew, could kill him in an instant. He had forty-six training jumps to his credit; he had the experience, he had the knowledge. He kept scanning the ground below. He recognised a railway line, a bridge spanning the Volga, a tongue of land cleaving the river. He knew exactly where he was. Numerous times he had jumped over this very spot in his training. He steered as best he could towards it, even singing to himself as he did so, his default reflex when things got hairy. The back-up parachute wallowed threateningly beneath him. There were just a few thousand feet left to go, four or five minutes in the air at most. He was so close – he might just avoid the river. The second parachute might not open. He had nearly made it.

He found himself briefly in cloud. A thick white mist, a few bumps – then he was through. There were people below. There were cars. There was a main road. 'I could tell everyone's eyes were on my two orange canopies.' He was steering away from the river, still struggling with the breathing tube, the back-up parachute trailing beneath him. He saw a deep ravine and next to it a field.

> I saw a woman there grazing a calf. I thought to myself, Well, I'm probably going to end up in that ravine, and you can't do a thing ... Then I saw that I would just land in the field. I thought to myself, It's time to get down. My feet touched the ground. It was a very soft landing. The field turned out to be freshly

> ploughed, very soft, it hadn't had time to dry. I didn't even feel the landing. I didn't understand at first that I was standing on my legs. The reserve parachute landed on top of me. The main parachute fell ahead. I slackened it off, and released my harness. I took a look: I was still whole. That meant I was alive and well.

It was 10.53 – although for fifty years his landing time would be incorrectly published as 10.55. It had taken him eleven minutes to parachute to the ground and one hundred and six minutes to circumnavigate the earth. He stood in his orange spacesuit and helmet in the middle of the freshly ploughed field and the only people he could see were an old woman and a little girl, and they were both running away.

SIX KILOMETRES NORTH at the missile battery in Podgornoye they had heard a bang in the sky, like a supersonic jet smashing through the sound barrier. In the radar room, Yuri Savchenko, a young army cadet, was watching the circular sweep of his screen when an object suddenly split in two at an altitude of seven kilometres and began descending within their area. Perhaps *this* was what they had been waiting for all morning? Quickly Savchenko and his colleagues calculated the critical parameters of the two targets, height, distance, azimuth, tracking their separate paths through the skies. Their commander, Major Gassiev, was also watching. The targets were minutes away from landing. They were both coming down on this side of the river, somewhere between the two villages of Smelovka and Uzmorie. Gassiev noted the co-ordinates, summoned between fifteen and eighteen men including Savchenko, commandeered a car and a truck and tore off down the dirt track to find them. Levitan's latest announcement was still ringing in his ears. He had stated that Yuri Gagarin was over Africa. Now, incredibly, it looked like the cosmonaut might already be here in Russia, and that Major Gassiev might be one of the first to meet him.

OTHERS HAD HEARD the bang too, for kilometres all around, but the villagers at Smelovka and Uzmorie most of all because the entire drama happened almost directly above their heads. The same adjectives fill the testimonies of witnesses, many of them lovingly preserved in recordings and transcripts to this day in local museums. There was an 'explosion', a 'crack', a 'thunderclap', a '*baam!*' – 'it was very very loud' – and people rushed outdoors to see what had just happened. Like the rest of the Soviet Union, almost everybody by now had heard Levitan's announcement, which meant that almost everybody was also wondering whether those two objects falling from the sky, a ball and a man, could really be *it*. And right here, in their quiet corner of Russia.

Viktor Solodkyi was one of those villagers. Sitting in Uzmorie's school almost sixty years later, he can still recall every detail of that morning. He was then twelve, studying at his parents' home when he switched on the radio and heard the news. He finished his homework, then slipped out into the garden to share the sensational event with his best friend Nikolai Pisarenko, a next-door neighbour. Their homes were on Uzmorie's Lenin Street, a collection of one-storey, corrugated-roofed wooden houses straddling a dusty thoroughfare where hens and geese clucked for pickings, a picture-book of rustic Russia. Nikolai was watering his garden. It was a beautiful, unusually warm day. Then, suddenly, '*Ba-boom!*' The noise ripped across the sky.

> 'What was that?' we wondered. We looked up. Then we saw a tiny black dot. It was dropping, dropping, dropping from the sky. It had a parachute. And it kept flying down. The dot was getting bigger and bigger as it came down. And so my friend and I ran after it.

They raced down Lenin Street and out into the countryside. The dot was resolving into a ball, a sphere suspended from its parachute and it was heading towards a ridge-line, perhaps a kilometre away. They dashed across the fields, skirting patches of low marshy ground flooded by the spring rains, then up to a newly paved road just below the ridge. A few cars were parked there and four men from those cars were also running up the ridge ahead of them, following the ball and its parachute. By the time the two boys got to the top the ball had just landed. It was barely metres away from the edge of the precipice, almost close enough to roll down the side. A lucky landing. Solodkyi thought it might even have bounced when it hit the ground. But the sight of it was so staggeringly strange.

> It was warm to the touch, still hot, and charred. It was lying on its side. The parachute was thirty metres away. There was a buzzing sound from inside, like the sound a transformer makes. There were no other sounds. One side was silver, the other was burned … well, melted. The men who were older than me said, 'We need to look inside, see if someone's in there or not.' I said, 'We don't need to. He's up there!' And I pointed to a man flying with two parachutes, carried by the wind, two kilometres away.

He began to chase after the descending man, but tripped up in a muddy ditch. He ran back to the sphere instead where Nikolai and the other men were pulling out tubes of food and chocolate from inside. Then Viktor looked back the way he and Nikolai had just come. 'People were running up the hill. They came from here, from there, there were schoolchildren too. They came on motorbikes, on cars, on tractors. They just kept coming. They kept on coming.' Some were on the road, some were crossing the fields, all of them heading towards the burned sphere.

AT THE COMMAND centre in Tyuratam, in the crowded room with the school map and the pin and its tiny red flag, everybody waited for the message that Gagarin was safely down. Even though all voice communications with him had long since disappeared, a telemetry station at Simferopol in the Crimea had managed to pick up a brief signal from a sensor on the spacecraft just before it re-entered the atmosphere. The signal was immediately passed along the chain: first, to the secret missile computation centre near Moscow and from there the two and a half thousand kilometres down the single telephone line to Korolev in the cosmodrome's command room. That was the only signal before re-entry. Once Gagarin began his fiery plunge, the rapidly ionised air surrounding the spacecraft created a total radio blackout. It lasted several tense minutes. The next signal would come, if it came at all, from a 'Peleng' beacon embedded in Gagarin's parachute straps, meaning that he had ejected successfully from the sphere – although whether alive, dead or injured could not yet be known. All eyes in the command room, including Korolev's, were riveted on the phone, willing it to ring.

Oleg Ivanovsky had been the last man to see Gagarin before his hatch was closed. Now he found himself holding his breath:

> I looked at my watch. Many others were doing the same …
>
> A few more minutes, and the Pelengs will appear.
>
> A minute …
>
> Two minutes …
>
> Then an ecstatic yell!
>
> 'I've got the Pelengs! Hurray!'
>
> Hurray! Hurray! And suddenly the tension is gone. Everybody's faces are different. Everybody is thumping each other. No one is left untouched.

Korolev called Khrushchev on his hotline, also waiting anxiously at his Black Sea resort in Pitsunda. 'The parachute has opened!' he told the premier. 'He's landing. The spacecraft seems OK!' Khrushchev had only one question, and he kept shouting it down the phone. 'Is he alive? Is he sending signals? Is he alive? Is he alive …?' But Korolev was not able to answer him.

THIRTY-THREE

GAGARIN'S FIELD

ONE HUNDRED AND SIX MINUTES AFTER LAUNCH

April 12, 1961: 10.53
Between Smelovka and Uzmorie
40 kilometres south of Saratov

WHEN ANNA TAKHTAROVA and her five-year-old granddaughter Rumia saw the orange man landing in the field, their one sensation was fear. Anna was the forester's wife and she lived in a hut close by. She and little Rumia had been planting potatoes when the man descended out of the sky on his parachute. 'I looked,' said Rumia, 'and something orange was flying above and right into our field. "Grandma, look, look!" I shouted, terrified. Grandma was scared too. She picked up the bucket, grabbed my hand – and we ran.' The man was waving at them, shouting at them to stop.

> I thought it was a monster – not a person – he stretched his arms out and was approaching us, stumbling a little ... My grandmother started praying – and when she started praying I got really scared because that meant something terrible was happening. But he kept shouting, 'Wait, wait!' And then I said to grandma, he's speaking Russian so he must be a human being! He was speaking with a deep voice. His voice sounded

> like it was coming from inside a barrel. My grandma stopped … She asked, 'Who are you? Where are you from?' And he said, 'I'm from a ship.' And she said, 'What ship? There's no sea here.' And he said, 'I'm from the sky.'

Clearly Anna and Rumia had not been listening to the radio. But presumably the letters CCCP – for USSR – painted at the last moment on Gagarin's helmet earlier that morning must have helped, even if the voice in the barrel was alarming. 'I'm a Russian, comrades!' he said. 'A friend!' At least he did not look like a spy. But he was still extremely odd. 'The man was dressed very strangely – not like one of ours at all,' recalled Anna. But then Gagarin took off his helmet and Anna, like millions to come, was instantly conquered.

> I looked again. The man was smiling. And he had such a heart-warming smile that all my fears vanished in a moment. Then he asked, 'Do you have any idea where I can get to a phone?' 'To a phone?' I said. 'It's a long way on foot. Better to ride than to walk. Do you know how to harness a horse and cart?' And he laughed. 'I can't do it myself but maybe you can give me a hand?'

Gagarin had just flown around the world. Now he needed a horse and cart. To those familiar with only slightly later TV footage of NASA spacecraft returning to earth – the Mercury, Gemini or Apollo splashdowns from the 1960s – Gagarin's return is in a league of its own, an exercise in the surreal with a uniquely Russian twist. Instead of banks of TV cameras broadcasting the event live across the globe, instead of steaming warships and giant aircraft carriers, chattering helicopters, leaping frogmen and non-stop commentary relayed in real time back to mission control, there is just Gagarin and Anna, her granddaughter and a bucket of potatoes. And no phone.

The phone was the priority. Having lost his emergency radio when it accidentally snapped off on his way down, Gagarin needed to report to *someone*. He had landed far enough off course for none of the search and rescue teams to have a clue where, exactly, he now was. Like Vitaly Volovich, the parachuting doctor, most of those teams were based four hundred kilometres away in Kuybyshev, closer to where Gagarin was supposed to end up. They were all now desperately trying to find him in their planes and helicopters. So was Major Gassiev, who was not even part of the official search and rescue effort but was nevertheless scouring the countryside for a man in an orange spacesuit.

Meanwhile four – and possibly as many as six – curious farmworkers from the local collective farm had joined Gagarin's little party. Unlike Anna they *had* heard Levitan's broadcast. They had spent a heavy night ploughing and were unwinding in their rest house when the news was announced. Even rest houses in Soviet collective farms were supplied with state-controlled radios; sometimes they got the radios before they got the tractors. Like everyone in the area, the farmworkers had heard the crack in the sky – like an 'explosion' – and had seen the orange parachutist drifting towards the field near Anna's forester hut a few hundred metres away. They had run off to find him. All of them belonged to the collective farm at Uzmorie. The field where Gagarin landed belonged to the neighbouring collective farm at Smelovka. As a result the two villages would each claim the honour of being the first to receive Gagarin on earth, a dispute that would continue, with some rancour, over the following six decades. It continues still.

Gagarin went up to the farmworkers. 'We introduced ourselves,' he wrote in his secret report. 'I told them who I am.' The men were stunned. 'We've just been hearing all about you on the news!' said one of them, Yakov Lysenko, a tractor driver. 'You're supposed to be over Africa!'

> He was very lively and happy … He was wearing a jump-suit, or whatever it's called, and he said, 'Boys, let's be acquainted. I am the first spaceman in the world, Yuri Alekseyevich Gagarin.' He shook hands with everyone. I introduced myself, and he said, 'Boys, don't leave. All the bosses will be here any minute

> now. They'll come by car, lots of people, but don't leave.' But of course everyone forgot about us. They came from a city or a military garrison. They took him into a car straight away … We've never seen him since.

The car came quickly and it was Gassiev's. He jumped out with his men. When Gagarin stated his rank, Gassiev interrupted him – he was a *major* now, not a first lieutenant, a double promotion in one skip. It was all on the radio. 'I blushed with embarrassment,' Gagarin would declare later. He then attempted to make his official report but Gassiev flung his arms around him and kissed him. Savchenko had brought along his camera, but 'the emotional shock' was too much for him; he forgot to take a picture. Everybody was suddenly kissing and embracing. But Gagarin still needed to get to that phone. Gassiev stuffed him in the car and off they went, leaving Anna, Rumia and the collective farmworkers standing nonplussed in their field. The whole encounter, from Gagarin's landing to Gassiev taking him away, had lasted approximately fifteen minutes.

On Gagarin's request, one of Gassiev's two vehicles was despatched with some soldiers to guard the capsule's landing site two kilometres away. Gagarin did not go there himself. Officially he was still not supposed to reveal any details about his mission and nobody was supposed to ask him. But they were all too excited and kept peppering him with questions in the car. When they got back to the missile base it was teeming with soldiers, all eager to get a glimpse of the man they had just been hearing about on the radio. By this time Gagarin was beginning to act rather oddly, as Savchenko observed:

> I don't know how the cosmonaut actually felt but he was not behaving entirely normally. At times he was reserved, withdrawing into himself, then he suddenly began to laugh, loudly and uncontrollably, and without any obvious reason. Clearly he simply could not comprehend that he had managed to return alive and unharmed, and that he was standing on earth surrounded by other people.

At least he got to make his phone call. He went inside and rang Major General Yuri Vovk, an air force divisional commander at Kuybyshev. His message to Vovk was brief and direct: 'I ask you to report to the Party and the government that the landing went well. I feel well, I have no injuries or bruises.' He was then asked if he wanted something to eat, but declined. He just wanted a cup of tea, he said – as hot and strong as possible. He was helped out of his spacesuit, leaving on his blue zipped thermal undergarment. In his classified report he is careful to make a point of this: 'I was not photographed wearing the orange … outer suit nor with the helmet on.' One can almost sense approving nods from the KGB. The secrets of the Soviet spacesuit would yet remain safe; but Gagarin would nevertheless be photographed in almost every other possible way, starting immediately.

He never got his cup of tea. The world was already descending on him. The instant he stepped back outside everybody crowded round him, wanting to touch him, kiss him, hug him. Savchenko snapped off an entire roll of film, the first photographs taken of Gagarin after his return from space. Others did the same. Very few of these earliest images were revealed to the Soviet public at the time. They are raw, immediate, messy and much too real; Gagarin's shock of hair is matted to his sweating forehead; his face is lined with exhaustion; he throws back his head laughing, rather as Savchenko described. He clasps an old woman in his arms, a sudden, spontaneous action. There is nothing here of the carefully triumphal pose of a Soviet hero – all that would come later. Gassiev even brought along his six-year-old son Sasha, lifting him up to Gagarin at the centre of a group shot of fifty-odd grinning soldiers that could have come straight out of the Russian Revolution.

The impromptu photo opportunity lasted perhaps forty minutes. From the moment of Gagarin's landing, timings are sometimes hazy; events too. The memories of eyewitnesses often collide, stories become embroidered, sometimes changing with each new re-telling. One has the sense of a Russian myth being created, or rather multiple myths like versions of the Gospel, except in this case with the man descending from heaven rather than the other way round. The picture is further complicated by the many half-truths and frequently outright lies perpetrated by Soviet news organs over the days – not to mention decades – that followed; lies that themselves are often contradictory, inevitable casualties of a state obsessed with secrets. A historian has to pick his or her way carefully. But like lampposts in a fog certain markers stand out, moments that seem to ring true.

One of them happened just before Gagarin finally left the base. In a touching gesture, Gassiev took off his army cap and gave it to Gagarin to keep as a memento. But Gagarin had nothing to give in exchange. So Gassiev took his Communist Party membership card out of his pocket and handed it to Gagarin with a fountain pen. On it Gagarin signed his first autograph after his space flight. 'Believe me, Major, I'll definitely find you after this!' Gagarin joked, handing back the card. The two then jumped back in the car and set off towards the Vostok sitting on its ridge. The spacesuit was packed into the boot where nobody could see it.

On the way, they saw a helicopter. It was one of the search teams looking for Gagarin. In one version of this story, the pilot had already spotted the orange parachute in Anna Takhtarova's field. He instantly landed only to be told by Anna and the tractor drivers that Gagarin had been and gone – but gone where they were not able to say, since the missile battery's location was top secret. The helicopter pilot took off and flew around in circles desperately searching for Gagarin until, by the purest luck, he saw a man waving madly at him from Gassiev's car who turned out to be Gagarin himself.

It is an irresistible story, a *Keystone Cops* moment as one historian puts it, but sadly almost certainly untrue. Far more likely, the pilot was intentionally heading for the missile base when Gagarin hailed him down, since Gagarin would have given its location in his telephone report. At any rate Gagarin climbed in to be met with yet another robust round of embraces and kisses. The helicopter then appears to have flown to the Vostok site, already crowded with hundreds of curious onlookers, perhaps even briefly touching down there. But Gagarin did not get out. Instead he was taken on to Engels, the nearest military air base, thirty kilometres away. The plan was to make a stopover there before a transport plane could fly him to Kuybyshev where he was to spend the next two nights with Korolev and the other cosmonauts. The greatest of all photo opportunities was thus missed: Gagarin would never be seen standing by the burned and battered sphere that had just carried him around the world. But the fiction that he had landed inside it would very soon be unleashed on that world.

THINGS MOVED QUICKLY back at the cosmodrome. As soon as Gagarin's report that he was safely on the ground reached the command room everyone, in Mark Gallai's words, 'exploded – literally exploded':

> It was difficult to make out what people were yelling and shouting. Some of them were shouting 'hurray', but the individual words were drowned in the noise. The decibel level of all this noise wasn't much less than a rocket launch. But the strength of human emotion – how can you measure it merely by decibels?

Korolev was impatient to get to the Vostok landing site, fifteen hundred kilometres north, before flying further north to Kuybyshev to meet up with Gagarin by evening. But this was Russia – or at least a piece of it transplanted to Soviet Kazakhstan. There were priorities to be observed. As Gallai remembers it, a bottle of champagne miraculously appeared:

> After the first toast – 'To success!' – Sergei Pavlovich Korolev, having drunk his champagne, slammed his beautiful crystal glass on the floor with a grand flourish, paying tribute to the old custom. Shards of glass flew in all directions like shrapnel. Many of the others present were about to follow the spectacular example of the Chief Designer, but one of the launch site's

> directors stopped us. 'The Chief Designer can do it,' he said, 'But, comrades, not you.'

After all the draining tension, Korolev's energy was suddenly overflowing. Presumably he found a moment to call Khrushchev still waiting anxiously in Pitsunda, but this is not recorded. Within minutes of the report of Gagarin's landing, he was corralling his key personnel, senior engineers and members of the state commission, ordering them to hurry to the cosmodrome's airport. Then he tore off there himself. He was back in his element, his fears forgotten, rushing ahead; he always liked to move fast. Gallai was told to join him. So was Oleg Ivanovsky:

> Someone came out of the hotel with a piece of paper shouting something. I heard a list of names – Feoktistov, Gallai, my surname too. 'Sergei Pavlovich has ordered all of you to get your things ready and be in the car in ten minutes ...' I ran and grabbed the first things that came in my hands, stuffed them in a small case and ran back.
>
> The steppe passed by with crazy speed. Our Gazik [car] was jumping up and down on the bitumen road making a terrible racket. We got to the airport. The IL-14 plane ... was already warming up its engines. Then we took off.

The party mood continued on board. Korolev and all the senior rocket designers were behaving, as Ivanovsky put it, 'like first-year students after a successful exam. They were close to breaking into a dance!' Korolev kept telling jokes, laughing himself to tears. He talked about space flights to come, about Titov's coming up next and the other 'great guys', about one day putting a permanent space station in orbit. With another four hours of this before they landed, Gallai decided the only thing to do was to go to sleep. He badly needed it too.

Theirs' was not the only party plane in the sky. Kamanin and the other five cosmonauts had also left the cosmodrome, a couple of

The complex instrument panel of the Mercury capsule.
The manual control handles are on the lower left.

A reconstruction of the Vostok version. The number pad to unlock the manual controls is on the left. The single manual control stick and 1960s car-style radio are to the right. The 'Vzor' porthole is centre below the miniature schoolroom-style globe.

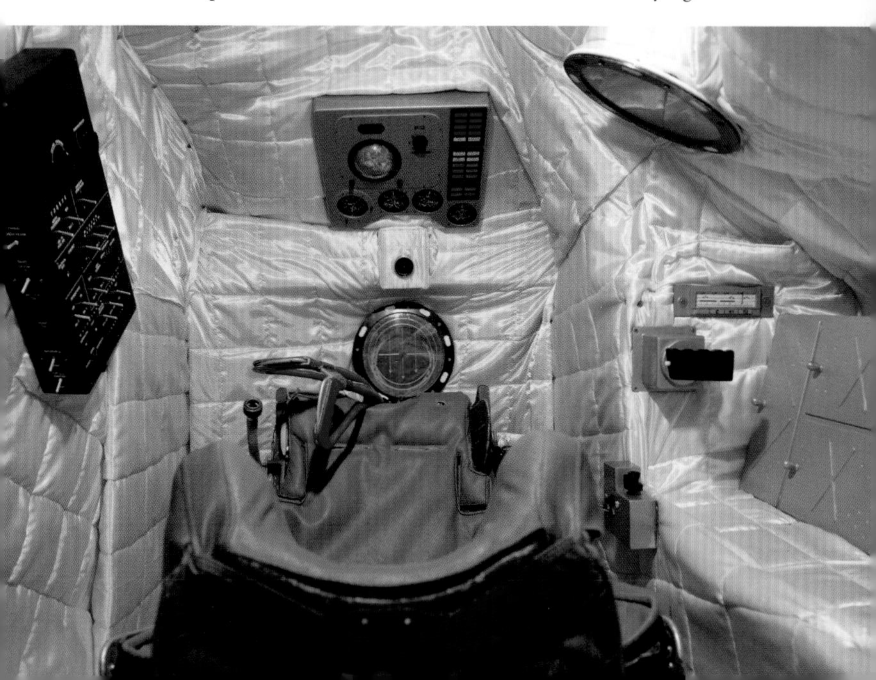

An artist's impression of the Vostok spacecraft in orbit, showing the sphere containing the cosmonaut. Beneath it is the instrument module with its braking engine. The two sections were supposed to separate before re-entry.

The much smaller Mercury capsule. 'You don't climb into the Mercury spacecraft,' said Glenn, 'you put it on.'

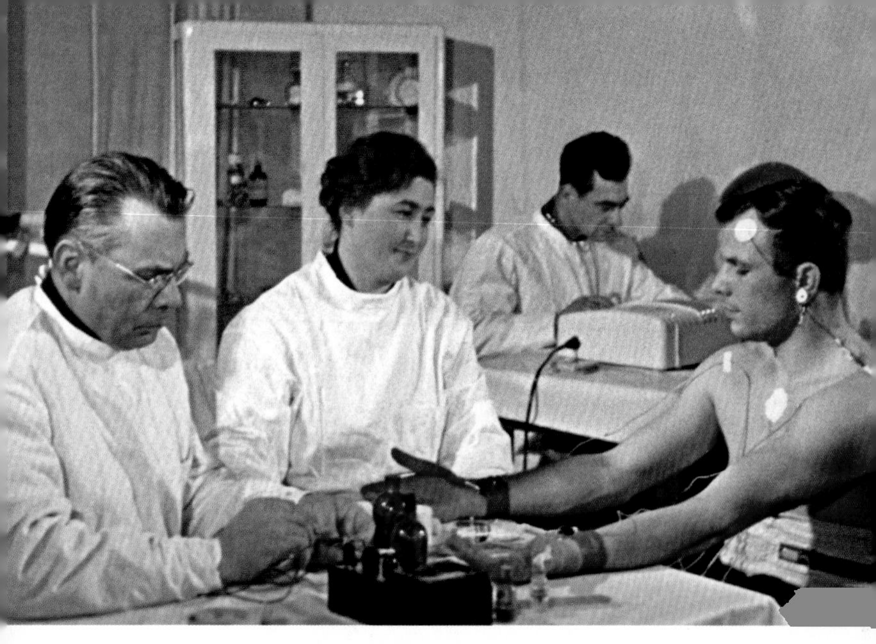

Gagarin's medical before his flight. Centre is Dr Adilya Kotovskaya. 'I did my job steadily, without fuss and with love.'

Korolev's masterpiece. The 260-tonne R-7 rocket showing its scale and four huge booster engines. A Soviet secret when Gagarin flew in 1961, this one was exhibited in 1967 in Communist Budapest.

A pensive Gagarin on the bus ride to the launch pad. Behind him a fully suited Titov is ready to take over.

'*Poyekhali!*' – 'Let's go!' A frame from Suvorov's movie footage as Gagarin blasts into the sky.

Mobbed at Major Gassiev's missile base. One of the first unofficial photos of Gagarin after landing.

The charred Vostok sphere after landing. 'This is history now,' said Korolev. 'This is the first one.'

Moscow April 14, 1961. Gagarin and Khrushchev at the airport. The Soviet premier 'looked like a child who had finally received a long-awaited present – and a very expensive one.'

The crowds go wild in Red Square. 'There were tears of joy,' said Galina Mishina. 'People kissed strangers in the streets.'

May 5, 1961. The tension builds as Alan Shepard waits for launch.

The Redstone finally lifts off carrying America's hopes. The tiny Mercury capsule is at the top.

An anxious President Kennedy and Jackie watch the launch on live TV.

Aftermath. The cameras are never far away. Gagarin at home with his wife Valentina and daughters Galina *(L)* and Elena *(R)*.

Parents of a Soviet icon. Gagarin's mother Anna and father Aleksey.

The icon today. A heroic Gagarin dominates the Moscow skyline.

hours earlier. They had driven to the airport once Gagarin had reached orbit, intending to meet him soon after he landed. Like every one of the search and rescue teams, their plane had been heading to the wrong landing site when the pilot received the message in mid-flight that Gagarin was safely down. Kamanin recorded the reaction in his diary entry for that day:

> After this wonderful message everyone – there were ten of us in the plane – began to kiss and dance, and Vasiliy Vasilyevich Parin [one of the doctors] took out the cherished bottle of brandy. I suggested that we drink it when we got to meet Yura.

They changed course for Kuybyshev, dancing and sleeping, but with still a long flight ahead. They would all meet up at a dacha there later that evening, a country house on a bank of the Volga belonging to the regional party authorities: Kamanin and the cosmonauts from this plane, Korolev and his team from the other. And of course Gagarin himself. Then the brandy would no doubt be opened.

AT HER HIGH school in Gzhatsk, Gagarin's fourteen-year-old niece, Tamara Filatova, Zoya's daughter, had spent most of the last two hours crying. When earlier that morning her teacher had announced the astonishing news to the class that her very own Uncle Yura was in space, she had not known what to think; she had barely any idea what 'space' even was, except that it was some sort of abyss from which people were unlikely ever to come back. She adored her uncle – he was fun, he was handsome, he was generous; he had bought her an expensive bicycle when she was little and he was a student – and now he might be dead. Then around lunchtime her teacher had come up to her and asked her why she was crying. Her uncle had landed safely, he said. It was just on the news.

Levitan had made the announcement at 12.25 – ninety-two minutes after Gagarin had, in fact, parachuted into his ploughed field. Tamara's teacher told her that all school was being stopped for the rest of the afternoon. She was to go straight home. Without further prompting she rushed off. But when she reached the house it was almost unrecognisable. There were big, important-looking cars – Volgas, Chaikas, Zils – pulled up outside. There were even helicopters clattering overhead. There was a crowd by the front door, almost all of them strangers, and she had to squeeze past to get inside. And there it was even worse. At least twenty people were jammed into the four rooms, a few of them friends and neighbours but a host of party officials and Russian reporters too. They were all clustering round her mother Zoya, her uncle Valentin and her little brother Yura, and asking them lots of questions about her uncle. Somehow they had also managed to set up three phones in the house where there was no phone before, and all of them were ringing non-stop. The noise was 'terrible' and it was getting worse by the minute. Some reporters were even trying to climb through the windows.

And then, into all this noise and confusion, her grandfather Aleksey appeared. Valentin saw him come in at the door:

> He froze on the threshold, blinded by camera flashes, confused, helpless. And when the photographers had shot off all their rolls of film, the other reporters rushed up to Father with their notebooks and their pencils and their portable tape recorders. To every question Father, utterly bewildered and unused to such a crowd of people, kept repeating the same thing. 'Thank you, thank you, I tried to bring up all my children to believe in hard work.'

He was exhausted. The journey back from Klushino had not been easy: first the horse, then the tractor across the boggy fields, finally the official car. He arrived with his axe and his toolbox just as he had left a few hours earlier that morning; but everything else was different. And the reporters would not leave him or the family alone. They kept asking for photographs of Yuri, as a baby, as a child, as a student, as a young air force officer. And they not only asked but also took, stripping the house bare of almost every framed or album photograph of Yuri there.

None of those photographs would ever be returned. Yuri was no longer just Aleksey's son, Zoya's brother or Tamara's favourite uncle. He had become state property. He was a Soviet son – soon to be an icon. And yet, standing in their little sitting room amid all these excited, voluble, interrogating strangers, Tamara could not help feeling thrilled by it all. 'My soul was singing,' she remembers, 'without my understanding what my uncle had actually done.'

ANNA WAS STILL on her train when Levitan broadcast the news of her son's safe return to the bosom of Mother Russia. By the time it pulled into Moscow's Belorussky station, she saw huge crowds in the square outside. They were shouting and laughing – '*He's landed! He's landed! Hurray!*' – the atmosphere, she remembers, was 'like a holiday'. Then she heard people shouting her son's name.

She burst into tears and went down into a subway. A concerned woman came up to her, as she describes in her memoir:

> 'Are you all right?' she asked. 'Has something upset you?'
>
> I smiled. My tears flowed like a river. I said, 'No, I'm feeling joy.'
>
> The woman laughed. 'Me too. A man has just flown in space! Did you know?'
>
> 'I know,' I nod, 'I know.'
>
> And she says, 'His name is Yuri Gagarin. Remember that.'
>
> 'I will remember, my dear. I will remember.'

She took the suburban train to Chkalovsky, walked to Valentina's apartment block and up to the fifth floor. The stairs were still crowded, the apartment packed. There were photographers there, reporters from *Pravda*, friends, neighbours, the other cosmonaut wives – and Valya herself. She looked drained. A military officer was asking everybody to leave – 'Comrades, we're asking you please to give Yuri Gagarin's family a rest!' The two children, little Elena and the baby Galina, were crying. All the noise and strangers were upsetting them. Anna went straight up to them and calmed them down. Then she put them to bed. 'Valya needed a break,' she wrote later. After all, that was why she had come.

THIRTY-FOUR

TRIUMPH AND DEFEAT

April 12, 1961: 12.30
Engels Air Base
Saratov District, USSR

NOTHING COULD HAVE prepared Gagarin for his reception at Engels the moment his helicopter landed after its short flight. However overwhelmed he had felt at Major Gassiev's missile unit half an hour earlier, this was altogether on a different scale. Already crowds of people were arriving at the base as news of his imminent arrival raced through the city, and their numbers were building every minute. Nobody could understand how the whole of Engels suddenly knew that Gagarin was *here*, by what mysterious operation of the bush telegraph the story of his arrival had spread, and spread so quickly. But it had, and masses of people were there to cheer and wave and yell themselves hoarse as soon as the helicopter door opened. Gagarin stepped out into the middle of it all and, by his own account, was 'moved to tears'.

Less than two hours earlier he had been alone in space, looking down on the earth; now he was back on that earth and nothing would be the same again. It was still only just after midday, but the physical and psychological distance he had travelled since breakfast was unfathomable and unique. As the crowds pressed around him he 'felt a rush of emotions'. A high-ranking general thrust a telegram of congratulations into his hands. It was from Khrushchev himself. A path was cleared for Gagarin to get to the terminal building. He went upstairs to the second floor and briefly collapsed into an armchair. Somebody snapped a photograph of him sitting there. He stares into space, dazed, like a soldier returned from combat. The photograph was not released by the authorities.

Officials would not leave him alone. They huddled round him, expressions of 'joy, wonder and admiration' glistening on their faces as the parachuting doctor, Vitaly Volovich, remembers. Volovich had arrived by plane himself a few minutes earlier to conduct Gagarin's first post-flight medical, but he was finding it a challenge even to get started. First the two men 'hugged and kissed' – on this day the hugging and kissing would never stop – then Colonel General Agaltsov, Deputy Commander of the Soviet Air Force,

> literally burst into the room to us, strode up to Yura, embraced him and said, 'Well, Major, congratulations!' Yura said, 'I'm not a major.' And the general said, 'You've been bumped up a rank!'

Thus Gagarin's new rank, first mentioned by Major Gassiev, was officially confirmed to him. Volovich opened his bag to begin the medical but Gagarin was called away to the phone. Now General Agaltsov's superior, Commander-in-Chief of the Air Force, Marshal Konstantin Vershinin, wanted to speak to him. Vershinin congratulated him and also confirmed his new rank. Gagarin thanked him. The call ended. Volovich tried again with the medical. But Gagarin was wanted on the phone again. This time it was Leonid Brezhnev, Chairman of the Presidium of the Supreme Soviet and one of the most powerful figures in the USSR. Gagarin reported that the mission had gone well and that he felt well. Brezhnev congratulated him. Gagarin put down the phone. Volovich tried a third time. Gagarin was called away again. A radio-telephone link had been set up connecting Engels to Pitsunda. Nikita Khrushchev wanted to speak to him.

The recording of that call has survived. It was also filmed and photographed. Standing up smartly, Gagarin spoke to the Soviet premier. Like everybody else, Khrushchev showered him with congratulations. 'You have made yourself immortal,' he said. 'With your feat you have glorified our Motherland!' He asked a few questions about what Gagarin had seen from space – 'I saw the earth from a great height,' said the cosmonaut helpfully, as if Khrushchev needed reminding. Then the premier got down to the point.

'I will be delighted to meet you in Moscow,' he said. 'Together with you, together with all our people, we will solemnly celebrate this great feat in space exploration. Let the whole world look and see what our country is capable of, what our great people and our Soviet science can do!'

'Let other countries try to catch up with us!' answered Gagarin, equally enthusiastically.

'Right!' said Khrushchev. 'I'm glad that your voice sounds so cheerful and confident, and that you're in such a wonderful mood! You're quite right. Let the capitalist countries catch up with our country which has paved the way into space. We are all proud of this great victory!'

Khrushchev was bubbling over with excitement. For a man who was able to get animated by the size of his aeroplanes compared to those of the Americans, Gagarin's triumphant flight presented a magnificent opportunity. He had already decided: he would have a monumental celebration in the Soviet capital on April 14, two days from now. It would be *the* biggest celebration since another Soviet victory – over the Germans – in 1945 at the end of the war; and very possibly much bigger. It would be the triumph of triumphs, at once an orgy of Soviet self-congratulation and a boast of Soviet supremacy. After all, those capitalist Americans had barely managed to get even a chimpanzee up in space; but the Soviets had placed a man in orbit around the earth. Even better, that man was clearly every bit as thrilled as the Soviet leader at the prospect of demonstrating just what his nation and his political creed could achieve. Gagarin had indeed proved the perfect choice. He unquestionably deserved that jump in rank. Here was a Soviet hero Khrushchev could flaunt to the world.

After asking about his parents and congratulating them too – 'They have the right to be proud of their son!' – Khrushchev ended the call. 'Goodbye, dear Nikita Sergeyevich,' said Gagarin, and he put the phone down. He had just made friends with the most powerful man in the USSR. Now he was smiling, and his smile really *was* terrific. There it is, in the movie footage and photos of the occasion. His thousand-yard stare had quite gone. So had Volovich, who had given up on the medical. It would have to wait until later.

In the meantime yet another dignitary had turned up on the second floor. Ivan Borisenko, the USSR's sports commissar, had come to sign the certificate confirming the lie that Gagarin had landed while still inside his spacecraft and thus claim the world altitude and a host of other world records:

> I, the undersigned, Sports Commissar of the Central USSR Aeroclub … Borisenko Ivan Grigoryevich, hereby witness that on April 12 1961 at 10.55 Moscow Time in the region of Smelovka village … a spaceship-sputnik landed with cosmonaut Gagarin Yuri Alekseyevich. The spaceship-sputnik bore the markings USSR-VOSTOK.

But Borisenko had witnessed nothing. He never went to the Vostok site; he filled out all the paperwork at Engels. The spacecraft landed at 10.48 not 10.55 – five minutes before its occupant had come down; it never bore the markings USSR-VOSTOK; and Gagarin landed two kilometres away in a ploughed field. But for the rest of his life and with a fidelity that is itself almost heroic, Borisenko stuck to his lie like glue, polishing and embellishing it to the point (twenty-two years later in 1983) where he manages to be at the landing site well before Gagarin even lands there:

> We never took our eyes off the sky where, from second to second, a dot should appear – the canopy of a giant orange parachute. Here he is, lower and lower. After 108 minutes Yuri Gagarin, the first person to orbit the earth, lands. We hurry to him. He stands happy, smiling among the jubilant collective farmers ... As required by the sports code, I checked the identification marks on the ship on which was written VOSTOK-USSR. Yuri Alekseyevich looked a little tired.

About the only thing that is true in that statement is that Gagarin was tired. But there was no time to rest. He remained at Engels for perhaps a couple of hours. Just before he left he gave an impromptu press conference over the phone to reporters from *Pravda*, *Izvestia* and the head of *Agitprop* – the USSR's Department of Propaganda and Agitation – duly declaring that his feat was not his alone but belonged to the entire Soviet people. No western reporters were invited. The press conference did not last long. His plane was waiting outside to take him to Kuybyshev. He had spent most of his time at Engels speaking to some of the most important and influential people in the land, but there was one person he had not yet spoken to and that was his own wife. She had no phone.

Somebody grabbed some flowers out of a vase and stuck them in his hands. Surrounded by officials, he stepped onto the airport apron and into pandemonium.

The hundreds who had been there when he first arrived had since swollen to thousands. People were breaking through the boundary fences, pushing past guards who were helpless to stop them. Some were climbing trees at the airport perimeter just to get a glimpse of the heroic young spaceman – in the rush and tumble one man fell down and broke his arm. Everybody was trying to reach Gagarin, touch him, embrace him, as he made his way to the plane. Later Gagarin would say that this sudden new-found celebrity was even more disorienting than re-entering the earth's atmosphere. It became so wild that Volovich had to wave his handgun to get Gagarin up to the boarding steps. At the top the cosmonaut turned around, beamed and thanked the crowds for their greeting. Then he ducked inside. The engines started, the plane taxied to the runway and took off.

By now it was approaching three in the afternoon. Gagarin was finally on his way to Kuybyshev, his next and last stop of the day; a day that never seemed to end. At least Volovich was able to perform his medical on board. Gagarin's fatigue was such that he was briefly sick. But when Volovich checked his blood pressure it was normal – 130/65. Indeed it was so normal that Gagarin joked that maybe he had never been to space at all.

Meanwhile the plane droned northwards. As he looked at Gagarin lying back in his seat, Volovich could not help experiencing a sense of awe. 'It was an amazing flight,' he told the author at the age of ninety in the last interview of his life, 'because it was difficult to imagine that sitting with us was someone who was the first person to have left the earth. It was an amazing feeling.' Gagarin, he noticed, sometimes closed his eyes. But he was not asleep. It was 'as if for a moment he had returned to the cabin of his spacecraft', alone above the world, watching the stars.

WHILE GAGARIN WAS flying to Kuybyshev, President Kennedy's valet glided down the central hall on the second floor of the White House and knocked on the president's bedroom door. George Thomas was a great favourite of both Jack and Jackie, famous for his good humour even if he was not always reliable about waking up the president on time. But today he did not need to worry. Kennedy was already up.

In Washington it was just before eight o'clock in the morning. As always Thomas handed the president a clutch of the morning papers. They were all full of Gagarin's triumph. In the six hours since Levitan's first announcement the front pages had been rapidly rewritten to accommodate the day's dramatic events. In bold capital letters every one of them rammed home the same story: '*SOVIET ORBITS MAN AND RECOVERS HIM*,' exclaimed the usually sober *New York Times* in the first of a three-line headline. '*SPACE PIONEER REPORTS: "I*

FEEL WELL'". The space pioneer was identified: 'Major Gagarin, 27 years old, is an industrial technician.' Amazingly 'Tyura Tam' was also identified as the launch site despite the best Soviet efforts to hide it, a tribute to the *Times*'s powerful network of US intelligence sources. Three entire pages inside the paper were also devoted to the flight, including a helpful guide on how to pronounce the space pioneer's name: 'He's "You-Ree Gah-GAH-Rin" with the accent on the second syllable of his last name.' The trial of the Nazi mass murderer Adolf Eichmann had moved to second billing. You-Ree Gah-GAH-Rin was everywhere.

He was also on TV. Or at least his accomplishment was. Only a few minutes before George Thomas's wake-up call, NASA's bullish administrator James Webb, only recently appointed to the top job by Kennedy, had actually been congratulating the Soviets on their achievement on all three major networks. What that also implied about NASA's own achievements – or lack of them – Webb left unsaid, which was not the case with most of the newspapers scattered about President Kennedy's bedroom. They all pointed out the obvious truth that America had lost the space race, and lost it spectacularly. The US had been *asleep*. And on Kennedy's watch.

Only the day before Webb had been complaining to the ongoing House Science and Astronautics Committee that Kennedy's proposed NASA budget was far too miserly if the Americans were ever to take on the Russians and win. And now look what had happened. Next to Gagarin's achievement, Project Mercury's proposed fifteen-minute manned *suborbital* flight suddenly looked embarrassing.

Before the president had much of a chance to take it all in, his press secretary Pierre Salinger was on the line. Salinger, who unlike his chief had been awake since Wiesner first called him at 1.35 a.m., filled him in. 'Do we have any details?' asked Kennedy, an odd question considering plenty of those details were already available to him in all those newspapers in his bedroom and on his own TV. Salinger read the pre-drafted statement, which Kennedy then approved. The call was over. Within minutes the president's words went out to the world:

> The achievement by the USSR of orbiting a man and returning him safely to the ground is an outstanding technical accomplishment. We congratulate the Soviet scientists and engineers who made this feat possible …

But if Kennedy ever thought that would do the trick he was wrong. However badly this day had started for both the United States and its president, it was soon to get a lot worse.

ALAN SHEPARD WAS at the Cape when he learned the news. For the past few days he and Glenn, twins in all but name by now, had been busily training on the simulator for Shepard's flight, whenever that would be. Now they were both staying at the Holiday Inn in Cocoa Beach, blissfully far from Hangar S with its trailers of chimpanzees. The Holiday Inn had a pool and a bar and not a chimpanzee within sight or smell.

Early in the morning a public affairs officer from NASA had come to the hotel and woken Shepard up. He told Shepard what had happened overnight. Then they turned on the TV. There it was on every channel. There *he* was, his competitor, his rival – at least in name. It had happened, exactly as Shepard had feared ever since Ham's flight. He had been beaten by a Russian just as his wife had once joked. Now he would never claim the cup of glory. He would never be first. He watched the TV in disbelief. Then he smashed his fist down on a table with such force that the NASA officer thought he might have broken it.

THIRTY-FIVE

THE BALL ON THE HILL

April 12, 1961: Afternoon
Capsule Landing Site
Between Uzmorie and Smelovka

BY MID-AFTERNOON THE enthralled crowds at the Vostok landing site had reached at least several hundred, and more were turning up all the time. Within hours of its arrival from space the strange, scorched, alien-looking sphere had become, in the words of one of those onlookers, 'a place of pilgrimage'. The road to the top of the ravine where it sat was choked with dust as people in cars, on motorbikes, on bicycles and on foot headed up for a closer look. Some of them had driven more than a hundred kilometres to see it. Major Gassiev's original modest guard of soldiers had since been increased by one of the Vostok search and rescue teams. Its chief was Arvid Pallo, whose story opens this book. But Pallo's spotless record of finding and retrieving space dogs and dummies in snowbound wastes, impressive though it was, was not much use to him now. The capsule was supposed to be a state secret, but the crowds were not having any of it. Tatiana Makarycheva was a twelve-year-old schoolgirl at the time:

> There were soldiers with machine-guns. When we ran up they said, 'No, you mustn't approach, because it may explode!' We were taken aback. Explode? But … they could not restrain us. As soon as they turned their backs we'd run up and tear off bits of it.

In desperation the guards drove wooden stakes into the ground around the capsule and covered it with a tarpaulin, but still the spectators would not be deterred. And they were not just twelve-year-old schoolgirls; they were grown-up school teachers too, good Soviet citizens who ought to have known better, like Margarita Butova:

> I couldn't hold myself back either … It was obvious that [the sphere] had been exposed to very high temperatures: it was blackened, melted, with solidified droplets of metal. There was a little oval hatch. I asked a soldier to raise the tarpaulin even just a little bit, so that people could take a look inside. They ordered everybody to stand back about twenty paces, and grasping the tarpaulin from both sides, lifted it up. Not being able to contain their emotions, people raced towards the capsule.

'Everybody was trying to break past the guards and get closer to the ship,' said Pallo, 'to look inside and snatch something as a souvenir.' Many succeeded. One boy grabbed the very last of Gagarin's tubes of chocolate. 'I remember tasting some and spitting it out,' said Tatiana Makarycheva. 'If you offered it to us today we wouldn't eat it!' Others ripped bits of burned foil from the sphere's cracked skin, tiny pieces of which are preserved to this day in a glass box in Uzmorie's village school. The pilfering was threatening to remove anything and everything of value in the spacecraft. In the end Pallo, acting wisely on the principle that it was better to join them than beat them, decided to give the crowd a spontaneous and what turned out to be 'noisy press conference' about Gagarin's 'flight, his well-being and the operation of the ship's systems'. Many of the questions, as Margarita Butova remembers, were sharper than anything *Pravda*'s tame reporters would ever have dared to ask:

> We asked the soldiers, 'But where did Gagarin himself land?' They told us. 'About two kilometres away.' We walked over there along the edge of a ploughed field … Gagarin had parachuted directly into that field.

So much for the fiction that he had landed inside his capsule. The consequences of this collision between state-sanctioned propaganda and the truth would explode in the state's face over the next two days, when even the official Soviet news agency TASS managed to get its stories muddled up and publish the real version until somebody spotted the error and promptly removed it – where it stayed removed for another thirty years. But for one brief moment that lifting of the tarpaulin before the people that afternoon was almost a symbolic act, a tantalising glimpse backstage that only a select few were normally permitted to see. By its sheer emotional power, Gagarin's flight had smashed past the gatekeepers whose task was to keep state secrets secret.

And then, suddenly, the state was everywhere, attempting to clean up. As the afternoon wore on military police and KGB officials in their black Volga cars began arriving in the area, and they brought dogs with them. There was that missing emergency radio to find, and Gagarin's inflatable dinghy which had snapped off as he parachuted down. There was his ejector seat too, another dangerous clue to the truth about his landing. The KGB men combed the fields, woods and villages for the missing items, questioning people who might know where they were, or who would inform on those who did. The ejector seat was quickly recovered a kilometre from the landing site and driven off in a military vehicle. The radio and the dinghy proved more of a challenge. These had been discovered by a couple of teenage shepherds who promptly spirited them away and divided the spoils. One took the radio and hid it under bales of straw. The other took the boat – once inflated, it might be useful for a spot of fishing on the Volga. The challenge was getting it to inflate. Tatiana Makarycheva knew both boys; although, even fifty-eight years later, she would still not reveal their names:

> One of them tried using a bicycle pump. It didn't work. Then he tried it with a car pump … That didn't work either. Nothing happened. Then he saw a small orange capsule attached to the boat, like a deflated orange balloon. He took it off and pierced it. And there was an explosion … The whole yard was suddenly covered in orange dye! … All the cows, the hens – everything was orange! Well, they had to confess then.

By late afternoon four helicopters thundered across the sky. They landed at the Vostok site just beyond Pallo's hastily erected wooden stakes, startling the crowds. Out jumped Korolev's engineers and state commissioners, and then Korolev himself, wearing his black overcoat and hat. They had just flown from the air base at Engels, transferring to the helicopters after their long plane ride from the cosmodrome. At Engels they had missed Gagarin by a couple of hours, but they would see him in the evening when they got to Kuybyshev: the cosmonaut would come later; now it was the capsule's turn.

Korolev ducked under a rope to get closer to the sphere. There was a flurry of excitement from the crowd as he did so. People were watching him, wondering who he could possibly be. Nobody recognised him, as nobody recognised any of the men with him: clearly he was too old to be Gagarin himself, but he was obviously somebody very important. He stood by the sphere. He stood like that, as Mark Gallai remembered, 'for several minutes'. Then he stroked it gently. He said, 'You could fly it into space again.' He seemed amazed by its condition, despite its ordeal by fire. 'This is history now,' he said. 'This belongs to all humanity. This is the first one.'

He ran his hand over his forehead, put his hat on, spoke a few words to his colleagues. And then he left.

AT FIRST GAGARIN'S aircraft was not able to land at the Kryazh air base near the city of Kuybyshev, his last stop of the day. If anything it was even *worse* than Engels down there. Just as in Engels it was a mystery how everyone seemed to know Gagarin was coming. But again they did know; and as the Ilyushin plane dipped its wings hundreds were swarming across the air base below. 'There were people all along the runway,' recalls the pilot, Viktor Malygin. 'And they kept running back and forth while we circled above it. They just kept running back and forth.' For several minutes the plane continued to circle until soldiers had cleared the runway and it could finally land. Malygin was then ordered to taxi to the furthest point from the terminal, away from the crowds.

Kamanin was waiting there, along with the other five cosmonauts. Gagarin stepped off the plane, still wearing the sky-blue undergarment he had put on at the cosmodrome a thousand hours ago, the same suit he had also worn in space; nobody had thought to provide him with a change of clothing. He was cold. Here in the centre of Russia spring had yet to arrive. Volovich lent him his precious leather flying helmet – the same helmet he had worn in 1949 parachuting over the North Pole. At least it would keep Gagarin's head warm, although Volovich would never see his helmet again – a casualty of the mania in the days that followed.

Kamanin went up to Gagarin. 'I had worried and worried [about him],' he recorded in his diary later that night, 'as if he were my own son.' Now they too hugged and kissed. But new crowds were beginning to press and swell around them, as they had earlier in Engels. 'There was a real danger he might be crushed,' Kamanin wrote, 'and Yura, although he was smiling, looked utterly drained.' It was time to be off. Kamanin led him towards the exit when Titov rushed up:

> [Gagarin] was surrounded by generals and I was just a first lieutenant with very small shoulder-straps, as they say. But I was interested to know: what was the weightlessness like? Yura was walking down the gangway and I pushed everyone aside.

> All of them looked at me. 'Who's this lunatic lieutenant?' they said … But I reached Yura. 'How was the weightlessness?' I asked. 'It's all right,' he said. That was our first meeting after the flight.

A hurried question, a hurried answer; but in that first encounter at the airport a gulf had already opened between Gagarin and Titov – between Gagarin and all the other cosmonauts. He was shielded, surrounded, he was *different*. It was not only that his experience had made him different; not only that he had seen and faced things that they had not; not only that he was now a major and Titov his junior. His essential status was different. In a matter of hours he had become enveloped with a strange celebrity, mantled with a kind of awe that set him apart – from Titov, from his friends, and from the rest of humanity. Titov saw it all, and it was very hard for him. 'I still feel jealous,' he confessed with remarkable honesty years later. 'Right up until now.'

Everybody piled into cars. Escorted by a police motorcade, they drove towards the dacha through Kuybyshev's streets, thronged with yet more cheering people. 'Someone in the crowd threw a bicycle under the car's wheels,' recalled Titov, 'because they wanted Yura to stop and say hello. The car swerved to avoid an accident.' The motorcade raced through the city and out at last into the countryside. The dacha sat on a hill overlooking the Volga. Despite the bare wintry trees and earth-brown landscape, it was a glorious setting with, as Kamanin noted, 'a beautiful view of the river' from its third-floor balcony. Here Gagarin and the other cosmonauts were to settle and unwind for the next couple of nights. Gagarin would give his secret briefing to the state commission in the morning. Now, finally, he could rest.

But rest was not easy. The house was rapidly filling with officials – local party bosses, military commanders, members of the ministry of internal affairs, KGB regional chiefs and a host of others who, wrote Kamanin, 'had absolutely nothing to do with the event'. But they had all come to look at Gagarin. Most were already drinking downstairs while he took a shower, finally getting rid of that thermal undergarment At some point it appears he managed to speak to Valentina and his mother on the phone. They had been moved to a hotel in Moscow because of all the crowds in the apartment. We do not know what was said; but Gagarin's sister Zoya learned some of it later. 'Of course,' said Zoya, 'we couldn't quite believe that everything was actually all right until we could actually see him. You know, we Russians have a saying. You have to touch it to believe it.'

Later that evening Gagarin played a quiet game of billiards with Kamanin and Titov. Then he went back to his room. He was there when Korolev arrived.

Gagarin's first biographer, Yaroslav Golovanov, interviewed Gagarin in 1968. In Golovanov's own account of this meeting, Korolev's 'eyes were wet' when he saw Gagarin. The two men embraced and kissed. Gagarin seemed to want to soothe him. 'Everything is fine, Sergei Pavlovich,' he said. 'Everything is good.' Korolev could not find the words to answer him. 'You need to rest,' the chief said at last. 'Tomorrow you can tell us everything.'

They went downstairs. A festive supper had been laid on the table but nobody was really in the mood. There were, according to the unfailingly frank Kamanin, a few 'very boring and solemn toasts' but everybody was far too tired. Most of them had been up for twenty hours or longer. They soon disappeared to their individual bedrooms. Gagarin went into his. Before midnight he was fast asleep, while on the other side of his bedroom door a KGB minder sat on a chair and kept guard through the night. This time there were no strain gauges underneath his mattress.

IN WASHINGTON IT was 4 p.m. President Kennedy stood up to face more than four hundred members of the press at the State Department Auditorium. Earlier that day, after sending a formal message of congratulations to Nikita Khrushchev, he had met NASA's chief James Webb and Senator Robert Kerr, a former governor of Oklahoma, in the Oval Office to discuss an upcoming conference on space in Tulsa. This meeting had been scheduled long before Yuri Gagarin's bombshell dropped into the president's lap, and inevitably Tulsa took a back seat. Gagarin's flight dominated the conversation. As a prop for Kennedy, perhaps to help him to visualise NASA's own manned space effort, Webb had brought along a little model of the Mercury capsule. It was a mistake. In the light of what had just happened it lent an element of farce to the meeting. Kennedy made a sour joke about the cartoon-like 'contraption' plonked on the Oval Office table. Webb, he said, 'might have bought it in a toy store ... that morning'.

Then came the afternoon press conference. Kennedy began by reminding the audience that this was the sixteenth anniversary of the death of President Roosevelt, as well as the anniversary – he did not say which one – of the announcement of a polio vaccine. After three more minutes of this he opened the floor. The first question was not about space; it was about Cuba. Had the president reached a decision on how far the country would go to help an anti-Castro uprising or invasion in Cuba? The president's response was evasive. 'I want to say there will not be under any conditions any intervention in Cuba by United States armed forces.' He did not say that a CIA-trained Cuban invasion force was just five days away from invading Cuba. Nor that a part of the US Atlantic Fleet was already on its way to the Caribbean to give them clandestine support.

The second question was about Gagarin. Could the president give his views on the Soviet achievement of putting a man in orbit, and what that meant for the US space programme? The president could; but the effort was manifestly painful. In the news footage he looks almost shell-shocked. His eyes are downcast, his face strained, his language stumbling, littered with hesitations and 'ers' – eight of them in the first two sentences:

> Well it is a most impressive ... er ... scientific accomplishment and also I think that ... er ... we ... er ... all of us ... er ... as members of the race have the greatest ... er ... admiration for the ... er ... Russian who participated in this extraordinary feat.

> I have already sent congratulations to Mr Khrushchev and … er … I … er … send congratulations to the man who was involved.

Aside from the odd, doubtless stress-induced, omission of the word 'human' before 'race', the truly glaring omission was the actual name of the 'man who was involved'. Not once in the thirty-minute press conference did Kennedy mention Gagarin. What he did instead was attempt, not very persuasively, to answer some tough questions about the lack of progress in the US space programme. The Soviets, he admitted, had bigger rockets, 'which were able to put up bigger weights and that advantage is going to be with them for a long time'. It was not exactly fighting talk. When asked to respond to the challenge that Americans were 'tired' of always coming in second behind the Soviets, he became noticeably irritated. 'However tired anybody may be, and no one is more tired than I am,' he countered, 'it is a fact that it is going to take some time and I think we have to recognise it.' But, he admitted, 'we are behind.'

It was not the president's finest hour. First the evasions over Cuba, now this. Wasn't there a real danger, asked one reporter, nailing the heart of the matter, that the Communists were once more proving that their 'system was more durable than ours'? Kennedy stumbled again. 'I do not regard the first man in space as a sign of the weakening of the er … of the … er … free world,' he said, before embarking on a rambling, semi-coherent discourse about American desalination technology which would, he pronounced, 'really dwarf any other scientific accomplishment, and I'm hopeful that we will intensify our efforts in the area'. Maybe so; but however worthy an ambition, turning saltwater into freshwater was never going to cut it with most patriotic Americans compared to beating the Russians in space. Essentially Kennedy was missing the point, just as his predecessor President Eisenhower had missed it after the Soviets launched Sputnik and he had gone off to play golf. There was too much second; what was needed was *first*. What was needed was action; action soon – action *now*.

But there was no action. Instead, and with the sort of grand ironic flourish that usually happens in novels, the crowning moment of this day was the long-awaited publication of Donald Hornig's panel of inquiry report on the Mercury manned space programme – commissioned by President Kennedy's science advisor, Jerome Wiesner, back in February. It landed in the White House, all sixty-five close-typed pages of it, and boiled down to one simple conclusion: Mercury was not ready. Caution was advised. More consultants were needed. Extra centrifuge test runs with humans and chimpanzees should be made before 'committing an astronaut'. 'We are not as sure as we would like to be,' its authors concluded, 'that a man will continue to function properly in orbital missions.'

Except, of course, one man just had. And now the leader of his country was planning to celebrate with the biggest party in Russian history, just to drive home the point.

THIRTY-SIX

PARTY TIME

April 14, 1961: 13.00
Above Moscow

THE VIEW FROM Gagarin's window was like nothing he or anybody on board had ever seen. Several hundred thousand people already jammed Moscow's streets as the big four-engine Ilyushin-18 thundered low over the capital just before one o'clock in the afternoon, escorted by seven MiG-17 jet fighters. At less than a thousand feet, hardly higher than the famous spired tip of Moscow's State University with its massive five-pointed Soviet star, the arrow-shaped formation swept over Leninsky Prospekt, over Red Square, over the Kremlin, the gleaming silver jets keeping perfect symmetry just a few metres from the big Ilyushin's wingtips. Sitting in the co-pilot's seat, Gagarin was entranced.

> They were lovely MiGs, just like the ones I used to fly. They came in so close to us that I could clearly make out the pilots' faces. They were smiling broadly and I smiled back. Then I looked below and gasped. The streets of Moscow were flooded with people. Human rivers seemed to be flowing in from every

> part of the city, and over them, sail-like red banners waved on their way towards the Kremlin.

'I am going to give him such a welcome!' Khrushchev had told his son Sergei, and here was its first act, a victory parade above the USSR's capital that nobody down there, nor the millions listening to the radio or watching on television across the Soviet Union and beyond, could possibly miss – this would also be the first ever live TV broadcast from the USSR to the countries of western Europe. Those excited Moscow crowds had been building all morning, waiting for Gagarin's plane to turn up from Kuybyshev. But he was nearly late. His departure from the air base there had almost been held up by massed crowds far bigger even than when he had arrived two days earlier. Such had been the mayhem that a decoy aircraft had to be parked elsewhere on the base to divert attention from the real thing, and Gagarin had only just managed to get on board before some of the people realised they had been duped. Their anger had been alarming, Gagarin's first brush with the dark side of celebrity, and perhaps too an intimation that his privacy from this point on was an illusion.

But now he was over Moscow, two hours and twenty minutes later, all eight aircraft announcing his arrival with a tremendous din. 'The weather was wonderful ... as if by divine grace,' said Aleksey Dubovitsky, the Ilyushin's navigator. 'We did our honorary lap of the city. And Gagarin started clapping.' Then the formation dipped its wings and screamed off towards the city's airport at Vnukovo, twenty-five kilometres south-west, to begin the second act of Khrushchev's welcome.

IN THE BARELY thirty-six hours since Gagarin had remained at the dacha in Kuybyshev, the clamour surrounding his space flight had swollen to epic proportions. While he was resting, walking, playing billiards with his fellow cosmonauts and disclosing truths about his flight behind closed doors to the state commission – many of which he would never be allowed to reveal in public – the world's media were busily turning the story into one of the biggest events of the century; indeed, *the* biggest event of the century according to London's mass-circulation *Daily Mirror*, whose April 13 front-page headline was exactly that: '*THE GREATEST STORY OF OUR LIFETIME, THE GREATEST STORY OF THE CENTURY*'. And the paper added: 'Today the *Mirror* salutes the Russians.' The same mantra was repeated everywhere, as were the inevitable parallels with Columbus and Charles Lindbergh, whose own sensational 1927 first solo flight across the Atlantic had aroused similar passions. Gagarin's orbital odyssey was 'The Number One Event of the Twentieth Century', it was 'a Universal Good', it was 'the Greatest Feat of Science in Man's History'.

The excitement spread like a forest fire. Many newspaper kiosks in major cities around the world ran out of stock within minutes as people snapped up every last newspaper left. People were enthralled by Gagarin's vivid descriptions of the earth from orbit, of the intense blues of the atmosphere and the brilliance of the dawn – '"I Can See Everything," Says First Spaceman'. Nobody could get enough of the smiling First Spaceman and his astonishing flight, even if some, like the celebrated British columnist Marjorie Proops, never one to mince her words, saw a different angle in the whole story: 'Why wasn't it a woman, eh? That's what I'd like to know.'

And the Soviet press had a field day. Gagarin's triumph was not only the biggest event of the century, it was the biggest of blows to capitalism and its arch-representative, the United States of America. That, after all, had been largely the point, at least for the Soviet leader. '*GLORIOUS EVENT IN THE HISTORY OF MANKIND*' ran *Pravda*'s April 13 headline, next to a drawing of a heroic figure streaking away from earth while holding up a banner of Lenin several times bigger than the globe. 'Honour and Adoration to the Russians!' declared another headline; 'Long Live the Party! Hail Our Motherland!' And on it went. A twelve-thousand-foot mountain was named after Gagarin, as was a glacier, a rare mineral – *Gagarinite* – and hundreds of newborn Russian babies. There were calls too to name one of

Moscow's major squares after him. Despite Khrushchev's banning of personality cults after Stalin's death, it was clear that an unstoppable full-scale version was already rapidly developing around the young hero. Ecstatic and very bad poems about him and his flight were printed too, not least a 'Hymn to the Space Ship Vostok' written by a poet from another Communist country – China – which duly appeared in *Pravda*. In fact the paper received so many hundreds of celebratory verses from its readers that one letter writer to the editor described April 12, 1961, as 'a record day for the production of poetry in the history of humankind', as well as for human space flight.

What of course was missing in all those poems and paeans of praise was very much, or frankly any, detail about some of the more technical aspects of the flight. Those TASS news agency 'confusions' over the precise nature of Gagarin's return to earth were soon cleaned up – or rather muddied up – to reflect the official fake version, and there was nothing at all about the launch site or the type of rocket Gagarin flew on or the Vostok itself. Indeed the only picture of the spacecraft that appeared in the Soviet press was copied directly from an American artist's impression in *Life* magazine, a cylindrical object that bore absolutely no resemblance whatsoever to the real thing. But to most Soviet citizens none of those details mattered compared to the central and ineluctable truth that, as *Pravda* spelled it out, 'America Is Stunned – They Are Consumed With Envy'.

And America *was* stunned and consumed with envy. Since Kennedy's lame press briefing on the afternoon of the flight, things had just gone from bad to worse. It was Sputnik all over again. '*HOW RUSS WON SPACE RACE*' headlined the *Chicago Daily News*, in case anybody still had doubts on that score. '*KHRUSHCHEV BOASTS OF SPACE LEAD – "TRY TO CATCH US" HE TELLS US*', ran the New York *Daily News*, and such was the paper's indignation that a Red had got into space before an American that it also scorned as fake news one of the few specific facts the Soviets really had told the truth about, namely Gagarin's working-class background:

> Russia's greatest hero since Lenin, spaceman Yuri (George) Gagarin, who according to the Commies is the son of a humble Russian carpenter and a product of the superior Russian state schools, is actually the grandson of a Russian prince who was shot by the Bolsheviks. He was chosen for the space flight because he was not too highly regarded as an officer and if he did not make it he would not be missed.

Along with all those acres of newsprint, a host of TV specials rammed home the same mixed messages of battered pride, awe and disbelief at the Soviet success, messages laced with a thick dose of national self-laceration. 'Russia has won a great scientific victory,' said NBC's *Man Into Space* report, 'and the US did not realise its rich scientific talent as it should have.' But this was not just a national humiliation – it was dangerous for the whole world. 'No contest in history,' opined New York's WPIX the day after the flight in *The Race for Space – A Startling TV Special*, 'has held such frightening and decisive consequences for free men as the contest between Russia and America.' While Kennedy appeared to be keeping his head down, in many of Washington's corridors of power reactions were equally enraged, if not downright panicked. The same House Science and Astronautics Committee that for months had been nit-picking its way through NASA's budget was suddenly screaming for action. 'Tell me,' Congressman James Fulton asked NASA's chief James Webb the day after Gagarin's flight, 'how much money you need and this committee will authorize *all* you need.' Congressman Vincent Anfuso was more blunt: 'I want to see our country mobilized for war because we are at war.'

And then of course there was NASA. '*"SO CLOSE YET SO FAR" SIGHS CAPE*' ran the resigned headline in the *Huntsville Times*, the local paper for the rocket builders at the Marshall Space Flight Center, but Marshall's chief Wernher von Braun made a special point of offering his 'heartiest congratulations' to the Russians, while on ABC News John Glenn displayed the characteristic political tact that would one day stand him in good stead as a US senator: 'They just beat the pants off us, that's all – there's no use kidding ourselves about that.' Perhaps the only NASA figure who got anywhere near to saying what he really felt was the one who felt it most. On the record Alan Shepard confessed to 'a deep sense of personal disappointment', although without mentioning the hand-smashing incident. But off the record he was still burning up with rage: 'We had them,' he kept repeating to anybody who would listen, 'we had them by the short hairs – and we gave it away.'

Such, then, was the impact the young, smiling cosmonaut had made in the two days since he had landed in his ploughed field and asked Anna Takhtarova where he might possibly find a phone. By the time his Ilyushin airliner touched down at Moscow's Vnukovo Airport at exactly one o'clock on that Friday, April 14, he had probably become the most famous man on the planet.

THE PLANE CAME to a stop a hundred metres from a flower-bedecked podium on which stood Khrushchev, Brezhnev, members of the top-level Presidium, ministers, marshals, ambassadors and Gagarin's family – his parents, his sister Zoya, his two brothers and his wife, Valentina, whom he was about to see for the first time since he had left home for the cosmodrome nine days earlier. All the family had been provided with new clothes for the occasion and all were still reeling. It was only two days since Aleksey Gagarin had picked up his toolbox and axe and set off across the fields to help to build a workers' club. 'Can you imagine,' remembers Gagarin's niece Tamara Filatova, 'these villagers from the country meeting the First Secretary of the Communist Party and the Minister of Defence – and everyone calling them by their names? All of them saying, *thank you* for your son! Can you imagine what that was like? It makes me exhilarated, even today!' It also made her terribly envious. She had to remain at home to look after her little brother and the piglets.

Inside the parked plane, Gagarin was almost overcome by an attack of last-second nerves. But Kamanin was with him, giving him a few comforting words of advice and casting a quick professional eye over his new major's uniform and smart greatcoat. In a touching passage in his diary, Kamanin describes how they had been rehearsing the next part of the proceedings the previous evening at the dacha, with Kamanin playing the part of Khrushchev and Gagarin delivering his official report. They had practised it twice. Now it was about to happen. The aircraft door opened to cheering from thousands of spectators who had gathered at the airport – all yelling *Ga-ga-rin! Ga-ga-rin!* – while live TV cameras pointed at Gagarin as he emerged. At the bottom of the stairs was a red carpet, a long one, which led all the way to the podium.

> Never – not even in the spacecraft – had I been so nervous as I was at that moment. The carpet seemed to stretch on for ever and ever. But while I was walking along it, I managed to keep my presence of mind. I knew that all eyes were on me.

Khrushchev's son Sergei was also watching:

> The rather short major walked and walked – it seemed as if he would never reach the end of it. A minor mishap occurred halfway: one of the garters holding up his high officer's socks came undone. The sock fell down, the garter slipped out of his pants and rose triumphantly up in the air with every step, then hit his leg painfully. The cosmonaut continued his ceremonial march without paying any attention.

Sergei Khrushchev has always remained convinced it was a sock garter; others, including Gagarin himself, remember it was shoelace. The controversy rolls on to this day. In the film footage shot by Vladimir Suvorov, *something* definitely comes undone, and Gagarin definitely ignores it. There were powerful pressures to remove the offending footage from public release – it was hardly compatible with orthodox notions of a Soviet hero – but Gagarin himself argued for its inclusion on the grounds that it made him more human and so there it stayed, forever to flap away on his shoe as he walks up the red carpet. At any rate, he made it to the podium without falling flat on his face in front of millions of viewers. There he saluted and made his official report:

> I am happy to report that the mission assigned to me by the Central Committee of the Communist Party and the Soviet government has been accomplished. The first flight in history has been made aboard a Soviet spaceship. I feel very well and ready to accomplish any other mission assigned to me by our Party and the government.

All that practice had paid off. He was word perfect. Everybody applauded. The crowds cheered. Khrushchev wiped a tear from his eye. Gagarin nimbly mounted the few steps up to the podium, where he found himself swept up in a full-scale bearhug from the portly premier who also kissed him repeatedly on both cheeks. The hugging went on and on before Khrushchev finally let others have a go. Gagarin threw his arms around his father and mother. Anna was crying – 'Please don't cry, Mamma. I won't do it again,' he said – and he turned to embrace Valentina, standing next to Khrushchev. She was not crying or smiling. She was biting her lips. Amid all the excitement she was the most uncomfortable person standing there. Naturally reserved and deeply shy, she had just been plunged into the worst of all possible places by her husband's feat, right in the centre of the world's gaze. And the airport was just the beginning. The third and biggest act of the day was yet to come.

FROM THE SKY the human rivers flooding into the city towards the Kremlin had made Gagarin gasp; now he was among them, sweeping into Moscow in an open-topped car with Khrushchev and Valentina beside him, with a string of cars carrying the nation's greatest dignitaries behind them, with escorts of police on motorbikes, with tens of thousands of people on all sides, on rooftops, on balconies, in windows, clinging to lampposts, lining every inch of every street along the route to Red Square, and all of them cheering, shouting, singing, crying, clapping, waving their hats and placards and banners for *him*. 'I doubt if anybody in the world was more thrilled than I was on that festive day,' he wrote later, a statement which, if unverifiable, certainly has the force of emotional truth – although the man sitting and smiling on his right might have run him a close second. Khrushchev's face, recalled his interpreter, Viktor Sukhodrev, 'was radiant. He looked like a child who had finally received a long-awaited present – and a very expensive one.'

Not since Moscow's Victory Day parade in June 1945 had the capital seen anything like it. Of course this being the USSR much of it had been carefully planned even *before* Gagarin's flight on April 12. A remarkable secret decree sent out to members of the Presidium on April 11 already has the schedule for the day's events mapped out; all that is missing is the name of 'the cosmonaut'. But once the flight had happened and Gagarin was safely back on earth the party preparations swung into action, and they swung rapidly. Loudspeakers were set up on Moscow's streets to relay live commentary of the parade, along with 'millions of yards of red bunting', and 'hundreds of thousands of photographs and portraits of the young hero', as the *New York Times*'s Moscow correspondent put it, all of which had to be mass-produced and mass-distributed within an astonishingly short space of time. Even as Gagarin was flying to Moscow helicopters were thudding backwards and forwards over the city's rooftops showering down leaflets about the day's coming parade. Like confetti they fluttered from the sky onto the crowds gathering below, tokens announcing the capital's greatest ever party.

But what made this party so radically different from others was not just the scale of its planning; it was the scale of the opposite, its *spontaneity*. Khrushchev was determined that it should be less controlled, less policed than any previous parades, a prospect that horrified KGB bosses, who were, according to the premier's son Sergei, 'panic-stricken at the thought of a crowd'. But Khrushchev overrode them all. 'Father would not hear of any objections ... This was going to be an unorganized crowd!' One can almost hear the amazement in that exclamation mark. *An unorganized crowd!* And Khrushchev did it because instinctively he grasped the overwhelming emotional power of this moment for his people. He grasped that for a nation that had lost some twenty-seven million of its citizens just sixteen years earlier in the Second World War, many of whose cities had been devastated, much of whose industry had been obliterated, Gagarin's space flight meant *everything*. It gave hope and colour to people's lives, purpose to their sacrifices, pride in their nationhood and their political creed. For a nation so brimming with historical insecurities in its relationship with the west, and especially with America – insecurities that Khrushchev understood only too well – it gave a massive, euphoric injection of confidence.

And it did something more. For one brief instant it also appeared to leave politics and all the mess and muddle of the earth behind. As *Life* magazine's reporter wrote at the time, Gagarin's voyage 'outsoared the shadow of the Cold War and touched the hope and imagination of all men'. A voyage in *space*! It was almost too fantastic. 'Here is a man,' announced the BBC's legendary Richard Dimbleby, his voice thick with awe over those live television images from Moscow, 'who has done and seen things that no other ... person has done and seen.' And here was that man now, riding through the gates of Red Square in the premier's flower-strewn limousine with his shy, pretty wife next to him, the son of an ordinary Russian carpenter who had risked his life to make this bold leap from earth. People could *connect* to him. 'There were tears of joy,' remembers Galina Mishina, a baker, who was among the crowds. 'People kissed strangers in the streets. And Gagarin – he was everybody's love. He and his smile.' It is a sentiment echoed in hundreds, if not thousands, of oral and written testimonies from that day.

The last time Gagarin had been in Red Square, on a visit with Titov and Nelyubov just two weeks before his flight, nobody had given him a second glance. Now the noise was deafening, an acclamation of triumph from the assembled thousands in which the one discernible constant was his name, repeated over and over. Framed by the golden domes and spires of the Kremlin, the message here was more vividly and overtly political: the whole world should see what the Soviets could do. The whole world should see where its best future lay. A sea of red flags marched from one end of the vast square to the other. Over most of the façade of GUM, Moscow's famous department store, hung a huge banner of Lenin with the slogan, '*Vpered k Pobede Kommunisma!*' – 'Forward to the Triumph of Communism!' In an ideology addicted to symbols surely nothing could be more symbolic than this climax of the day's third act, the presentation of the young Soviet hero atop the mausoleum containing the preserved corpse of the Soviet Union's founding father, Vladimir Lenin.

Gagarin stood on Khrushchev's right, flanked by the nation's leading political figures in their hats and heavy overcoats. The national anthem blasted from loudspeakers and then, to a volcanic eruption of applause, Gagarin stepped up to the microphone:

> I am boundlessly happy that my beloved Motherland has been the first to achieve this flight, the first in the world to penetrate into outer space … The first Sputnik, the first space ship, and the first space flight – such are the stages on my motherland's great road to penetrating the secrets of nature … Long live our Soviet Motherland! Long live our great and powerful Soviet people! Long live our Communist Party!

Perhaps, like his speech at the airport, this one had also been practised with Kamanin back at the dacha; we shall never know. But in this moment Gagarin's identification with the Soviet leader standing right next to him was complete, an identification that would cost him dear when that leader fell from power three years later. All that, however, was yet to come; now Khrushchev, beaming from the mausoleum's balcony, praised the mighty strength of the USSR and its scientists, declaring that this 'greatest of man's dreams' had all been undertaken for the sake of peace and that it was the 'most noble, most radiant victory' for Marxism-Leninism. 'We say there is no force in the world capable of turning us off this path … Victory will be ours!' He then announced that the 'pilot-cosmonaut' beside him would be decorated as a Hero of the Soviet Union, the nation's highest honour. 'His name,' Khrushchev boomed, his voice echoing from the loudspeakers around Red Square and across the USSR, 'will be immortal in the history of humanity.'

Among the masses below, Gagarin's fellow cosmonauts, wearing civilian clothes, unrecognised and unknown, looked on in utter amazement. They had been driven to Red Square earlier in a bus, away from inquisitive eyes. Nothing prepared them for what was now happening to their friend. 'It was astonishing to see him there,' said Titov. 'It was only then that I realised the importance of the event that had moved all the people ... The whole world was glad because a man had gone into space. It was extraordinary.' This was the accolade and the glory, the immortality that so nearly might have been his with just one simple stroke of Kamanin's pen. But Titov was down here, not up there. He tried to attract Gagarin's attention, to gain some mark of recognition; they all did – 'We shouted, "*Yura! Yura!*"', says Titov's wife Tamara who was present too, 'but there was so much noise, he didn't see us.' Then some of the cosmonauts lifted Titov up and threw him into the air above the heads of the crowd. Titov, Tamara remembers, was wearing a conspicuous yellow coat:

> Yura then saw him. You could see him asking, what's going on over there? Something unusual. Who's that being thrown into the air? He looked, he peered at the crowd. And we yelled at him. And then he saw Gherman, he saw us all, he grinned and waved at us. And then he turned to Nikita Khrushchev and we saw that he too raised his hands at us in greeting. And we all felt happy.

Amid the clamour and press of people it was a brief but poignant connection between friends. Titov had first noticed the change two days before at the Kryazh air base near Kuybyshev. Now it was inescapable. Gagarin had crossed his frontier: not only into space but into another life. In time, Titov and some of the other cosmonauts would follow him there; but none of them would experience quite what he had experienced and none with the same force, because none of them did it first.

But there was one man for whom there was no such moment of recognition that afternoon in Red Square, because he was not in Red Square; and that was Sergei Korolev. He had travelled near the very back of the motorcade from the airport in his own car, an old, nondescript Chaika. He wore no marks of distinction, no medals pinned to his lapel in case he was identified by western intelligence agents. He was as anonymous as he was always required to be. Before he ever got to Red Square his car was held up by the huge crowds. It was impossible to go any further. So he drove back home where he watched the whole event on television with Nina. Later that night he returned to the city for a grand reception held to honour Gagarin in the glittering Georgievsky Hall at the Grand Kremlin Palace. Under brilliant chandeliers the chosen hundreds of guests, among them Gagarin's parents and siblings in their new suits and dresses and his anxious, uncomfortable wife, toasted the young hero and watched Brezhnev pin the Order of Lenin and Gold Star medals on his chest, the first of scores of medals and decorations to come. But there were no such public honours and no public toasts for Korolev. He had done more than anybody else in that marbled hall to make Gagarin's flight happen. But apart from a select few, nobody there had a clue who he was.

'We work underground,' he had once said, 'like miners, unknown and unseen to the world.' It was the bargain he had struck in order to realise those childhood dreams and launch humanity's journey beyond the earth. It was never easy for him, as it was never easy for those who loved him most. But sitting in her Moscow flat filled with her father's memorabilia and photographs nearly sixty years later, his daughter Natalya, then a young doctor living in the capital, still cherishes her own special memory of that day:

> I walked with the crowds and one of my colleagues asked, 'I wonder who's the Chief Designer who launched Gagarin into space?' I wanted so badly to say that this was my father. But I could not. He had absolutely forbidden me to tell anyone what he was doing ... But of course my heart was bursting with pride.

THE PARTY WENT on late into the night, and not just inside the Kremlin palace. Outside in Moscow's packed streets the singing and dancing continued until dawn. There were spectacular fireworks too, dazzling displays of colours exploding in the night sky. In the days, months and years to come Gagarin would go on to be fêted, celebrated and adored in cities around the world; and even though there would be grand parties and singing and fireworks in those places too, there would never be another day like this.

But for Gagarin and for Valentina, that day was ending. Earlier at the reception Anna had noticed how 'from time to time Valya touched Yuri's arm, as if to reassure herself that he was actually alive. I understood her.' Now at last, for the first time since they had said goodbye nine days before, they could be alone together. And together they faced a future in which the only certainty was that nothing in their lives would ever be the same again.

It had so nearly not been this way. Somewhere in his suitcase, still sealed inside its envelope, was the letter he had written to Valentina in case he never came back. It would not be needed now. He had encircled the earth and he had seen the stars. He had seen the night and then the new day racing towards his little ship. He had seen the impossible beauty of the atmosphere that enables life to exist; and he had seen its impossible thinness too. He had seen it all.

And he had come back.

EPILOGUE

ENDGAME

MID-APRIL–MAY 1961

If somebody can just tell me how to catch up. Let's find somebody – anybody. I don't care if it's the janitor over there, if he knows how.

President John F. Kennedy, April 14, 1961

ON APRIL 14, 1961, the very day that Moscow was celebrating Gagarin's flight with the biggest party in its history, a crisis meeting was also taking place five thousand miles away in Washington, DC. As Gagarin and Valentina were returning from their splendid reception at the Kremlin and the streets were still jammed with revellers, Kennedy sat down with his advisors in the White House's cabinet room. He was markedly different from the figure who had faced the press conference two days earlier: not hesitant or evasive, but openly alarmed. There was no talk of desalination projects this time. 'If somebody can just tell me how to catch up,' he said. 'Let's find somebody – anybody. I don't care if it's the janitor over there, if he knows how.' And he kept on saying it: 'Can we leapfrog them? Is there any place we can catch them? What can we *do*?'

One by one the men sitting around the coffin-shaped table began setting out the options. Among them were NASA's chief James Webb and his deputy Hugh Dryden, both now caught at the centre of a storm of public anger and accusation after Gagarin's triumph. Another was Jerome Wiesner, the Mercury-sceptic science advisor whom Kennedy had appointed. Yet another was David Bell, an ex-marine who ran the Bureau of the Budget with a fanatical eye for detail. Various leapfrogging alternatives were explored: building bigger rockets, building more satellites, pushing faster to get a Mercury astronaut into space and into orbit – even the chastened Dr Wiesner, 'slumped down in his chair so far that his head seemed to be at table level', according to the *Time* reporter Hugh Sidey who was also present, glumly accepted that Gagarin's feat had finally removed many of his fears about manned space flight. Kennedy listened – he was a renowned listener; but none of what he was listening to was enough. Something more was needed, something big enough to capture the world's attention and restore America's political and ideological pre-eminence.

Something like going to the moon.

NASA already had a long-term future programme for a manned moon mission – named Apollo, after the Greek god who rode his golden chariot across the heavens; but this tentative Apollo programme was a distant and uncertain adventure that Kennedy himself had effectively quashed in the latest round of NASA's budget cuts. Now here it was again, suddenly back on the table. But how much, asked Kennedy, would it *cost*? Dryden gave him some rough estimates. It could cost possibly $40 billion – approximately $345 billion in 2021 money – an unfathomably enormous sum of money that, according to Sidey, left the budget director David Bell in a state of 'muted horror'. Dryden added that it would also require a crash programme similar to the Manhattan Project that had built the atomic bomb during the Second World War. He did not say that it might end up costing more than ten times as much as the bomb. Nor that having spent all that money it might not even work. It would take years to develop the technology. The moon was a quarter of a million miles from earth – at least a week's journey there and back. The furthest anybody had been in space so far was once round the planet in a hundred and six minutes. And he was a Russian.

As the shadows lengthened across the lawns outside the cabinet room, the men batted the arguments back and forth. *A journey to the moon!* It was fantastic, but its ambition was spectacular. Kennedy had never shown any great enthusiasm for space, at least since becoming president. But times were different. This was now a national emergency. Sidey could see that the president was for the first time 'beginning to get the feel of a challenge'. And yet – *$40 billion*? Kennedy could not stop tapping his teeth with his fingernails, a well-known sign of stress with him. He rubbed his hands through his hair – 'agonisingly', says Sidey. Over and again he returned to the same theme. 'The cost,' he said. 'That's what gets me.' He stood up and went to the window. After a long pause, he turned back to the table. 'When we know more, I can decide if it's worth it or not,' he said.

It was an opening, even if it was not a green light. But perhaps there was more to the president's anxious teeth-tapping and agonised hair-rubbing than he was letting on to. Because what he did not say at that meeting was that even as it was taking place – even as the crowds were also singing and celebrating in the streets of Moscow – six ships carrying the CIA-trained men of Brigade 2506 had just departed Puerto Cabezas in Nicaragua for the Bay of Pigs in Cuba with the objective of invading the island and removing its president, Fidel Castro. Everything was suddenly coming together at once.

THE INVASION WAS a disaster.

The bombing began within hours of the White House meeting at dawn on Saturday, April 15. On April 17 the brigade began its assault on the beaches; and by Wednesday evening, April 19, less than five days after it had started, the operation had collapsed. Despite Kennedy's public denials, news media had been flagging an imminent CIA-backed invasion for days, if not weeks, before the first Cuban rebel hit those beaches. The result, in the words of the president's press secretary Pierre Salinger, was 'the least covert military operation in history'. Trapped on the sands they landed on, the brigade fighters were repeatedly raked with shells and gunfire from Castro's rapidly mobilised planes, tanks and surrounding troops. As the brigade's position deteriorated, CIA chiefs urged President Kennedy to authorise US air strikes as the only way to save the battle. But Kennedy refused. There were no air strikes. The dangers of escalation were too great; Khrushchev himself, fresh from his victory with Gagarin, informed the president bluntly that if American forces intervened, the USSR would 'provide the Cuban people and their government with all necessary assistance'. The consequences, warned Khrushchev, could 'bring the world to a military catastrophe'.

That threat was enough. A hundred and fourteen Cuban fighters were killed, the remainder taken prisoner. Over the following months hundreds of the regime's political opponents were also executed. Far from removing Castro, the failed invasion only secured his grip on power; and it also drew him even more tightly into Khrushchev's embrace, an alliance that would one day lead the superpowers to the brink of nuclear war. But for now Khrushchev's triumph was complete. In the space of just one week he had shown the whole world what Soviet power, know-how and influence could do: it could put a human being in space; it could stop an invasion; it could humiliate America.

And it could humiliate America's president. Those days in mid-April, wrote Pierre Salinger, were 'the grimmest I can remember in the White House'. Only a week earlier he had witnessed Kennedy's stumbling press conference on the day of Gagarin's flight. Now he witnessed his chief's despair after the Bay of Pigs. Kennedy was haunted by the disaster. He felt the weight on his conscience at leaving men exposed to die on the beaches. Aides noticed that his face had become drawn and thin, his clothes uncharacteristically dishevelled, his hair not properly brushed. The morning after the invasion, Salinger found him crying in his bedroom. He was, the press secretary wrote, 'in the most emotional, self-critical state I had ever seen him'. Although he took full responsibility for the operation, Kennedy could not forgive the CIA chiefs who had planned and pushed for it. He called them 'sons of bitches'. And more than once in the days that followed he would blurt out in the middle of a meeting, 'How could I have been so stupid?'

Ted Sorensen, one of Kennedy's intimate circle and his principal speechwriter, wrote that the president 'was not accustomed to failure in politics or in life'. But now he *had* failed; moreover his brazen lies about American involvement in the Cuban operation had been brutally exposed. He suffered, as Salinger described it, the 'scorn' of the Communist world while, critically, 'neutral nations were now more receptive to overtures from the Kremlin'. On that Wednesday, April 19, when everything was disintegrating at the Bay of Pigs, he was discovered pacing endlessly back and forth in the Rose Garden in the hours before dawn, the loneliest of figures. This was the ninetieth day of Kennedy's term as president. The blazing optimism of his inauguration speech in January, the bright hopes and vigour of his vision, had suffered two major blows in quick succession.

But a solution, the possible way out first tentatively explored in that cabinet room meeting on April 14, was still at hand. Kennedy did not need the White House janitor to tell him what he needed to do next. It was time to turn the corner. And on the ninety-first day of his term, he turned it.

On April 20, he sent a memorandum to his vice president, Lyndon Johnson, a single page of questions whose impact on the coming decade – and long beyond – would turn out to be seismic. Among all the options was there, asked Kennedy, a 'chance of beating the Soviets' by going to the moon? 'Are we making maximum effort?' he demanded. 'Are we achieving necessary results?' The ball was rolling. And it kept on rolling. Four days later, on April 24, Wernher von Braun was summoned at short notice to Washington to give his own views. Could a moon landing be done? Was it even *possible*? By April 29 the rocket-builder had sent his answer to Johnson. Not only could it be done, but 'with an all-out crash program' it could even be done by 1967–8. 'We have,' wrote von Braun, 'an excellent chance of beating the Soviets to the *first landing of a crew on the moon*' – and that underlining emphasis is his. Despite his sober tone, the German rocket-builder's excitement beats beneath it. Here was von Braun's dream of a lifetime; suddenly within a finger's length of his grasp.

Even the cost of the adventure, the part that had so alarmed Kennedy, was coming down: not $40 billion now but possibly half that, $20 billion; still a colossal sum yet under the circumstances justifiable and perhaps necessary. And just in case Kennedy might still be having second thoughts, his vice president, a noted space champion, rammed the point home for him: 'Dramatic achievements in space,' Johnson wrote, 'are being increasingly identified as a major indicator of world leadership.' The president only had to look at the latest edition of *Life* from April 21. Within the magazine's nineteen pages devoted to space were telling reactions from young people 'overseas' to Gagarin's triumph: 'The Americans are licked,' said an Egyptian; 'I knew Russia would do it first,' said a Japanese student; 'The Americans talked a lot. Russia kept silent until success came,' declared an African student. And a German secretary: 'This makes one realize Soviet boasts of ultimate superiority may not be groundless at all.' If Kennedy wanted to demonstrate, and demonstrate decisively, his own leadership to the world then he was going to have to act.

And so, memo by memo, meeting by meeting, Kennedy inched closer to the momentous project he had contemplated on the day Gagarin was being adored in Moscow. There was, however, one more crucial step before such a project could be approved. An American actually had to fly in space and he had to get back alive.

ON THE MORNING of May 5, 1961, twenty-three days after Yuri Gagarin circled the earth, Alan Shepard launched into Florida's skies from Cape Canaveral.

Watched by an estimated forty-five million people on television and another half million packed on nearby beaches, roads and causeways, he lifted off from Pad 5 at 09.34 Eastern Time. He left the earth three days late. His flight had originally been scheduled for May 2, but poor weather had got in the way. At least his identity as America's first astronaut could finally be revealed to the public. Then, on May 5, NASA tried again – and again his launch was delayed, this time with seven separate holds for weather and technical problems, the last of them coming just two minutes before lift-off. The tension was almost unbearable. Would he *ever* get off the ground? By then Shepard had been strapped inside his cramped Mercury capsule Freedom 7 at the top of the Redstone rocket for well over four hours

– twice as long as anybody had anticipated. Desperate to relieve himself he had ended up urinating inside his spacesuit. 'I'm a wetback now,' he had joked. But his legendary test-pilot nerve and cool held fast as the clock ticked down towards zero, even as millions of Americans held their breath. 'Everybody,' remembers the Cape reporter Jay Barbree, 'hit their knees and got down and prayed.' 'I felt a shiver,' said Gene Kranz in Mercury Control. 'This was history.'

At her home in Bay Colony, Virginia, Shepard's wife Louise watched the countdown with her youngest daughter Julie and niece Alice, while scores of TV news crews swamped her front lawn, scaring the kids and poking their cameras at her living-room windows. 'Be sure to wave when you lift off,' she had told her husband earlier. 'I'll open the hatch and stick my arm out,' he had laughed right back. At her boarding school in St Louis, their eldest daughter Laura, who had once seen her father get so mad at the Russians after Sputnik, was watching in the principal's office. Ever since he had been selected back in January she had known that her father would be going first, but she still had to keep it secret from everybody at school – the hardest secret she had ever had to keep. Now she found herself praying, 'Please Daddy, don't mess up.' Up there on that rocket, according to his friend Jay Barbree, Shepard was praying the same thing. 'But,' said Barbree, 'he didn't say mess.'

In his secretary's office at the White House President Kennedy, his wife Jackie, his vice president and other members of his National Security Council were also watching the TV. In the last seconds before lift-off the room was enveloped, as the *Time* reporter Hugh Sidey wrote, in an 'ominous' silence. A photographer snapped a worried Kennedy staring intently at the screen. The possibility that Shepard might be blown up on live television was all too real. Over the three weeks since Gagarin's orbital triumph there had been calls from powerful political figures to postpone this flight or at least conduct it in secret in case of an accident. With the added humiliation of the Bay of Pigs, the last thing America or its president needed was a real-time rocket explosion witnessed by tens of millions. But Kennedy had chosen to go ahead in the open. This, in his view, was what elevated a free democracy over a Communist dictatorship. It was a source of pride, and not just for Kennedy. 'We took our chances in front of the world's cameras,' said Mercury Control's flight director, Chris Kraft. 'They took theirs in secret.' 'This,' wrote Sidey, 'was how free men did things.'

The gamble paid off. Shepard's flight was flawless. It was also short, lasting fifteen minutes and twenty-two seconds. In that time he reached a maximum altitude of 116 miles – compared to Gagarin's maximum of at least 188 miles. His ballistic up-and-down trajectory meant that he was weightless for only five minutes, and although he briefly reached space he could never reach orbit. When he splashed down in the Atlantic Ocean he was just 302 miles from the Cape, where he had started. Gagarin had travelled some 26,000 miles around the planet. Like Gagarin – and perhaps because of him – Shepard also commented on the 'beautiful view' from space, in his case of the Bahamas. But in truth he had forgotten to remove the grey filter from his periscope and the view would have been in black and white. 'Shit,' he kidded to his fellow astronaut Wally Schirra soon after he landed, 'I had to say something for the people.'

And the people loved him for it. Across the nation ecstatic newspaper headlines fell over themselves with superlatives. '*OUR SHEP DOES IT!*', '*SHEP DID IT!*', '*SHEPARD'S TRIP A-OKAY ALL THE WAY*' were typical offerings while the *Detroit Times* ran with '*BOY WHAT A RIDE! SPACEMAN SHOUTS*'. At last, said Sidey, 'the country had a hero'. Shepard had shown tremendous bravery and skill, and unlike Gagarin he had actually *controlled* his capsule for a short period of his flight. But behind all those golden superlatives there was no avoiding the fact that his flight was, as even CBS's Walter Cronkite conceded in his live broadcast, a less spectacular feat. In the Soviet Union, *Pravda* predictably stressed its 'inferiority' and wrapped up the whole story in a few lines on its back page. In four words the New York *Daily News* nailed the contrast: '*MILLIONS CHEER, IVAN JEERS*'.

Yet Shepard had survived, and that was the real point. He had stepped forward to reclaim America's honour, and moments after he was helicoptered onto the deck of the giant aircraft carrier USS *Lake Champlain* – there were no ploughed fields, buckets of potatoes or terrified peasants to greet his return to earth – he received an effusively congratulatory phone call from the president himself. Kennedy, wrote Hugh Sidey, felt like 'a million pounds had just been lifted from his back'. Three days later on May 8, the president demonstrated his gratitude with a big parade for the astronaut in the capital, a display of public adulation with obvious parallels to Gagarin's reception in Moscow, even if the whole occasion was on a rather smaller scale. The fêting nevertheless filled the day: there was a celebration for Shepard in Congress, a standing ovation from five hundred members of the press at the State Department, and at the White House Shepard was awarded the Distinguished Service Medal, a ceremony only slightly undermined by the fact that the president also managed to drop the medal on the floor.

But the way was now clear. 'The future of our entire manned space flight program,' wrote von Braun, had 'hinged on the success of this flight'. It was *the* key test, the last great hurdle that had to be overcome before the lunar enterprise could be finally approved. Shepard, and NASA, had passed it with flying colours. There were points of detail yet to be finalised, costs yet to be agreed and it would all take a little longer to get there; but on May 25, 1961, exactly forty-three days after Yuri Gagarin had first stunned the world, President Kennedy made the televised address before Congress that would forever afterwards be identified with his name:

> I believe that this nation should commit itself to achieving the goal, before this decade is out, of landing a man on the moon and returning him safely to earth. No single space project in this period will be more impressive to mankind or more important for the long-range exploration of space; and none will be so difficult or expensive to accomplish ... In a very real sense, it will not be one man going to the moon – it will be an entire nation. For all of us must work to put him there.

And so the president who had never shown much interest in human space exploration, who had barely mentioned space in his inauguration speech, had just committed himself and his country to the boldest, most uncertain and most costly of adventures. It was, said his science advisor Jerome Wiesner, a decision he had made not from passion but 'cold bloodedly'. It was made because it had to be made. The president himself, in that same speech to Congress, had framed the terms: as a battle 'between freedom and tyranny', a battle which, at such a pivotal moment in history, *had* to be won. Kennedy had just lost the first round in the space race; but in losing it, he had started a new, and bigger, race.

And though he himself would not live to see it, this time the United States would win.

ALL SEVEN OF the Mercury astronauts would go on to fly in space, but only one of them would ever get to the moon. On January 31, 1971, almost a decade after his short suborbital flight, Alan Shepard commanded Apollo 14 on its nine-day mission to the lunar surface and back. Along with astronauts Stuart Roosa and Edgar Mitchell, Shepard launched from Pad 39-A at what had been renamed Kennedy Space Center aboard one of Wernher von Braun's mighty Saturn V rockets – the tallest, heaviest and most powerful rocket ever built. On February 5, their lunar module Antares touched down near Cone Crater in the boulder-strewn Fra Mauro region after a challenging descent that Shepard characteristically described as 'a piece of cake' compared to landing a jet on an aircraft carrier. 'It's a long way,' he said when he stepped foot on the moon the following day, 'but we're here.' In the thirty-six hours he and Mitchell were there, Shepard made history by becoming the first human to hit a golf ball on the moon, using a six-iron club head attached to a rock collecting tool. Thanks to the moon's low gravity, the second ball, he said, went 'miles and miles and miles'.

He never flew in space again. In a 1991 filmed interview he was asked if he had any ambitions left. 'Mount Everest might be a neat place to go,' he said. 'If I didn't have to climb up there.' In 1996 he was diagnosed with leukaemia. He died two years later on July 21, 1998, at the age of seventy-four. His widow Louise died just a month later on August 25 at a little before five o'clock in the afternoon, the time her husband always used to call her when he was away at work. His eldest daughter Laura, who had prayed for him in the principal's office on the day he first left the earth, now has a home in Cocoa Beach only a few miles from Cape Canaveral. From her balcony it is possible to see the launch pad from where her father lifted off into space on that bright May morning in 1961, carrying America's hopes and dreams with him. She sometimes calls it 'Daddy's pad'.

John Glenn, Alan Shepard's back-up, friend and rival, became the first American to fly in orbit. In February 1962 his Mercury capsule Friendship 7 circled the earth three times. During much of his flight he controlled his spacecraft manually, splashing down in the Atlantic after nearly five hours in space. It was, he said, 'the best day of my life'. In many respects his orbital flight finally levelled the playing field with the Soviet Union. He was given a tremendous reception by President Kennedy and a wildly enthusiastic ticker-tape parade in New York City.

In 1974 Glenn's skills as a communicator which had so dumbfounded the other six Mercury astronauts at their first press conference in 1958 came into their own when he was elected a Democratic senator for the state of Ohio. He remained in the Senate for a quarter of a century. In 1983 he announced his candidacy for president but was beaten for the Democratic nomination by Walter Mondale. Fifteen years later, in October 1998, Glenn returned to space at the age of seventy-seven. After passing the same medical that younger astronauts were required to pass, he flew aboard the space shuttle Discovery in an attempt, among other research projects, to determine the effects of space on age. For nearly ten days he and his six crew members orbited the earth on a journey of some 3,600,000 miles. 'I realised how much I'd missed being in space all those years,' he said. He remains the oldest man or woman ever to have flown in space.

Of the Mercury Seven Glenn was the last to pass away. The jogging and dedication to physical fitness that had impressed and sometimes exasperated his fellow astronauts continued into later life. At ninety he still retained his pilot's licence. He died on December 8, 2016, at the age of ninety-five. Such was his status as an American icon that on learning of his death President Barack Obama ordered official flags to be flown at half-mast across the nation.

There were no such official national commemorations for Ham. After his flight in January 1961, the chimpanzee spent the next two years undergoing further observations back at the Holloman animal laboratory. He later began training for a second space mission until his evident lack of enthusiasm at the prospect disqualified him, and in 1963 he was transferred to the National Zoological Park in Washington. His fame went ahead of him: he appeared as a 'guest' on a few TV shows and even had a cameo role in a movie featuring the stunt performer Evel Knievel; but he found it difficult to adjust to life at the zoo. In 1980, now seriously overweight, he was loaned by the US Air Force to another zoo in North Carolina where he died in January 1983 at the comparatively young age of twenty-six. A plan to have him stuffed and placed on display at the Smithsonian Institute was rejected after a negative reaction from the public. His skeleton is preserved at the National Museum of Health and Medicine in Maryland while the rest of his remains are buried at the International Space Hall of Fame in Alamogordo, New Mexico. A memorial plaque there describes him as the 'World's First Astrochimp' who 'proved that mankind could live and work in space'.

Unlike Ham, the Soviet dogs Belka and Strelka who orbited the earth eighteen times in August 1960, *were* both stuffed after their deaths and are still displayed in glass boxes near the entrance to the Museum of Cosmonautics in Moscow. The same fate awaited Chernushka, the first dog to fly a dress rehearsal of Gagarin's flight in March 1961; she is preserved at Moscow's Institute of Space and Aviation Medicine where she was once trained. The second dog, Zvezdochka, was donated to a zoo but her subsequent destiny is unknown. It appears she either ran away or was possibly stolen by a trophy hunter aware of her history as a space pioneer. Their fellow traveller, the dummy Ivan Ivanovich, was auctioned at Sotheby's in New York in 1993 and sold to the American businessman Ross Perot for $189,500. Perot loaned Ivan, along with a substantial treasure trove of Soviet space memorabilia, to the Smithsonian Air and Space Museum in Washington, where his eerily human-like face with its creepy eyelashes unnerves visitors today just as it always did. As well as the Smithsonian example, another Ivan, wearing a spacesuit and strapped to his cosmonaut couch, exists at the Baikonur Cosmodrome Museum, complete with a sign claiming that this Ivan flew on March 25, 1961, and that it was donated to the museum by Korolev himself. As so often with the early Soviet space programme, the truth can be frustratingly elusive, even after sixty years. One only hopes that Perot got the right one.

Wernher von Braun's career reached its astonishing apogee with the Apollo 11 moon landing in 1969. The Saturn V rocket that he and his team designed and built at the Marshall Space Flight Center in Huntsville performed superbly – unlike its Soviet counterpart, the very slightly smaller N-1, which failed spectacularly on all four of its test launches, thus destroying any Soviet hopes of putting a Russian on the moon. Today there are three complete Saturn Vs with original parts on display. So enormous is the rocket that all three are exhibited on their side and not vertically, where they continue to inspire awe. Their creator has fared less well. In the immediate aftermath of the Apollo moon programme public interest in space rapidly waned, and with it collapsed von Braun's great dream of sending humans to Mars. In 1972 he left NASA to work in the commercial sector, but by then his past was catching up with him. A war crimes trial in Germany raised questions about his knowledge of the Dora concentration camp inmates used to assemble his V-2 missiles in horrific conditions. In 1976 a proposal to award him a Presidential Medal of Freedom was rejected because of his Nazi background. 'He has given valuable service to the US since, but frankly he has gotten as good as he has given,' wrote one of President Gerald Ford's advisors. The following year, in 1977, he died of cancer, just before some of his worst offences – his membership of the SS and use of slave labour – finally became public. His Faustian pact with Hitler had brought him huge fame, adulation and success. Without him and his German rocket engineers, the US space programme would undoubtedly have looked very different. But the moral cost remains a stain on his name. Perhaps the final irony is that the man whose V-2 missiles once rained down destruction on London and Antwerp has a crater on the moon named after him.

Von Braun's great rival, Sergei Korolev, died eleven years before him in 1966. His last years had also not been kind. After Gagarin's sensational success the pressure to keep on producing space 'spectaculars' with which to thrill the world was relentless, first from Khrushchev and after 1964 from his successor, Brezhnev. The result was that instead of a co-ordinated programme to compete with the Americans in the drive for the moon, there was a series of disconnected, if newsworthy, firsts: the first two-man spacecraft, the first three-man spacecraft, the first woman in space, the first spacewalk. Some of these were achieved at enormous risk. The three-man spacecraft, called Voskhod or Sunrise, was essentially the same Vostok that had carried Gagarin, except with three seats. It was so cramped that its occupants were not even able to wear spacesuits.

And things continued to go from bad to worse. In the whole of 1965 only a single Voskhod, this time a two-seater, was launched, while NASA successfully placed five two-seat Gemini spacecraft in orbit. Hampered by the death of some of his best and most loyal colleagues, repeatedly attacked and undermined by longstanding enemies like his former denouncer Glushko, exhausted and increasingly ill with his heart problems, Korolev had by then become a shadow of his former self, a man who had 'aged ten years' in the words of the diarist Kamanin. 'I am in a constant state of utter exhaustion and stress,' Korolev wrote to his wife Nina in 1965, 'but I can under no conditions show that these things are getting to me. I am holding myself together using all the strength at my command.'

But still the pressure continued, with Brezhnev demanding in December 1965 that Korolev launch two further Voskhods in time for the Twenty-Third Congress of the Communist Party in March 1966, an almost impossible deadline. In that same December Korolev was diagnosed with a polyp in his intestine. A routine operation was scheduled in the new year to remove it. On the morning he left for hospital, according to his daughter Natalya, he was 'greatly upset' when he could not find his lucky one-kopeck coins. The operation was performed by Boris Petrovsky, the Minister of Health, on January 14. It very quickly ran into complications. As soon as Petrovsky tried to remove the polyp Korolev began to bleed profusely. Several attempts to administer a deeper anaesthetic to stop the bleeding failed. Korolev's jaw, smashed by his NKVD torturers in 1938, made it impossible for three separate anaesthetists to intubate him. Petrovsky then discovered a malignant tumour twice the size of a man's fist in the wall of Korolev's rectum. What should have been a short operation lasted for more than eight hours as Petrovsky attempted to remove the tumour while Nina waited in a next-door room. In the end Korolev's weak and battered heart gave up. He had a cardiac arrest and could not be revived. He was pronounced dead at 4.30 p.m. He was two days past his fifty-ninth birthday.

In life Korolev was unknown to the world; in death the veil of secrecy surrounding his name and his accomplishments was finally lifted. Two days after he died *Pravda* published his obituary along with his photograph and a list of his medals. Yet so well hidden had been the secret that the outside world was slow to recognise his importance, with the *New York Times* only briefly noting his death on page eighty-two of its Sunday edition on the same day as *Pravda*'s obituary. Before his cremation Korolev's coffin was placed in Moscow's famous Hall of Columns in the House of Unions, where thousands came to pay their last respects. Gagarin, Titov and many of the other cosmonauts stood as guards of honour. On January 18, a day of severe frost, Korolev was given a state funeral in Moscow attended by enormous crowds. The urn containing his ashes was interred in the Kremlin's wall in Red Square, the nation's highest honour.

His death, wrote Kamanin, was 'an avalanche ... The country has lost one of its most outstanding sons and our cosmonautics has been orphaned.' His second wife Nina never married again. She devoted the rest of her life to organising Korolev's archives. She died in 1999. His first wife Kseniya died in 1991. His daughter Natalya, who saw so little of her father when he was alive, preserves her memories of him in her Moscow apartment, its shelves filled with his memorabilia and its walls hung with his photographs. 'All I wanted,' she told the author in 2019 at the age of eighty-four, 'was to hug and kiss him. Hug and kiss him. That's all I wanted.'

NONE OF THE Vanguard Six cosmonauts are still alive. The last to pass away was Valery Bykovsky in 2019. Five of the six went on to fly in space. The one that never did was Grigory Nelyubov, the man who had once been favoured by some to fly first. But the brash egotism that his instructors had noticed in him proved to be his downfall. In 1963, after a drinking session at a railway station bar, he ended up in an argument with a police officer. Despite repeated efforts by his superiors to make him apologise, he refused. As a result he was ejected from the cosmonaut corps. He was exiled to a remote air base in the far east of the USSR, where he returned to flying jets. He tried desperately to get back his place in the cosmonaut corps but failed. In photographs of the cosmonaut corps that were later published by the Soviet authorities his face and figure were airbrushed out. Even the fellow pilots in his squadron did not believe that he had once been a cosmonaut. 'I no longer exist,' he told his wife Zinaida. By then he was drinking heavily and had fallen into a deep depression. One day in February 1966 he disappeared from his house and was killed by a passing train in a snowstorm. He was thirty-one. Officially it was never established whether he had committed suicide, but he left a note for Zinaida. 'You were always better than anyone else,' he wrote. 'Forgive me.'

Gherman Titov flew into space on August 6, 1961, the second Russian and fourth person in the world to do so after Gagarin, Shepard and Grissom. His flight was in every respect far more challenging than Gagarin's. He flew seventeen orbits, a gruelling mission lasting one day, one hour and eleven minutes. He brought a camera with him, filming the earth and its atmosphere for the first time from space. He also took along a photograph of himself which he signed for his wife Tamara while in orbit: 'To my little Tomochka, nearly all the stars are asking me to send their warmest greetings to you.' His flight, however, was not without complications. For at least half the time he suffered badly from space sickness induced by prolonged weightlessness and vomited more than once. None of this was reported in the Soviet press after he landed, nor the fact that the two sections of his Vostok failed initially to separate, just as they had failed with Gagarin. His final descent was even more fraught: he nearly parachuted into an incoming train, which was forced to screech to a halt. Then a driver who raced off the road towards him hit a pothole and he had to use his own first-aid kit to bandage her head. As with Gagarin, Titov received a hero's welcome in Red Square on August 9, when once again Khrushchev boasted of Communism's triumphs from atop Lenin's tomb. Just four days later, on Sunday, August 13, East German Communist troops closed the border between East and West Berlin. In a move that dangerously escalated tensions across the Iron Curtain, construction workers began building the Berlin Wall.

Titov never flew in space again. Like Gagarin and sometimes alongside him, he was sent off on exhausting trips around the world to advertise the glories of the Soviet space programme and the system that had spawned it. Both men suffered from the experience, both sometimes turning to drink and occasional womanising, but Titov did not possess Gagarin's audience-pleasing skills nor his smile, and a less than triumphant tour of the United States in 1962 earned him the enmity of the American press. He did, however, manage to enjoy an impromptu barbecue with John Glenn that nearly went up in smoke when the grill caught fire.

By 1968 it was not only space that was closed to him. He was also forbidden to fly aeroplanes again, being considered too valuable a national asset to lose. The decision hit him hard. He nevertheless rose to a very senior position in the Soviet space programme before retiring in 1992. He died at his home of a heart attack in 2000, at the age of sixty-five. By then his fame had begun to disappear as he had always known it would. In the four decades since his flight he had been forced to carry the burden of being 'Cosmonaut Number Two', history's runner-up. It was the issue that came up in almost every interview he ever did, in almost every article, book or television documentary in which he ever featured – just as it would for Buzz Aldrin, the second man on the moon.

AND THEN THERE is Cosmonaut Number One.

In the immediate aftermath of his flight the Soviet government showered Yuri Gagarin and his family with rewards. A secret decree from the Council of Ministers, issued just four days after the Red Square celebrations and only published fifty years later, offers a revealing glimpse into the state's power and reach over all of their lives; and perhaps too, the relative material poverty of those lives beforehand. One by one the decree lists the gifts each member of the family would now be entitled to receive. Gagarin's parents would get an entire new house – 'with television, radio receiver, furniture' – while Valentina would, among many items, be provided with '3 dresses, a black suit, 2 hats, 2 women's handbags and 3 pairs of shoes'. There were toys, dolls and a pram for the two children, gifts of money for the three siblings, and, for Yuri Gagarin himself, a lengthy inventory of goods including '6 sets of silk underwear', a 'light summer coat' and an electric razor. Gagarin's Chkalovsky apartment, now provided by the state with luxuries like a refrigerator and a washing machine, was doubled in size; later, in 1966, the family was moved to an elite brand-new apartment block in the complex that shortly afterwards became known as Star City, the growing and well-guarded Cosmonaut Training Centre outside Moscow.

And Gagarin travelled everywhere, month after month, year after year. His brilliant charm and magnetism won him millions of fans in every country he visited. 'I was reminded,' wrote Kamanin about one trip to India, 'of Christ meeting his people.' Adored on both sides of the Iron Curtain, Gagarin was Khrushchev's – and the USSR's – perfect poster child. In July 1961 he had a highly successful tea with the Queen at Buckingham Palace, to where he was driven past cheering crowds in an open-top Rolls-Royce sporting the number plate YG1. And although he never toured the United States like Titov, he did address the United Nations in New York in 1963. But his rock star status created its own problems, just as it did for Titov. The drinking, fast cars and episodes of womanising caused enormous anguish to those who loved him most. An incident in an exclusive resort in the Crimea where he appears to have leaped or fallen from a second-floor balcony under, as Kamanin wrote, 'alcoholic intoxication' – and possibly to avoid being caught with another woman – nearly caused a major scandal. As with Titov's scandals, it was also hushed up; but the scar above his left eye from an operation as a result of his injury was a living reminder of how close he was sailing to the edge.

Perhaps it was a reminder too of just how deeply the pressures of living so viscerally in the public eye were taking their toll: on his health, his marriage and his career. Khrushchev's fall from power in 1964 and his replacement by Brezhnev in what was effectively a palace coup added to that toll, since Gagarin was inevitably regarded by the new guard as a Khrushchev favourite. The death of Korolev two years later in 1966 was a devastating blow. The two men truly loved each other. Korolev very rarely spoke about his terrible experiences in Stalin's labour camps. But he bared his soul about them to Gagarin just two days before he left for the hospital where he died.

By then Gagarin was desperate to get back to space. He finally persuaded Kamanin to allow him to begin training to fly a new spacecraft called Soyuz, or 'Union'. He worked hard. He became fit, committed and self-disciplined, much more like his former self. But the Soyuz was beset by hundreds of problems, a casualty of the increasingly directionless Soviet space programme since Korolev's death. In a dangerous attempt to compete with the Americans – now far ahead – it was rushed into service long before it was ready. One of Gagarin's closest friends, Vladimir Komarov, was selected to fly it in the spring of 1967 in time to celebrate the fiftieth anniversary of the Russian Revolution. Komarov knew he would very likely die in the attempt. Gagarin was forceful in his criticism of that decision, and even tried to go in Komarov's place, but he was overruled. The only time Gagarin's daughter Elena saw her father cry was the day he said goodbye to Komarov. Komarov was killed on April 24, 1967, when, after a host of problems in orbit, his parachute system failed on his return and he plunged at a hundred and twenty miles an hour into the ground.

'Father was so angry,' says Elena today, not least because 'the government did not want to announce what had really happened'. Gagarin forced them to tell the truth; such was his iconic power still. But on the ostensible grounds that he was too valuable to lose, he was removed from cosmonaut training. He tried to get back to flying jets, training with experienced instructors. And it was on one such flight on the morning of March 27, 1968, that he was killed when his two-seat MiG-15 smashed into snowbound woods near Moscow. He was just thirty-four. His instructor Vladimir Seryogin also died in the crash. The snow was so thick that a search team led by Kamanin only found the bodies the next day. Inside Gagarin's wallet was his driving licence and a photograph of Korolev.

It is probably fair to say that to this day the causes of that crash have never been satisfactorily established. Matters were not helped by the Soviet government's obsessive secrecy in its original investigation; there appear to have been at least three official investigations at once, including one by the KGB, but many of the files were not released for decades. Their findings were also inconclusive, suggesting that the aircraft may have lost control in bad weather attempting to avoid a bird or a weather balloon. A subsequent examination of the files by Gagarin's close friend Aleksey Leonov nearly twenty years after the accident pointed to a quite different cause: the possibility of at least one other military aircraft in the vicinity that should not have been there, perhaps even a supersonic fighter which got too close to Gagarin's MiG in the clouds and whose blast tossed their little jet into an unrecoverable spin. Over the years a number of alternative causes of the accident have been advanced, among them entirely unsubstantiated rumours that the pilots were ill-prepared or drunk or even attempting to hunt elk from their plane. As with the sudden and shocking deaths of other world figures like President Kennedy or Princess Diana, a potent cocktail of conspiracy theories still swirls around the tragedy. Inevitably there was, and remains, talk of foul play, especially given Gagarin's anger at the Soviet leadership both before and after Komarov's death; but there is no hard evidence for it. According to Gagarin's niece Tamara Filatova, it appears that his father Aleksey believed his son may have been deliberately killed, 'but without proof – who knows?' 'Nobody knows what happened,' says Gagarin's daughter Elena today, a reality that carries its own pain down the years. 'There are no final answers,' says Tamara. 'No peace.'

Yuri Gagarin's state funeral on March 30, 1968, was an epic outpouring of national and personal grief. Tens of thousands lined Moscow's streets in solemn silence as his coffin was escorted to Red Square in a strange and sad echo of that other joyous occasion seven years earlier. His ashes, and those of his flying instructor, were buried in the Kremlin Wall, close to Korolev's. His wife Valentina returned home to Star City with her two children, Elena then nearly nine, and Galina aged just seven. They rarely, if ever, spoke afterwards about what had happened. 'That was my mother,' says Elena. 'She never talks about her feelings.'

As the years passed, the gulf between Gagarin the husband, father, son and brother, and Gagarin the icon grew ever wider. The personality cult that had already begun to envelop him while he was alive grew ever deeper roots in Soviet, and Russian, society. Today there are numerous statues of him in cities across the country, in almost all of which his figure strikes a grand, heroic pose. One in Moscow towers over forty-two metres, or a hundred and forty feet, above what is now Gagarin Square, a monumental column on top of which a titanium Gagarin surveys the city he once conquered. There are museums too in every place associated with him across Russia and at the cosmodrome in Kazakhstan; museums that have become almost shrines. In one of them his Vostok sphere, still scarred by its trip in space, is approached along a red carpet. His landing site, graced by another statue, is a national monument, even if it is probably not exactly in the ploughed field where he landed. In the cosmodrome now formally called Baikonur, the cottage where he spent his last night before his flight is preserved precisely as it was, with his shirt and tie hanging at the end of his bed. Elsewhere artefacts are lovingly kept under glass: his gym shoes, his uniforms, his football and hunting knife, the towel he used on the day of his flight, the whistle he once blew to judge a basketball competition in college. Even his car, yet another gift from a grateful state, is preserved in mint condition inside a sealed glass case at his home town of Gzhatsk – today called Gagarin.

In the end even his own mother Anna lived in one of those museums dedicated to her son's memory, a lonely widow of eighty who, in 1983, was obliged by the state authorities to inhabit the top floor of one of eight Gagarin museums in Gagarin. Sometimes tourists would encounter her in its rooms filled with his medals or articles of his clothing, and she would offer them a personal tour, perhaps to relieve her loneliness; and sometimes too she would steal back at night to her old home, also transformed into a museum, to sleep there. She died just six months after moving into her museum.

Valentina remained in Star City, never abandoning the apartment she had once shared with her husband. Her children Elena and Galina grew up and left to have successful lives and careers of their own. Shy and always shunning the limelight, Valentina lived out the following years quietly in her home while the world outside, and especially her own country, changed dramatically around her. She shared fifty of those years with a pet parrot called Lora, a gift from Fidel Castro. From her window on the sixth floor she could look down on yet another statue of her husband, this one an effigy in stone standing proudly on its plinth in a park in front of her apartment block. But as the years went by the trees grew thicker and taller; and by the time she died in March 2020, her husband would only have been visible from her window in the depths of winter.

Still famous and revered as he was in his homeland, by then Gagarin's name had largely faded from the rest of the world's consciousness. Other names, and especially Neil Armstrong's, eclipsed his, epitomising the great adventure of human space travel with those footprints on the moon. But if that adventure appeared in the decades afterwards to lose its grip on the imagination it has since come back, and come back with force. Today there are more, and ever greater, adventures planned than ever before: back to the moon, to Mars, and one day perhaps to the other planets. The adventure is there, waiting for humanity. But the way was opened by Yuri Gagarin, the man who did it first, when he strapped into his little sphere on top of his rocket sixty years ago, and stepped into the beyond.

NOTES

Details of works and interviews cited are in Bibliography, Filmography and Original Interviews.

PROLOGUE: FIFTEEN MINUTES BEFORE LUNCH

1 'Mankind will not stay on Earth': Konstantin E. Tsiolkovsky letter August 12, 1911, to B.N. Vorobiev, Editor of the *Herald of Aeronautics Journal*. Russian Academy of Sciences Archive F.1528 D.173 p.3

1 'Control of space': Senator Lyndon Johnson addressing the Senate's Democratic caucus January 7, 1958, in Logsdon *John F. Kennedy and the Race to the Moon* 30

4 Towering more than 38 metres: According to Siddiqi in a note to SW the height of this R-7 (type 8K72K) was 38.36 metres or 125.8 feet

4 destructive power of two hundred Hiroshima bombs: The R-7 could carry a 3–5 megaton nuclear warhead, equivalent to 3,000–5,000 kilotons. The Hiroshima 'Little Boy' bomb had a destructive power of approximately 15 kilotons. To say the former was two hundred times more destructive than the latter is therefore a cautious 'least-case' estimate. It was almost certainly substantially higher

ACT I: FOUR MONTHS EARLIER

7 'The day will come': Scott and Leonov *Two Sides of the Moon* 35

7 'I'll be damned if I sleep': Koppel *The Astronaut Wives Club* 15

CHAPTER 1: PALLO'S DOGS

9 Arvid Vladimirovich Pallo: Most of the material and quotations regarding Pallo's rescue are from his own account in Rhea (ed.) *Roads*

to Space, 197ff. Additional material from four interviews Pallo gave to Alexander Loktev in *Herald* no.18 (225), August 31, 1999. Also Grahn *Sputnik 6 and the Failure of December 22*

12 'non-calculated trajectory': TASS quoted in *New York Times* December 3, 1960. Pchelka and Mushka launched aboard Korabl-Sputnik 3 on December 1, 1960. They and their Vostok were destroyed by the A.P.O. on December 2

12 a back-up sixty-hour timer: Siddiqi *Challenge* 260 and also in an email to SW December 16, 2019. Additional material on A.P.O. in Chertok *Rockets and People vol. III* 53 and Yury Mozzhorin in *Roads to Space* 414. Also Grahn *Sputnik 6 and the Failure of December 22*

16 At least seventy-four people: This figure is from Siddiqi in personal communication with SW. The precise number of deaths is unknown but likely to be much higher, possibly up to 150 according to the Russian space agency Roscosmos: 'Russia Remembers Horrific Space Accident' *Sydney Morning Herald* October 25, 2010. To this day there are by tradition no launches in the Baikonur cosmodrome on October 24

17 the story splashed across the American press: For example the *Aviation Week* edition of December 26, 1960, forecasts a possible manned US flight in March

CHAPTER 2: WHO LET A RUSSIAN IN HERE?

19 'nightmare': New York *Daily News* January 20, 1961. Additional details of pre-inauguration events and gala in *Washington Post* January 20, 1961, Sidey *John F. Kennedy* 43–7, Salinger *With Kennedy* 108–9 which includes the size of the TV audience

21 Building 60: Ehrenfried *The Birth of NASA* 39

22 *Life* magazine: The exclusive deal was negotiated (at no charge) by a celebrity Washington lawyer, Leo DeOrsey. It was reportedly worth $25,000 a year for each astronaut over three years – perhaps over $210,000 a year in 2021. That was in addition to the astronauts' ordinary salaries – another $7,500 – and their bonus flight pay. The deal allowed the astronauts control over the printed material which was also often presented under their own by-lines. See also Garber 'Astro Mad Men: NASA's 1960s Campaign to Win America's Heart' *The Atlantic* July 31, 2013

23 'Cosmonaut Get Ready to Travel' and 'Space, Expect a Visit from Soviet Man': Brantz (ed.) *Beastly Natures* 211

24 'Anyone who doesn't want to be': *We Seven* 21

25 'He could fly anything': Thompson *Light This Candle* 129
25 'I want to be first': Ibid. xx
25 'cool customer': *Life* March 3, 1961
25 'Icy Commander' and 'crash through that barrier': Brinkley *American Moonshot* 238
27 'little bear of a man': *Life* March 3, 1961
27 'If we wait any longer' and 'most difficult decision': Thompson *Light This Candle* 270
28 'You got it!': Ibid. 271

CHAPTER 3: THE HOUSE IN THE FOREST

32 'should quickly aim for space': Siddiqi *Challenge* 237
32 Dr Adilya Kotovskaya … claimed: SP and SW interview with Kotovskaya, July 2, 2013
32 Aleksey Leonov … recalls: SP and SW interview with Leonov, July 5, 2013
33 'The blood vessels': SW interview with Zaikin, April 27, 2012
34 'He made you want to live': SP interview with Natalya Popovich, December 16, 2019
34 'We all wanted to be first': Titov interview in TV documentary *Nelyubov: He Could Have Been the First* (Russia 2007)
35 'gives the impression of being tough': Kamanin *The Hidden Cosmos*, diary entry January 18, 1961
35 'He was a fighter pilot': SP and SW interview with Tamara Titova, October 3, 2019
35 'He would hear the doctors': Burgess and Hall *First Soviet Cosmonaut Team (FSC)* 82
35 Karpov also admired: Golovanov *Korolev: Facts and Myths* 683
36 'Without being mawkish': Burgess and Hall *FSC* 87
36 'All six cosmonauts': Kamanin diary entry January 18, 1961
36 a hidden diary: In correspondence with SW, Siddiqi writes: Kamanin's 'diary continued into the late 1970s but his official involvement with the space program ended in 1971 … Censored excerpts were published as early as 1966 (in the magazine *Ogonyok*) and then later in other journals. The first truly uncensored excerpts came out in 1988 during Glasnost. Further excerpts were published from 1988 to 1995 before the first volume came out'
37 'Who among these six': Kamanin diary entry January 18, 1961
37 vote on which of their peers: Seventeen votes for Gagarin in Siddiqi *Challenge* 262; Burgess and Hall *FSC* 138 gives twelve votes

37 'tentatively': 'Results of the Examination Carried Out with Cosmonaut Cadets at the Cosmonaut Training Centre of the Air Force' January 25, 1961. Baikonur Cosmodrome Museum Archive

38 comprehensive clinical and training report: 'Service Description of the Cosmonaut Cadet of the First Division of the Centre for Preparation of Cosmonauts of the Air Force Senior Lieutenant GAGARIN Yuri Alekseyevich, dated January 9, 1961'. Baikonur Cosmodrome Museum Archive

38 'He could do everything': Jenks *The Cosmonaut Who Couldn't Stop Smiling* 32

39 'He smiled less frequently': Doran and Bizony *Starman* 13

40 'He grew up very quickly': SW and SP interview with Elena Gagarina September 28, 2019

41 He moved his entire house: The actual house exists in Gzhatsk, now called Gagarin, as a museum. Another museum in the village of Klushino is a reconstruction of the house in its original location with period furnishings. The original dugout is at the back. There are still apple trees in the garden, perhaps descendants of the tree from which Boris was hanged

41 hit musical: Jenks *Cosmonaut* 56. According to Jenks 55, one of Gagarin's earliest essays was entitled 'The Class Struggle in the Countryside and Collectivization'. Gagarin was in Saratov between 1951 and 1955. The college is still flourishing today. The former canteen is now a People's Museum dedicated to Gagarin. Among its many treasures is Gagarin's July 1951 application letter: 'I would like to improve my knowledge of foundry work and be as useful as possible to my Motherland.'

41 'a new illness': Gagarin *Road to the Stars* 36

41 'He's crazy about flying' and *Young Stalinist*: Jenks *Cosmonaut* 72

42 'could talk to anybody': Transcript of Titov interview for the 1998 BBC / CNN series *Cold War*, King's College, London

CHAPTER 4: INAUGURATION

43 'felt fresh and promising': Sidey *John F. Kennedy* 46

44 'Let every nation know': Transcript and video of inaugural address in John F. Kennedy Presidential Library (www.jkflibrary.org)

44 so-called 'missile gap': Logsdon *John F. Kennedy* 12–13; Brinkley *American Moonshot* 166 states that Kennedy used the missile gap during his campaign as a 'calling card' even though 'its contention … proved to be a fiction'. See also *National Intelligence Estimate Number 11-5-61* (April 25, 1961) for the CIA's own fairly accurate intelligence about both Soviet

missile numbers and capability at the time. Also Greg Thielmann 'The Missile Gap Myth and its Progeny', Arms Control Association (https://www.armscontrol.org/act/2011-05/missile-gap-myth-its-progeny)

44 'duck and cover': A Civil Defense film (1952) directed by Anthony Rizzo

45 Imre Nagy: For a revealing account see David Pryce-Jones 'What the Hungarians Wrought: The Meaning of October 1956' *National Review* October 23, 2006

45 'bury': Khrushchev said this on November 18, 1956, at a reception at the Polish embassy in Moscow. 'Whether you like it or not, history is on our side. We will bury you!' *Time* magazine November 26, 1956. A CIA document from February 7, 1962 (approved for release on January 22, 2002) questions whether Khrushchev did actually make that remark: https://www.cia.gov/library/readingroom/docs/CIA-RDP73B00296R000200040087-1.pdf

45 'like sausages': Thielmann 'The Missile Gap Myth and its Progeny'

45 take off his shoe: This supposedly happened during the 902nd Plenary meeting of the UN on October 12, 1960. In one version Khrushchev was protesting against a speech by the Philippine delegate attacking the Soviet Union. Khrushchev's son Sergei also describes the incident in *Nikita Khrushchev* 414–15, where the delegate is Spanish. Photographs of the episode are almost certainly fake and it is quite possible there was no actual banging of a shoe at all. Like so many of colourful Khrushchev's stories the line between reality and myth can sometimes be blurred

45 'bringing the aggressive-minded imperialists': Schlesinger *A Thousand Days* 302

46 'testicles of the West': Jenks *Cosmonaut* 127

47 'He realised': Chertok in BBC *Cold War* episode 9

47 'approximately $174 billion in 2021': inevitably a very approximate figure, based on Bureau of Statistics CPI figures and the average inflation rate since 1961: https://www.in2013dollars.com/us/inflation/1961?amount=100 The same multiplier is used in all US$ currency adjustments

48 'look up and see the Soviet flag': Brinkley *American Moonshot* 204

48 'serious problems within NASA': Logsdon *John F. Kennedy* 32–5. An excellent contemporary profile of Wiesner is in *Aviation Week* January 16, 1961

CHAPTER 5: SHAME AND DANGER

50 took his ten-year-old daughter Laura: Details from SW interview with Laura Shepard Churchley November 18, 2019

52 'like kids dropping rocks': Koppel *Astronaut Wives Club* 15
52 'flashing down at hypersonic': Mieczkowski *Eisenhower's Sputnik Moment* 11
52 'a life and death matter': BBC *Cold War* episode 9
52 a second-rate power: McQuaid 'Sputnik Reconsidered' *Canadian Review of American Studies* 37 (2007)
53 'Sputnik does not raise': Mieczkowski *Eisenhower's Sputnik Moment* 11
53 'One small ball': 'NASA's Origins and the Dawn of the Space Age' *Monographs in Aerospace History* 10 (NASA) https://history.nasa.gov/monograph10/onesmlbl.html
53 'Muttnik': Brinkley *American Moonshot* 137
54 'made our potential enemies': Khrushchev *Khrushchev Remembers* 47
54 'Arguing the Case': *Life* November 18, 1957
54 'relocate': SW interview with Barbree, December 17, 2019
55 166 rocket launches: According to Barbree's Wikipedia page. The figure is almost certainly higher now (2021)
55 'into hiding': Barbree *Live from Cape Canaveral* 11
55 'Why doesn't someone go out': Ibid.
55 'Flopnik': For instance UK's *Daily Herald* December 7, 1957; 'Stayputnik': Brinkley *American Moonshot* 147; 'Oopsnik' and 'Dudnik': *Time* magazine December 16, 1957
57 'War is war': Neufeld *Von Braun* 161–2
57 And if challenged: Von Braun was probably arrested on March 22, 1944 and released on April 1-2 (Neufeld *Von Braun* 170f). In addition to his supposedly defeatist remarks overheard at a party, he was also accused of being more interested in building spaceships rather than weapons. He was released on the intervention of Albert Speer, Hitler's minister of armaments, who successfully pleaded von Braun's cause with Hitler himself
57 'He feels no guilt': Ibid. 208
57 'Once the rockets are up': Lehrer's 1965 song concludes: *You too may be a big hero, Once you've learned to count backwards to zero. 'In German, oder Englisch, I know how to count down, Und I'm learning Chinese!' says Wernher von Braun*
58 'The secret of rocketry': Neufeld *Von Braun* 290
59 'even the enemies of Socialism': Harford *Korolev* 142
59 'Just make sure': Sergei Khrushchev *Nikita Khrushchev* 349–50
60 'We have beaten you': Thompson *Light This Candle* 216
60 'we are not going to attempt': Harford *Korolev* 148

CHAPTER 6: THE AMERICAN TEAM

64 'There is the aggressive response': Burgess *Selecting the Mercury Seven* 33
64 'real men': Swenson, Grimwood, Alexander *This New Ocean* 144
64 'We had to ensure': Dr Robert B. Voas, interviewed by Summer Chick Bergen, Vienna, Virginia May 19, 2002 Johnson Space Center Oral History Project (JSCOHP)
65 William Randolph Lovelace: Dr Lovelace's daughters Jackie Lovelace Johnson and Sharon Lovelace generously shared their memories of their father and mother in telephone interviews with SW. Both were small children when the Project Mercury welcoming cocktail parties took place at their home. In a private vote the children decided that their favourite astronaut was Scott Carpenter; their least favourite John Glenn. In contradiction to the 'real men' spirit of most of the Mercury selectors, Dr Lovelace actively championed the selection of female astronauts. In 1960 he invited and funded up to twenty-five experienced female pilots to undergo the same Phase I Lovelace tests as the men. Thirteen women passed. Unofficially they were known as the 'Mercury Thirteen'. But before they were able to take further jet orientation and centrifuge tests NASA intervened. The tests were cancelled. The women took their case to Congress in 1962 but failed to secure support. One of those who testified against them was Glenn. 'They did not want this program pure and simple,' says Jackie Lovelace. 'It had to be the biggest slap in the face.' See the excellent documentary *Mercury 13* directed by David Sington and Heather Walsh (2018)
65 'I didn't know the human body': Burgess *Selecting the Mercury Seven* 220
65 'It was your worst nightmare': Ibid. 210
65 'I just hope they never give *me* a medical': Ibid. 221
65 'I found it fascinating': Ibid.
65 'psyched out': Ibid. 214
66 'We were well patients': Ibid. 220
66 'a whole new host of torture machines': Ibid. 238
66 'These guys loved to press': Ibid.
67 'a beauty pageant': Kranz *Failure Is Not an Option* 36
67 'This is the worst': Mercury Seven press conference April 9, 1959
67 'I have no problems at home': For video of complete astronauts' press conference (in three parts): https://www.youtube.com/watch?v=FXj5lc_QUOM
68 'Who is this boy scout?': Thompson *Light This Candle* 202
68 'It looked like an atomic bomb': *NASA's Greatest Missions*, episode 1, Discovery Channel (2009)

68 'I saw a lot of rockets launched': Ibid.
69 'Boy those things were really scary!': Lunney interview 1998 JSCOHP
69 'We fried a lot of rattlesnakes': SW interview with Rigell
69 'We got a big kick': *NASA's Greatest Missions*, episode 1
70 A remarkable document from 1960: Jenks *Cosmonaut* 16–17
70 'they were garbage buckets': SW interview with Smirnov April 28, 2012
71 fishing trawlers: Rigell *Ike* 247
71 'the original vomit machine': Glenn interview 1997 JSCOHP
72 for Glenn it was the design of the instrument panel: The other four astronauts' specialities were: Carpenter – communications and navigation aids, Cooper – the Redstone rocket, Grissom – the capsule's automatic and manual controls, Slayton – the Atlas rocket (that would eventually take one of them into orbit)
73 'Are we having fun yet?': Barbree *Live from Cape Canaveral* 34
73 'crazy indications': SW interview with Rigell December 18, 2019
74 'Even to a rookie like me': Kranz *Failure Is Not an Option* 131
74 The cause was finally traced: SW is indebted to the late Ed Fannin, William 'Curly' Chandler, Ike Rigell and especially Terry Greenfield, all launch blockhouse engineers on MR-1 and other Mercury-Redstone missions, for their superb descriptions and explanations of this and other flights
75 first manned space flight was postponed: See the *New York Times* November 22, 1960, the day after MR-1, for the new schedule

CHAPTER 7: THE SOVIET TEAM

76 Luostari: For an evocative portrait of the air base and area today, including haunting images of the rotting former Soviet air base, see 'Cold War Airbase Turns Ghost Town' by Thomas Nilsen in *Barents Observer* September 5, 2013. Incredibly the control tower is (or was in 2013) still manned even though there are no aircraft: https://barentsobserver.com/en/security/2013/09/cold-war-airbase-turns-ghost-town-05-09
77 share her kitchen: SP and SW interview with Elena Gagarina September 28, 2019
78 'They asked me to sit down': Burgess and Hall *FSC* 18–20
78 middle of August 1959: As with many aspects of Soviet space history, the documentary evidence is contradictory with some accounts saying that the doctors visited the military units as early as June and July (these two dates are from the recollections of Dr Yazdovsky and Dr Kasyan). But

the air force order to form commissions to go out to the units was not signed until August 13, 1959

79 'The conversation' and 'asked permission': Siddiqi *Challenge* 244–5

79 initial pool of 3,461: The figures are from Pervushin *Yuri Gagarin: One Flight and a Whole Life* 259

79 role of cosmonaut: First uses of the word 'Cosmonaut' from 1955 according to *Merriam-Webster Dictionary*: https: //www.merriam-webster.com/dictionary/cosmonaut

79 Moscow's Institute of Aviation Medicine: It was renamed the Institute of Aviation and Space Medicine in late 1959

79 'He was unbelievably harsh': SP and SW interview with Valentina Bykovskaya October 10, 2019

79 'charming dog': Yazdovsky *Along the Paths of the Universe* 67–8

80 piles of letters from citizens: Ibid. 116

80 'members of other physically challenging': Ibid. 109

82 'He was asked how many cosmonauts': Burgess and Hall *FSC* 17

82 'In a short time, they were supposed': Yazdovsky *Along the Paths* 110

82 GMVK: *Gosudarstvennaia MezVedomstvennaia Komissia*. The 'State Inter-Agency (Departmental) Commission' was officially in charge of selecting the cosmonauts. It consisted of representatives of different interested agencies and parties such as Defence, Medicine/Biology, Production Plants and Space

82 'It was done by the KGB': Pavel Popovich from interview with Bert Vis September 30, 1996

82 in early October: Gagarin was certainly already at the hospital by October 11, 1959, and remained there until November 4. Siddiqi email to SW May 15, 2020

83 'We were instructed not to talk': Scott and Leonov *Two Sides* 40

83 'After forty minutes the face': Yazdovsky *Along the Paths* 130

83 'Many people fainted': SW interview with Zaikin April 27, 2012

83 'You sit on a seat' and subsequent quotations: SP and SW interview with Volynov July 1, 2013

84 'There were lots of doctors': Burgess and Hall *FSC* 23

84 'get a comparative estimate': SW interview with Bogdashevsky April 27, 2012

84 'you have to walk three hours a day': SW is indebted to Tamara Titova for permission to quote from her husband's letters

85 'I would stand by my post box': Titova interview in TV documentary *Gherman Titov: First after Yuri* (Russia 2010)

85 'He addressed us in a fatherly way': Scott and Leonov *Two Sides* 40

86 'We knew that the Americans': SW interview with Zaikin
87 'We were united by one thing': Titova interview in TV documentary *Nelyubov*
87 tiny shared apartments … in Moscow: Married members of the cosmonaut group were moved from the sports club at Frunze airfield to shared apartments at 95 Leninsky Prospekt in June 1960. The building still exists. Between September and October 1960 they were moved again to Chkalovsky near the Chkalovsky military air base north-east of Moscow. Bachelors were housed in their own quarters while the married cosmonauts and their families lived at 4 Tsiolkovsky Street (now 25 Lenina Street). This apartment block also still exists. SP interview with Igor Lissov and subsequent emails
87 'We had to drape': Scott and Leonov *Two Sides* 47–8
87 'We didn't bring anything': Popovich interview in TV documentary *Separated by the Sky* (Russia 2010)
87 'Since I'd already married': SP and SW interview with Tamara Titova
87 Central Aerodrome's former meteorological office: This two-storey building also exists, although the Frunze airfield no longer does. Its address today is Leningradsky Prospekt, house 39/building 17
88 'He came inside': *The Red Stuff*, documentary directed by Leo de Baer (Netherlands 2000)
88 'He was a man of average height': SP and SW interview with Leonov July 1, 2018

CHAPTER 8: THE KING

89 'My loveliest, dearest girl': Letter to Nina January 27, 1961, quoted in *Korolev: The Horizon of Events. Tender Letters of a Stern Man 1947–1965* 321
90 single-storey cottage: Today preserved in its original condition when Korolev used it, as is the very similar cottage next door where Gagarin and Titov spent the night of April 11, 1961
91 'Take a good look': SP and SW interview with Kotovskaya
91 'Mankind will not stay on Earth for ever': Koroleva *Father* 343. The quotation is from Konstantin E. Tsiolkovsky letter August 12, 1911, to B.N. Vorobiev, Editor of the *Herald of Aeronautics Journal*. Russian Academy of Sciences Archive F.1528 D.173 p.3
92 'He liked my stories': Harford *Korolev* 17
92 'We should be friends': Ibid. 1
92 June 27, 1938: Details of arrest in Koroleva *Father* 10

93 1.9 million prisoners: Applebaum *Gulag* 516. For a penetrating assessment of prisoner and death numbers in the Gulag camps see also her Appendix: 'How Many?' 515–22

93 Two were shot: They were Ivan Kleymenov and Georgy Langemak

93 smashed over the head: Harford *Korolev* 52. This account is taken from Leonov to whom, along with Gagarin and Titov, Korolev described his appalling experiences just a few days before he died

93 'I beg you to save': Facsimile of telegram in Koroleva *Father* 23

93 Huge numbers would die: Harford *Korolev* 51

94 'brilliant': Ibid. 4

95 'We developed a kind': Ibid. 56

96 'epic': Koroleva *Father* 441

96 'He had lively black eyes': Harford *Korolev* 70

96 'We were standing near the gates': Ibid. 95

96 'For Korolev rage': Rhea *Roads to Space* 239. The shaken witness was engineer Anatoly Abramov

96 'toilet sinks or frying pans': Jenks *Cosmonaut* 96

98 'There could be only one conclusion' and other quotations: Sergei Khrushchev *Nikita Khrushchev* 329f

98 Disneyland: After being denied the chance to visit it, Khrushchev reportedly said, 'What, do you have rocket pads there? Is there an epidemic of cholera?' The 340 members of the press corps following him lapped this sort of thing up. Elder 'Cold War Roads' *Newsweek* November 15, 2014

99 'We were entering the sanctum': Sergei Khrushchev *Nikita Khrushchev* 106–7

100 'How dare any country': Khrushchev speech to Presidium of Central Committee of the Communist Party December 8, 1959. Nikita Khrushchev *Reformer* 655

100 'I don't want to exaggerate': Siddiqi *Challenge* 117

100 'thumbing his nose' and subsequent quotation: Sergei Khrushchev *Nikita Khrushchev* 111f

101 6,700 square kilometres: Hall and Shayler *The Rocket Men* 45

101 In time the secrets of both the R-7 and the launch complex would be … penetrated: By 1961 the CIA's intelligence regarding both Soviet missiles, including the R-7, and Soviet missile launch sites was in many respects remarkably accurate. A good example is to be found in the CIA's National Intelligence Estimate (NIE) *Soviet Technical Capabilities in Guided Missiles and Space Vehicles*, dated April 26, 1961, which not only details the Soviet ICBM programme, including the test launches

and capabilities of the (unnamed) R-7, but also contains an impressively accurate drawing ('from photographic and other sources') of the R-7 launch pad at the secret missile complex in Tyuratam

101 'kidnapping': Finer *The Kidnapping of Lunik* (CIA document sanitised for release September 1995)

102 hugely successful Disney TV show: *Man in Space* (first broadcast March 1955), *Man on the Moon* (December 1955), *Mars and Beyond* (December 1957)

102 'Father never complained': Koroleva *Father* 355

103 'At last we were successful': 'Here Are the Men and Formulas that Launched Sputnik' CIA March 18, 1958 (approved for release 1996)

104 'Here I am': Letter to Nina October 11, 1960. *Korolev* 312–14

104 'This is not some military toy': Rhea *Roads to Space* 55

105 'a Sputnik for *human* flight': Siddiqi *Challenge* 193

105 'You are the chosen one': Scott and Leonov *Two Sides* 55

106 'We're Going Places': *Life* January 27, 1961

CHAPTER 9: SUBJECT 65

108 Psychomotor: Full technical details in *Results of the Project Mercury Ballistic and Orbital Chimpanzee Flights* (NASA 1963). Left-hand lever was an 'intermittent or discrete avoidance procedure', right-hand lever was a 'continuous avoidance procedure'

109 'only animals that are lawfully acquired': 'Animal Handling Statement' in NASA press release January 28, 1961 (no. 61-14-3)

109 'Usually we try to get them as babies': Burgess and Dubbs *Animals in Space* 117

110 'set them about four or five feet apart' and subsequent Dittmer quotations: Maybury *A Tribute to Edward C. Dittmer* New Mexico Museum of Space History, Curation Paper 8 Summer 2012

110 Enos … faeces: Gunther Wendt interview February 25, 1999 JSCOHP. Enos was famously difficult to handle. He subsequently went on to orbit the earth on November 29, 1961, in advance of Glenn's first US manned orbital flight the following February

111 Ham would not be the first animal: By far the best and most authoritative account of animals used for rocket and space research both in the US and USSR is Burgess and Dubbs *Animals in Space*. SW acknowledges his huge debt both to the book and to Colin Burgess over numerous email communications

111 'We got no information or data': Burgess and Dubbs *Animals in Space* 45

113 'We didn't force a monkey': Ibid. 134

113 'it pretty much smashed the brain': *One Small Step* film documentary by David Cassidy and Kristin Davy (2008)
114 'P. Varin' and quotation: *Aviation Week* March 20, 1961
115 'respond well to training': Yazdovsky *Along the Paths* 20
115 'trust in the person flying': Gallai *With a Man on Board* 36
115 one estimate suggests: Asif Siddiqi in an email to SW (May 14, 2020) confirms that 'there are so many conflicting numbers on this issue! From a document from the Russian State Archive for Scientific-Technical Documentation the numbers appear to be: 31 launch attempts with dogs (of which 29 were successfully launched), 41 dogs, 22 deaths. Again, these numbers vary from source to source and are complicated by the fact they renamed dogs with the same name (sometimes if a dog died, they would simply name a new dog with the same name)'
116 'Perhaps you'd also like them to have blue eyes': Siddiqi *Challenge* 94
117 'to wish Ham luck': The NASA film is at: https://www.youtube.com/watch?v=uDyfDKqsNGsandfeature=youtu.be
117 'Al would gladly have traded': Kraft *Flight* 3
118 'It was quite a thrill': Maybury *Tribute to Edward C. Dittmer*
120 'You don't climb into the Mercury': *We Seven* 104

CHAPTER 10: A HELL OF A RIDE

123 'It was bad for the family': SW interview with William 'Curly' Chandler December 18, 2019
123 Juno missile U-turn: July 16, 1959, Juno II disaster at Launch Complex 5/6
124 'I always cry thinking': SW interview with Kathy and Ike Rigell December 20, 2019
124 a very lengthy 640 hours: The full countdown is itemised in *The Mercury Redstone Project (MRP)* 7–11
125 'We literally wired the world': Kraft *Flight* 90
126 'I'm Flight and Flight is God': Ibid. 2
128 'If I could send an abort signal': Ibid. 134
129 'a hell of a ride': Kranz *Failure Is Not an Option* 32
129 Within the first few seconds: Details of Ham's flight in *Results of Project Mercury Ballistic and Orbital Chimpanzee Flights*; Kraft *Flight* 125–6; Catchpole *Project Mercury* 259–264; Burgess and Dubbs *Animals in Space* 249–55; Swenson et al. *This New Ocean* 297–304; Ehrenfried *Birth of NASA* 121f
130 94 beats per minute: *Results of Project Mercury Ballistic and Orbital Chimpanzee Flights* 28

130 'The onboard movie camera': Kraft *Flight* 126

130 Stuck-open fuel valve: *MRP* 8–9. Specifically, the 'mixture ratio servo control valve stuck in the open position'. This was one reason why too much liquid oxygen was pouring into the combustion chamber. 'The propellant consumption rate was also increased by hydrogen peroxide pressure which drove the turbo pump faster.' Hydrogen peroxide gas spun the turbo pumps which drove the propellants – ethyl-alcohol and liquid oxygen – into the combustion chamber where they were ignited. These two issues together caused the engine to stop early, having run through its liquid oxygen supply

131 set to abort the mission *automatically*: The automatic abort system on the Redstone was known as the Automatic Inflight Abort Sensing System and is described in detail in *MRP* 5.2f. Essentially, along with both the RSO's abort and an in-cockpit manual abort option in the astronaut flights ('the chicken switch'), the Huntsville team had designed an automatic abort if the Redstone transgressed certain pre-set parameters (for example, an incorrect trajectory), the advantage being that the system would spot errors before a human in a fast-moving rocket flight. One parameter was the rocket's fuel running out too soon at which point a chamber pressure sensor would detect the anomaly and initiate an abort. In MR-2, the system was flying 'closed loop' for the first time, meaning it would actually operate if anything *did* go wrong. On this flight, the system had been set to disarm 137.5 seconds after lift-off, after which time the rocket would be high enough not to require the abort mechanism

130 'We've never had a Redstone': Kraft *Flight* 4

131 'A 17-G kick in the ass': Ibid. 5

132 'I look around mission control': Ibid. 6

133 Traces of vomit: In *Results of Project Mercury Ballistic and Orbital Chimpanzee Flights* 30. Other physiological data 29–32. According to *MRP* 8–9, Ham was supposed to fly up to a maximum height of 114 miles but actually flew to 156 miles. Instead of a planned distance of 291 miles, he covered 415 miles

134 'Looking far less perturbed' and 'cuddly astronaut': New York *Daily News* February 1, 1961

134 'balked and screeched': *El Paso Times* February 4, 1961, which also has the story of Joe Pace's finger

134 'grinning hero': *Life*, February 10, 1961

134 'a disaster': Barbree *Live from Cape Canaveral* 46

134 'the most extreme fear': Jane Goodall in *One Small Step: The Story of the Space Chimps*

134 5 per cent of bodyweight: *Results of Project Mercury Ballistic and Orbital Chimpanzee Flights* 31

135 'physiologically, an Astronaut': Quoting Lt Col. James Henry NASA, *El Paso Times* February 4, 1961

135 Marshall Space Flight Center: The Army Ballistic Missile Center in Huntsville, where von Braun and his colleagues had designed missiles and space vehicles for the army was transferred to NASA in July 1960, and thus from military to civilian control. The full name of the new body was the George C. Marshall Space Flight Center, often shortened to MSFC

135 'The damn Germans': Kranz *Failure Is Not an Option* 31

135 Joachim Kuettner: A German engineer designated *Booster* who operated a panel in Mercury Control Center linking with the launch pad blockhouse. During the Four-Inch Flight fiasco of November 21, 1960, Kuettner was speaking in German to the German blockhouse engineers when this incident occurred. Ehrenfried *Birth of NASA* 121

ACT II: DECISION

137 'What would happen to anyone': SP and SW interview with Volynov

137 'Sure I knew all seven': SW interview with Barbree

CHAPTER 11: THE RISK EQUATION

139 Achille and Gian Battista Judica-Cordiglia: Perhaps the best account is Grahn *Notes on the Space Tracking Activities and Sensational Claims made by the Judica-Cordiglia Brothers*: http://www.svengrahn.pp.se/trackind/Torre/TorreB.html

140 'We are very concerned': Press conference February 8, 1961, in John F. Kennedy Presidential Library (www.jkflibrary.org)

141 'Was Mercury ready to fly?': From an interview with Hornig in Logsdon *John F. Kennedy* 49

141 'impressive scientific achievement': Kennedy made this statement on February 14, 1961. See *Aeronautical and Astronautical Events of 1961* NASA Report to Committee on Science and Astronautics US House of Representatives (US Government Printing Office 1962) 6

141 'Mercury Chimp Test Might Be Repeated': *Aviation Week* February 13, 1961

142 'At least one unmanned shot': McClesky and Christensen *Dr Kurt H. Debus: Launching a Vision* (52nd International Astronautical Congress, Toulouse 1–5 October 2001) 15

143 'man or no man': Swenson et al. *This New Ocean* 309

143 'their feelings are quite bitter': Neufeld *Von Braun* 359
144 'Some of those people were just awful': Gilruth in a recorded interview with Linda Ezell, Howard Wolko and Martin Collins, National Air and Space Museum, 1986
144 'brushed aside our comments': Kraft *Flight* 127
144 'We were never a team': SW interview with Greenfield December 20, 2019
144 'doesn't care what flag he fights for': Neufeld *Von Braun* 337
144 most expensive public funeral in history: Gilruth refers to this in his interview. As director of the Office of Science and Technology Policy from 1957–61, George Kistiakowsky had been President Eisenhower's chief science advisor. A brilliant explosives expert, he had helped to develop the plutonium (atom) bomb in the Manhattan Project during the Second World War. Interestingly Donald Hornig, who chaired the sceptical Project Mercury panel of inquiry, had worked on the atom bomb under Kistiakowsky
145 'They are all big boys': *We Seven* 20
146 'boomtown': Barbree *Live from Cape Canaveral* 14
146 'executive drive': Koppel *Astronaut Wives Club* 49
146 'a terrifying driver': SW interview with Julie Shepard Jenkins November 22, 2019
146 'Definition of a sports car': Thompson *Light This Candle* 259
146 'Cape cookies': Apparently the astronauts' own phrase, Koppel *Astronaut Wives Club* 47
147 'because I'm the one he really loves': Ibid. 55
147 'his dick would have fallen off': Thompson *Light This Candle* 255
147 'pants zipped': Glenn *Glenn* 294
147 'He'd give you that famous stare': Thompson *Light This Candle* 179
148 'Absolutely' and 'I think the answer is Yes': NASA press conference, Cape Canaveral February 22, 1961
148 'I don't think it will be fruitful' and 'tests': NASA press conference, Cape Canaveral February 21, 1961

CHAPTER 12: AT OUR HEELS

149 'undoubted': Kamanin diary entry February 12, 1961
149 'I really want to sleep': Ibid. February 4, 1961
150 'was apparently more interested': Sergei Khrushchev *Nikita Khrushchev* 282
150 'unbearable' and 'distracting': Note from R.D. Malinovsky and M. Zakharov to F. Kozlov, L. Brezhnev and D. Ustinov February 15, 1961, in *Soviet Space* 304

151 'the devil's venom': Siddiqi *Challenge* 214
151 'He would tell us that the Americans: Harford *Korolev* 3
152 'My personal opinion is this': Kamanin diary entry February 24, 1961
153 'Valya is feeling well': Unpublished letter from Gagarin to his family February 13, 1961, supplied by Tamara Filatova
153 'hot-blooded': SP interview with Natalya Popovich
153 'He invited all the women': SP and SW interview with Elena Gagarina
153 'You must subordinate': SP and SW interview with Tamara Titova
154 'frightening': Ibid.
154 50 and 60 roubles a month: This figure is quoted in Scott and Leonov *Two Sides* 85
155 'He was not afraid': SP and SW interview with Tamara Titova
156 'The men were doing': Burgess and Hall *FSC* 88–9
156 'No one was supposed to know': SP and SW interview with Volynov
157 'Sometimes people didn't get through it': Ibid.
157 'Although I had never': Burgess and Hall *FSC* 121–2
159 98 per cent probability: *Mercury-Redstone Project* ref. 5.3.4, which also contains the relevant tables
159 'We're ready to go. Let's go!' and 'For God's sake': Thompson *Light This Candle* 279

CHAPTER 13: IVAN IVANOVICH

161 'He was dressed extraordinarily': Suvorov *The First Manned Spaceflight* 49
162 'in some fantastic movie': Ibid. 8
162 'It was a top-secret assignment': Ibid. 4
163 'What a time!': Ibid. 2
163 'mysterious personality': Ibid. 24
163 'Keep shooting!': Ibid. 52
163 'his slightly jutting chin' and 'By his mood, gesture': Ibid. 25
164 engraved watch: Vladimir Yaropolov in *First Forever* 311
165 'I was dressed in some flimsy beret' and 'lashes': SP and SW interview with Kotovskaya
165 'We all knew that it was categorically forbidden': Yaropolov in *First Forever* 309
166 'The main purpose was to ensure': Siddiqi *Challenge* 265
166 'Noah's Ark': Yazdovsky *Along the Paths* 92
167 after the images were finally extracted by CIA: Details of demodulating Vostok's TV signal in Henry G. Plaster *Snooping on Space Pictures* Vol. 8 CIA Studies in Intelligence, Fall 1964 (approved for release 1994)
168 'Ivan Ivanovich is not injured': Suvorov *First Manned Spaceflight* 52

169 'a large crowd of collective farmers': Kamanin diary entry March 9, 1961
169 'main objective': *Pravda* March 10, 1961
169 'Yes, another brilliant victory': Kamanin diary entry March 10, 1961
170 like two boots tied: The metaphor is in Doran and Bizony *Starman* 111
170 'Final separation': Chertok *Rockets and People Vol III* 64

CHAPTER 14: THE BIGGEST MISSILE SITE IN THE WORLD

171 'the time is not far off': *New York Times* March 15, 1961
171 'crazy, noisy games': SP and SW interview with Elena Gagarina
172 'Yuri Gagarin is the first candidate for the flight': Kamanin diary entry March 16, 1961
172 'They are all rather short': Suvorov *First Manned Spaceflight* 56
172 'one to two technical questions' and subsequent entries for the day: Kamanin diary entry March 18, 1961
173 *Instructions from the Central Party Committee to the Cosmonaut on the Use and Control of the Spacecraft Vostok 3A*: Original in the Russian State Archive of Scientific-Technical Documentation (RGANTD). SW is grateful to Asif Siddiqi for making his copy available
174 'cosmonauts were pleased': Kamanin diary entry March 21, 1961
175 'fairly well prepared': Ibid. Diary entry March 23, 1961
175 'The cosmodrome astounded the cosmonauts': Yazdovsky *Along the Paths* 138
175 Leninsky: In 1969 the name was officially changed to Leninsk, in recognition of its status as a substantial town
176 'to get the cosmonauts acquainted': Vladimir Solodukhin in *First Forever* 279
176 'his extensive establishment': Kamanin diary entry March 22, 1961
176 'No photograph, no piece of cinema': Gallai *With a Man on Board* 30
177 The total area of the complex … Ohio: Hall and Shayler *The Rocket Men: The First Soviet Manned Spaceflights* 45
177 'every stone, brick, board and nail': *Unknown Baikonur: Collection of Memoirs by Veterans of Baikonur* 54
178 'Our first impressions were depressing': Rhea *Roads to Space* 322
178 'You look around at all this': *Unknown Baikonur* 10
178 'hung all over the area': Siddiqi *Challenge* 136
178 'We were sweating and greedily sucking': *Unknown Baikonur* 10
178 'the Home of the Black Death': Rhea *Roads to Space* 336
179 'Did they really do it?': Ibid. 324
179 'we had no days off': SP and SW interview with Khionia and Vladimir Kraskin January 30, 2015

179 'romantic': From interview with Khionia Kraskina in BBC *Cold War* 1998
180 'After the devastating war': Chertok *Rockets and People* 12
180 The only maps CIA analysts: Hall and Shayler *The Rocket Men* 54
180 'crown jewel of Soviet space technology' Burgess and Dubbs *Animals in Space* 143
181 'and now just look at how many silly things': Khrushchev speech May 7, 1960. Transcript in CIA Information Report June 9, 1960 'Soviet Version of U-2 Incident'
181 sentenced to ten years: Powers did not serve all his sentence. On February 10, 1962, he was exchanged in a prisoner swap with KGB agent Colonel William Fisher at the Glienicke Bridge in Berlin. In 1977 he was killed piloting a helicopter for a news station in Los Angeles
181 The CIA was immediately suspicious: Details of CIA surveillance of Tyuratam from *Resolving the Missile Gap with Technology* (CIA Historical Collections)
182 the Corona satellite: Brugioni *Eyes in the Sky* 378
182 'It was horrible to see the funerals': This and following quotations from SP and SW interview with Khionia and Vladimir Kraskin, and interview with Khionia Kraskina in BBC *Cold War* 1998

CHAPTER 15: THE PRICE OF PROGRESS

184 'Stand by to administer emergency aid': Golyakhovsky *Russian Doctor* 129
185 'The patient on the stretcher': Ibid. 129
185 'He was a very good-natured': Burgess and Hall *FSC* 133
186 'I still tremble even now': Ibid.
186 the chamber's soundproofed steel walls: The isolation chamber is today in the Museum of the First Flight in Gagarin's home town of Gzhatsk, now called Gagarin
186 40 per cent of the total atmosphere: Details of chamber pressure and oxygen ratio from Igor Lissov in an email March 3, 2020, and SP interview with Irina Ponomareva December 17, 2019. Ponomareva worked with the cosmonauts on the isolation chamber experiments. Burgess and Hall *FSC* 126 quotes Leonov who claims that the oxygen enrichment was 68 per cent
187 For the next ten days: Some accounts (for example, Kamanin, Aleksander Seryapin, see next note) suggest that Bondarenko was on day ten of a thirteen-day schedule. Others (Pervushin *Yuri Gagarin* 353) that it was a fifteen-day schedule. In Burgess and Hall *FSC* 126, Bondarenko's schedule was due to *end* on the tenth day and he ripped

off his sensors with understandable relief before throwing the cotton wool pad onto the electric hotplate

187 What happened next is still not entirely clear: The version in the text is largely based on SP interviews with Ponomareva and Volynov. According to his own testimony, Volynov was part of the subsequent investigation into the fire. Additional material is from a detailed account by Boris Yesin, Head of Cosmonaut Correspondence at the Cosmonaut Training Centre, in an interview with *Novosti Kosmonavtiki* (issue 5, 2001). A different version is given by Alexander Seryapin, then a military doctor at the Institute of Aviation and Space Medicine, also in an interview with *Novosti Kosmonavtiki* (issue 1, 2009). Seryapin recalls that on March 23 Bondarenko was on day ten of his thirteen-day schedule in the chamber. He says that Bondarenko had just finished using the toilet and was performing a programmed experiment in which he pricked his thumb for a blood sample and wiped it clean with an alcohol-soaked ball of cotton wool. He then dropped the ball by accident onto the live electric hotplate, starting the fire. The observers outside had not been watching him closely because he had previously been in the toilet. In truth, as so often with the Soviet space programme's history from this period, it is impossible to say which version of the story is correct. But in every version the alcohol-soaked ball of cotton wool results in the same tragic outcome

188 'It's my fault ... I'm so sorry': Burgess and Hall *FSC* 127. These words, or words very like them, are also repeated in SP and SW interviews with cosmonauts Zaikin and Volynov

188 'the horrible news': Kamanin diary entry March 23, 1961

188 Their colleagues back in Moscow were certainly told: This is confirmed in SP and SW's interviews with Volynov, Zaikin and Ponomareva

188 'I got there and the first thing': SP and SW interview with Volynov

188 'We were speechless': SW interview with Smirnov

189 'Valentin violated fire safety instructions': Yazdovsky *Along the Paths* 141

189 'Everything was taken away': SP interview with Ponomareva

189 'With fond memories from your pilot friends': Burgess and Hall *FSC* 130. In 1991 the International Astronomical Union named a crater on the moon after Bondarenko

189 Six years later: The Apollo 1 accident happened on January 27, 1967

190 'a whole puddle of saline solution': Quotations and other details of the March 24 State Commission meeting in Kamanin diary entry March 24, 1961

192 'In that spring everybody involved': Gallai *With a Man on Board* 33

192 'All the time': Kamanin diary entry March 20, 1961
192 'the Americans are at our heels': Harford *Korolev* 3
192 It would lift off from Pad 5: The Redstone test flight (see next chapter) lifted off at 12.30 EST (17.30 GMT) on March 24. Ivan Ivanovich lifted off at 05.54 GMT on March 25, 12 hours and 24 minutes later

CHAPTER 16: BOOSTERS AND DUMMIES

194 'we refused to waste a Mercury capsule': Kraft *Flight* 129
195 'It was, c'mon – how many times': SW interview with Lunney December 14, 2019
195 Information Plan: Redstone Development Test MR-BD March 21, 1960: In NARA Fort Worth, Texas
196 'Conclusion: All booster corrections': *Memorandum on Mercury-Redstone Booster Development*, Patrick Air Force Base, Florida March 26, 1961 in NARA Fort Worth, Texas
196 'Whatever the Germans had done': Kraft *Flight* 129
196 'The March 24 Redstone flight was an absolute beauty': Burgess *Freedom 7* 99
197 'It is still too early': New York *Daily News* March 25, 1961
197 'However, a close look at the tapes': Burgess *Freedom 7* 63
198 'the Soviets have demonstrated': Logsdon *John F. Kennedy* 64. Webb's quotation is part of a six-page 'talking paper' he wrote the night before the meeting with the president. One can assume he made the same point as forcefully in person the next day
198 *Marines, Let's Go!*: Details of this and the Laos crisis in Schlesinger *A Thousand Days* 329–34
199 'You may not feel he has the time': Logsdon *John F. Kennedy* 67
199 cabbage soup: Doran and Bizony *Starman* 76–7. The story of the recipe appears to be confirmed in the authors' interview with OKB-1 engineer Oleg Ivanovsky
200 'We worried that if anybody': SP and SW interview with Kotovskaya
200 'we are going to need that luck for ourselves': Pervushin *Yuri Gagarin* 355
201 'He went up to the railings': Burgess and Hall *FSC* 137
202 'the helicopter would not capsize': SW is grateful to space journalist Igor Lissov for this online account of the rescue operation by Pallo: http: // epizodsspace.airbase.ru/bibl/stati/pallo/pallo4.html
202 'the rubbery, cold face of the dummy': This account by Vladimir Yefimov, a member of the rescue team, in Burgess and Hall *FSC* 137
202 'Probably already anticipating': Suvorov *First Manned Spaceflight* 53

203 'preliminary investigation': TASS quotations in *New York Times* March 26, 1961
203 'We don't consider [it] to be interesting': Topchiyev quoted in *Aviation Week* April 3, 1961
204 '*REDS ORBIT, LAND DOG*': New York *Daily News* March 26, 1961

CHAPTER 17: THE MILITARY-INDUSTRIAL COMMISSION

205 'no confidence in the cosmonaut's life support' and subsequent quotations: Kamanin diary entry March 27, 1961
206 'Neither I nor any of the individuals': Chertok *Rockets and People* 65
206 a specially shaped key to his office: Korolev's office at OKB-1 has been preserved as a museum. The key sits on his desk. OKB-1 is today the S.P. Korolev RKK (Rocket and Space) Energia Corporation
207 'I told him immediately': SP and SW interview with Koroleva
207 'Who is for a human flight next?': Kamanin diary entry March 29, 1961
208 a comprehensive document: *Note from D.F. Ustinov, K.N. Rudnev, V.D. Kalmykov and Others in the Central Committee of the Communist Party of the Soviet Union*, March 30, 1961 in *Soviet Space* 333–6
209 Korolev's handwriting listing eighteen: RKK *Energia* Museum exhibit
210 'write a few lines about the six brave cosmonauts': Kamanin diary entry March 20, 1961
211 'the cameramen filmed Yura': Golovanov *Korolev* 13
212 Decree of the Presidium of the Central Committee: *First Forever* 165
212 'speed up the departure': Kamanin diary entry April 3, 1961

ACT III: FINAL COUNTDOWN

CHAPTER 18: BRINKMANSHIP

216 'a man-eating shark': Rasenberger *The Brilliant Disaster* 122
217 'Cuba is Warned': *New York Times* April 4, 1961
217 'the silent secretary': Rasenberger *Brilliant Disaster* 160
217 'a brave, old-fashioned American speech': Schlesinger *A Thousand Days* 252
217 'I'd say, let 'er rip!': Rasenberger *Brilliant Disaster* 159
217 'You're the only one': Ibid. 162
218 'enormous confidence in his luck': Schlesinger *A Thousand Days* 259
218 'The cause for which I am parting' and all subsequent quotations in the chapter from Kamanin diary entry April 5, 1961

219 'I guess I was more aware than any of the other wives': SP and SW interview with Titova
220 'Take care of the girls, Valyusha': Golovanov *Korolev* 689
221 raw colour footage: This is held at RGANTD in Moscow (Russian State Archive of Scientific and Technical Documentation)
223 'Yevgeny Anatolyevich Karpov': Golovanov *Korolev* 688
223 'one of those fairy tales': Sergei Khrushchev *Nikita Krushchev* 430

CHAPTER 19: JUST IN CASE

226 'If a man fails to return to earth because of me?': Chertok *Rockets and People* 27
227 'Even the seatbelts': The engineer was Vladimir Yaropolov quoted in *First Forever* 309
227 'just in case': Chertok *Rockets and People* 62
228 'thoroughly': Kamanin diary entry April 5, 1961
228 'We worked fifteen days a week': Harford *Korolev* 166
228 'Negligence was a crime': Yaropolov *First Forever* 311
228 'in much the same way as an opponent's breath': Gallai *With a Man on Board* 32
229 'Of course everyone understood what this meant': Ibid. 51
229 'I like that brat': Burgess and Hall *FSC* 116
229 'Korolev was like a father to Gagarin': SW interview with Bogdashevsky
230 'Sergei Pavlovich loved Gagarin as a son': SP and SW interview with Koroleva
230 'My dearest Kotya': Letter from Nina to Korolev April 6, 1961 in *Korolev: The Horizon of Events* 326

CHAPTER 20: THE EGG TIMER

232 'The cosmonauts sent off reports to base': Vladimir Soludukhin in *First Forever* 281
234 'Gagarin, Titov and Nelyubov know manual descents': This and other Kamanin quotations for this day from diary entry April 7, 1961
234 original footage: in RGANTD Moscow
234 'joyful and healthy': Suvorov *First Manned Spaceflight* 56
235 'They had a real men's talk': Ivanovsky *Rockets and Space in the USSR* 128
235 even 'kamikaze'-style attacks: One of the most famous of these was Nikolai Gastello, a Red Army pilot who deliberately aimed his stricken DB-3 bomber at a German panzer column on June 26, 1941, just four days after the German invasion of the USSR. Gastello was posthumously made a Hero of the Soviet Union. Generations of Soviet schoolchildren

were taught about his deed. The cosmonauts would have been fully aware of his story, with its underlying moral of self-sacrifice and the willingness to take enormous risks for love of the Motherland

235 'You have to push that fear away': Popovich interviewed in Russian documentary *Separated by the Sky*

CHAPTER 21: CHIMP BARBECUE

239 'I don't think two people could have worked': Thompson *Light This Candle* 286

239 'Al's alter ego': Glenn *Glenn* 315

239 'my back-up': Thompson *Light This Candle* 285

241 'What the hell can we tell': Ibid. 281

241 '... then, at 900,000 feet': The cartoon was by the British cartoonist 'Emmwood' (John Musgrave-Wood). A framed copy exists in the John Glenn archives at Ohio State University

241 'Maybe we should get somebody': Thompson *Light This Candle* 280

241 'tentatively scheduled for this Spring': NASA press release March 27, 1961

242 'We don't even know what we're racing against': Glenn in the documentary *Mercury 13*

CHAPTER 22: CHOSEN

245 fake launch pad: Pervushin *Yuri Gagarin* 401–4

246 'Man's mental state is dependent': Swenson et al. *This New Ocean* 204

246 'invisible shadow': Gallai *With a Man on Board* 56

247 'The thing they most feared': In the Russian TV documentary *Yuri Gagarin: Three Days and a Whole Life* (2011)

247 'might still sail away': Gallai *With a Man on Board* 56

247 'That's it, the case is settled!': Kamanin diary entry April 8, 1961

248 'the sealed envelope would be glued to the inner lining': Ibid.

248 'On behalf of the Air Force': Ibid.

249 'I wanted to make my flight into space': Gagarin *Road to the Stars* 115

250 Even Khrushchev worried about that: In the documentary *The Red Stuff* Titov says of the Soviet premier: 'When he heard the name Gherman Titov, Khrushchev supposedly said, "Gherman, what kind of name is that? Is he German?"'

250 'I have never seen such a smile anywhere': SP interview with Natalya Popovich

251 'He was like a sphinx': SP and SW interview with Leonov

251 'up to that last minute I thought': Burgess and Hall *FSC* 145

251 'The joy of Gagarin': Kamanin diary entry April 10, 1961
251 'We all warmly supported': Titov *700,000 Kilometres Through Space* 81
252 'All the journalists': From documentary *The Red Stuff*
252 'nonsense!': Doran and Bizony *Starman* 85
252 'I'd say that he departed his life': From Russian TV documentary *Nelyubov*
252 'That wound was there': Ibid.
252 'Yura turned out to be the man': Doran and Bizony *Starman* 86
253 'hotly discussing': Vladimir Khilchenko in Rhea *Roads to Space* 385. Khilchenko also claims the R-9 exploded on the launch pad itself, but in an email to SW (February 24, 2020) Siddiqi states that this was not the case. Chertok also confirms the second-stage malfunction (*Rockets and People* 68) and he was assigned to the investigation by Korolev

CHAPTER 23: OPENING PITCH

255 'a nation not only of spectators': Interview with WGN-TV April 10, 1961, in John F. Kennedy Presidential Library
255 The aircraft carrier USS *Essex*: SW is grateful to Rasenberger *Brilliant Disaster* 176ff for these details
255 'anti-Cuban plotting': *Pravda* April 6, 1961
257 possibly by as much as fourteen kilos: Pervushin gives this figure, *Yuri Gagarin* 361. The engineer Vladimir Yaropolov puts the figure at 6–8 kilos. The account of Ivanovsky's removal of items of equipment to reduce the weight and Korolev's anger is from Yaropolov in *First Forever* 313. As so often in narratives of the Soviet space programme, there are alternative versions of this story. Pervushin suggests that Korolev *instructed* the Vostok technicians the night before to remove the surplus weight. In this version there was also even a discussion about replacing Gagarin with Titov, since the latter was approximately 4 kilos lighter. The idea was rejected on the grounds that Gagarin had already been selected
258 In the footage: Archives of RGANTD
258 'Fly, dear Yuri Alekseyevich': Kamanin diary entry April 10, 1961
258 'twin': Gagarin *Road to the Stars* 141
259 'There will be many speakers': Suvorov *First Manned Spaceflight* 57
259 'concert': Ibid. 55
259 'The ship is ready' and other quotations from this meeting: Kamanin diary entry April 10, 1961
259 'the Air Force command recommends': From Gagarinday.ru
259 'How young he looks!': Suvorov *First Manned Spaceflight* 58
259 'Permit me, comrades': Pervushin *Yuri Gagarin* 360

CHAPTER 24: ROLL OUT

261 'A Man is Already in Space' and 'the whole world is engulfed in rumours': Jenks *Cosmonaut* 125
261 'Man in Space Alert': *Daily Mirror* April 11, 1961
261 'unofficial sources': New York *Daily News* April 11, 1961
261 Dennis Ogden: Some modern accounts mistakenly date Ogden's front-page story for the *Daily Worker* on April 10, 1961, but it did not appear until April 12, the day of Gagarin's flight, complete with a drawing of the 'Rossiya' looking almost exactly like the American Mercury capsule
263 'Talking about a man up there': House of Representatives, Science and Astronautics Committee, April 11, 1961, from Hathi Trust Digital Library
264 'Those walking alongside it': Vladimir Solodukhin diary April 11, 1961, in *First Forever* 282
264 'He pulled over at the side of road': Gallai *With a Man on Board* 60
265 'I looked at [Gagarin] and in my mind': Siddiqi *Challenge* 273
266 'with thunderous applause': Solodukhin diary April 11, 1961, in *First Forever* 283
266 'I don't know, I must be rather frivolous': Golovanov *Korolev* 694
267 'just in case': Solodukhin diary, in *First Forever* 282
267 'Many years later at an informal dinner': Ponomarev in *First Forever* 553
268 'We have nothing more to do here': Gallai *With a Man on Board* 57

CHAPTER 25: NIGHT

270 'filling' and all Kamanin quotations: Kamanin diary entry April 11, 1961
270 'liveliness and self-confidence' and details of medical: *Data from Psychoneurological Observations of Behaviour* April 11, 1961
270 'like gymnasts': Gallai *With a Man on Board* 61
270 'They came in like mice': SP and SW interview with Kotovskaya
271 'The yard is hopelessly dark': Suvorov *First Manned Spaceflight* 59
271 strain gauges: 'In accordance to my instructions Ivan Shadrintsev (a radio electronics engineer) secretly glued strain gauges into the cosmonauts' mattresses. Wires from the sensors led across the road to another building, where he and the psychologist, Fyodor Gorbov, monitored the data.' Yazdovsky *Along the Paths* 140
272 'My sweet and much loved Valechka': Gagarin letter to Valentina from April 10, 1961. Valentina found the letter in her husband's suitcase after his return
272 'Sergei Pavlovich went off silently': Burgess and Hall *FSC* 153
273 'Night': Suvorov *First Manned Spaceflight* 60

ACT IV: LAUNCH

275 'Earth is the cradle of humanity': Konstantin Tsiolkovsky *The Investigation of Universal Space by Means of Reactive Devices* 16
275 'Can a man dream for more?': Gagarin's recorded speech April 3, 1961

CHAPTER 26: A MOTHER'S LOVE

277 'It's time' and other Suvorov quotations: Suvorov *First Manned Spaceflight* 61ff
277 Gagarin had concentrated hard: Doran and Bizony *Starman* 88
278 'Instead of advice': Ibid. 89–90
278 'Standing before the line-up': Vladimir Yaropolov in *First Forever* 314
279 'Readiness for launch': Ibid. 314
280 291-tonne mass: The R-7 missile specifically modified for Vostok missions was known as the 8K72. Its total lift-off mass was 291 tonnes. The total dry mass (that is, without its fuel and liquid oxygen propellants) was 28 tonnes. Thus approximately 90 per cent of its lift-off mass was propellant. Anatoly Zak *Vostok Launch Vehicle*: http://www.russianspaceweb.com/vostok_lv.html
280 'Decision: Healthy' and 'Usually he was a smiley': SP and SW interview with Kotovskaya
281 'he kept silent': SP and SW interview with Kotovskaya
281 'Why did he do this?': Gallai *With a Man on Board* 62
282 'Clearly he'd had a sleepless night': Doran and Bizony *Starman* 89
282 'Everything will be all right': Gagarin *Road to the Stars* 144
282 attacked by sharks: This and other details in Jenks *Cosmonaut* 132
282 'The ground staff who helped me': Gagarin *Road to the Stars* 144
283 'For the first time since his arrival': Doran and Bizony *Starman* 90
283 'good hand' and all reported speech on painting helmet: Gallai *With a Man on Board* 63f
284 'In the event of a Soviet manned shot': Logsdon *John F. Kennedy* 69–70
284 Salinger's meeting and Kennedy's evening including quotations: Sidey *John F. Kennedy* 110–11
285 'of masterminding the Nazi massacre': New York *Daily News* April 11, 1961
285 'I rather hold my breath': Ibid. The show is described thus in the *Daily News* TV schedules: '10 p.m. – 11 p.m. NBC (Channel 4) *JFK – Report No 2: President and Mrs John Kennedy.* The President describes how the White House functions under his administration and Mrs Kennedy discusses her role as First Lady.'

CHAPTER 27: KEY TO START

286 And so the little blue and white bus: There is an unverified story that Gagarin stopped the bus on the journey to urinate on the back-wheel, a time-honoured tradition among pilots before a mission. Like so many stories surrounding the events of this day this one is almost certainly a myth

287 'The closer we got to the launching-pad': Burgess and Hall *FSC* 154

287 'They're coming! They're coming!': Solodukhin in *First Forever* 284

287 'happy overcoat': Koroleva *Father* 439

287 'So they simply clang': Suvorov *First Manned Spaceflight* 62

288 'I still feel his hands on my shoulders': Gallai *With a Man on Board* 64

288 'It was difficult keeping any kind of planned': Kamanin diary entry April 12, 1961

288 'What could happen at this late stage?': Doran and Bizony *Starman* 92

289 'Of course he was hoping that when Gagarin': Ibid. 92

289 'I watched Gherman': SP and SW interview with Kotovskaya

289 'Gagarin looks in my direction': Suvorov *First Manned Spaceflight* 63

291 'Do you hear me?': This and all subsequent communications in VHF, HF and also Gagarin's own on-board tape recording transcribed in *First Forever* 42–64

291 'Yura, the numbers are 1-2-5': Doran and Bizony *Starman* 94, except Doran almost certainly gives the numbers incorrectly as 3-2-5. The space journalist Igor Lissov believes they were either 1-2-5 or 1-4-5. Gallai, who placed the envelope in the Vostok, says 1-2-5 in *With a Man on Board* 57. Kamanin says 1-4-5 in his diary entry for April 12, 1961. Take your pick!

292 'light touch': Chekunov and all subsequent quotations in *First Forever* 302

293 'Was Gagarin frightened?': *Novaya Gazeta* December 23, 2013

294 five thousand kilometres: Kamanin diary entry March 6, 1961: 'The range of HF (high frequency) communication is 5,000 kilometres, the VHF (very high frequency) is 1,500 kilometres, but these radii have so far been practically unchecked, and there is no confidence in the reliability of the connection'

294 a secret missile computation centre: It would take several years for the CIA to penetrate NII-4. An agency briefing describing NII-4's functions was only compiled in 1971, and when the briefing was declassified in 2003 over half the text was blanked out – as it still is. See *Moskva Scientific Research Institute NII Bolshevo 4 – Strategic Weapons Industrial Facilities USSR* CIA 1971 (approved for release 2003/08/05). For details of the Soviet missile tracking system, SW is indebted to Professor

Siddiqi in emails February 4 and 24, 2020, and also to Igor Lissov (email January 24, 2020). For the chemically treated paper and masks, Viktor Blagov, who worked at Bolshevo and later as a 'flight director' for the Soviet Mir and Salyut missions, wrote a fascinating online account (in Russian): *Korolev's Breakthrough into the Universe: Memories of a Space Industry Veteran*, November 21, 2018: https://www.i-podmoskovie.ru/history/korolyevskiy-proryv-vo-vselennuyu-vospominaniya-veterana-kosmicheskoy-otrasli/

295 'Suddenly we hear, "No KP-3!"': Solodukhin and all subsequent quotations in *First Forever* 284

295 'I went cold' and additional quotations: Ivanovsky *Rockets and Space* 146

297 'We loved our Vostok': Ibid. 143

297 'shining in all its futuristic beauty' and subsequent quotations: Suvorov *First Manned Spaceflight* 64

298 making everybody say all the same words: It is not exactly certain when Suvorov filmed the reconstruction in the bunker but it appears to have happened almost as soon as Gagarin was in orbit, presumably while everybody's memories of the words they had just spoken were still fresh. Suvorov also says that Korolev was filmed at the same time (Suvorov *First Manned Spaceflight* 65) but this is unlikely given he was needed at the command centre to track the flight. According to his daughter Natalya, Korolev's contribution was actually filmed later in his study in Moscow, complete with a reconstructed cloth-covered table and microphone. Under the circumstances his acting is quite convincing. Koroleva *Father* 352

299 'There wasn't a single person': Yaropolov in *First Forever* 315

299 'Our nerves were strained to breaking point': Rhea *Roads to Space* 387

299 'I remember his unblinking eyes': Kirillov and all subsequent quotations in *First Forever* 535

299 Bogdashevsky: SW interview Bogdashevsky

299 later analysis would put it at 46 per cent: Golovanov *Korolev* 697

299 'our dear one': SP and SW interview with Khionia and Vladimir Kraskin

302 'It hammers your ribcage': Doran and Bizony *Starman* 98

302 157 beats: Siddiqi *Challenge* 276

CHAPTER 28: THE FIRST SEVENTEEN MINUTES

304 'That's my boy!': Gallai *With a Man on Board* 66

305 '*Don't let anything happen to him*': From Gagarinday.ru http: // gagarinday.ru/articles/17-polet-yu-a-gagarina-v-kosmos-vzgljad-na-realnye-sobytija-teh-let-iz-xxi-veka.htmlp. 9

305 'Please God nothing goes wrong': SP and SW interview Khionia Kraskina
306 'like that of being ripped away': In Gagarin's (secret) Report to the State Commission (GRSC) after the space flight on April 13, 1961, *First Forever* 88
308 'The numbers were being repeated': Chekunov in *First Forever* 302
309 'I don't know how I looked at that moment': Kamanin diary entry April 12, 1961
310 'The Earth was moving to the left': GRSC April 13, 1961
311 'as the old tradition goes': Ivanovsky *Rockets and Space* 149–53
311 'and shook everyone's hand': Chekunov in *First Forever* 302
312 'We could hear Yuri's voice': Titov interview in Russian documentary *Yuri Gagarin Remembered* (1969)
312 'in nervous expectation': Sergei Khrushchev *Nikita Khrushchev* 432
312 'a small bright red flag': Gallai *With a Man on Board* 67
313 The culprit was … the R-7's second, mainstage engine: SW is indebted to Igor Lissov for his explanation of the causes of Gagarin's higher than nominal orbit
313 he would land in the wrong place: Perhaps counter-intuitively, the higher the orbit the *slower* the speed. A slower speed meant Gagarin would actually land short of his intended target once the Granit fixed timer activated the re-entry procedure

CHAPTER 29: THE NEXT THIRTY-TWO MINUTES

314 Before he left, his wife Anna: Anna Gagarina *A Word About My Son* and Valentin Gagarin *My Brother Yuri* for details of Gagarin's parents and family on day of flight
315 'How far away?': Zoya Gagarina in Doran and Bizony *Starman* 106
316 'The quality was poor' and 'The radio operator by my side': Scott and Leonov *Two Sides* 60
317 'Even if I'd known his orbit': SP and SW interview with Leonov
318 ELINT: Details in Plaster *Snooping on Space* and Gran *TV from Vostok*
318 Pierre Salinger: Sidey *John F. Kennedy* 111
319 'Amur Waves': The Vesna HF (high frequency) station at Khabarovsk kept broadcasting the song from 09.42–09.52 Moscow time. Three other Vesna HF stations also broadcast music for the same reason: Novosibirsk (music unknown), Alma-Ata (songs by Roza Baglanova, a popular Soviet soprano) and Moscow (songs about Moscow)
319 'Swallowing': GRSC April 13, 1961
319 'You watch them as in a dream': Gagarin *Road to the Stars* 154

320 'a beautiful blue halo' and other adjectives from Gagarin tape recording transcript *First Forever* 62

CHAPTER 30: MOSCOW CALLING

322 Yuri Levitan: Details from Levitan's own account in Adolf Dikhtyar *Before We Heard the Word 'Poyekhali' ('Let's Go')*. Reproduced in Russian Library of Astronautics and Cosmonautics: http://12apr.su/books/item/f00/s00/z0000055/st002.shtml
323 'Nikita Khrushchev called Malinovsky': In Russian TV documentary *Yuri Gagarin: Three Days and a Whole Life*
324 'like a whirlwind': Levitan in Dikhtyar *Before We Heard*
324 'Suddenly I saw Levitan without his jacket and tie': *Yuri Gagarin: Three Days*
325 '*Govorit Moskva!*': Russian Library of Astronautics and Cosmonautics
326 'came to a halt': Burgess and Hall *FSC* 158
326 'I froze': SP and SW interview with Titova
326 'My heart disappeared': Titova in Russian documentary *Gherman Titov: First After Yuri*
326 'Lord, thank goodness' and subsequent quotations: SP and SW interview with Titova
326 'She was filled with emotions': SP and SW interview with Elena Gagarina
327 'She says, "Mama, why aren't you listening"' and subsequent quotations: In *Yuri Gagarin: Three Days*
327 'I'm going to Valya': Doran and Bizony *Starman* 107
328 'Yes, everything is fine': Anna Gagarina *A Word About My Son* 46
328 'We listened to the radio': Doran and Bizony *Starman* 107
328 'Alekesy Ivanovich!' and following dialogue: Anna Gagarina *A Word About My Son* 46–8
329 the telephone next to Pierre Salinger's bed: Sidey *John F. Kennedy* 111
330 'You're kidding' and following dialogue: Barbree *Live from Cape Canaveral* 52
330 '*SOVIETS PUT MAN*': Cadbury *Race to Space* 246; also *The [London] Times* April 13, 1961: 'We're All Asleep'

CHAPTER 31: A BEAUTIFUL HALO

331 'weakly audible': GRSC April 13, 1961
334 Major Akhmed Gassiev: Details from Zverev and Oksyuta, *Yuri Gagarin on Saratov Soil*
334 Vitaly Volovich: Details of rescue team in SP and SW interview with Volovich

334 'A man is shortly to fly into space': Doran and Bizony *Starman* 87
335 'very abruptly' and following excerpts: GRSC
335 'approximately forty seconds': There is some uncertainty concerning exact timings, with figures variously given between 40 and 44 seconds. My source is Lissov in a series of emails 2020/21
338 Twelve minutes: Igor Lissov in a series of emails 2019/2020 to SP for all these timings and for other technical details of Gagarin's re-entry
339 'trapped in a house on fire' and 'pounding heart': Gagarin told these specific details to Golovanov in an interview. Golovanov *Korolev* 708
341 'Everybody was smiling': Anna Gagarina *A Word About My Son* 47

CHAPTER 32: BA-BOOM!

342 'very comfortably': GRSC April 13, 1961
343 'The silvery Volga': Gagarin *Road to the Stars* 160
343 'black and burned out' and 'a powerful wrench': GRSC
344 'I could tell everyone's eyes' and subsequent Gagarin quotations: Ibid.
345 10.53: Gagarin's landing time is often incorrectly published as 10.55, giving him a supposed flight time of 108 minutes. However secret official documents from the period give 10.53. The Vostok sphere landed five minutes earlier at 10.48. See 'OKB-1 report on the results of the launch of korabl'-sputnik with pilot Yu. A. Gagarin onboard' dated May 3, 1961 (*First Manned Flight* Vol II 42, document No. 154). In an email to SP (November 30, 2020) Igor Lissov, who has spent many years researching Gagarin's flight, writes: 'It is worth mentioning that nobody knows where the landing time originally announced of 10:55 am Moscow time comes from. No sports commissars were present at the landing … It is most likely that this time was roughly estimated by someone's personal watch with an unknown margin of error and with an unknown length of delay in relation to the actual event'
346 'explosion' and other quotations: Katz *The Dawn of Gagarin's Spring* 26–7
346 'What was that?' and subsequent quotations: SP and SW interview with Viktor Solodkyi October 1, 2019
348 'I looked at my watch': Ivanovsky *Rockets and Space* 153
348 'The parachute has opened!': Siddiqi *Challenge* 281

CHAPTER 33: GAGARIN'S FIELD

349 'I looked': Rita Takhtarova in the Russian Library of Astronautics and Cosmonautics: http://12apr.su/books/item/f00/s00/z0000055/st003.shtml

349 'I thought it was a monster': Rumia Nurskanova (aka Rita Takhtarova) in Videokanal IA 'Vzglyad-Info': https://www.youtube.com/watch?v=29Xetlm2ug0
350 'I'm a Russian, comrades': Gagarin *Road to the Stars* 161
350 'A friend': Doran and Bizony *Starman* 119
350 'I looked again': From the Russian Library of Astronautics and Cosmonautics
351 'four curious farmworkers': Some accounts say five or six. However SP and SW interviews in the area (October 2019) suggest four is more accurate. Two were tractor drivers (Yakov Lysenko and Vasily Kazachenko), one was a clerk (Ivan Rudenko) and one a fuel service man (surname Veryovka)
351 'explosion': Lysenko in Doran and Bizony *Starman* 118
351 'We've just been hearing all about you': SP interview with Tatiana Makarycheva and Viktor Solodkyi October 1, 2019
351 'He was very lively and happy': Doran and Bizony *Starman* 118
352 'I blushed with embarrassment': From the Russian Library of Astronautics and Cosmonautics
352 'the emotional shock': Yuri Savchenko quoted in Katz *Dawn of Gagarin's Spring* 7
352 'I don't know how the cosmonaut actually felt': Ibid.
353 'I ask you to report': The Russian Library of Astronautics and Cosmonautics
354 'Believe me, Major': Katz *Dawn of Gagarin's Spring* 9
354 *Keystone Cops*: The reference is in Jenks *Cosmonaut* 137
355 'exploded': Gallai *With a Man on Board* 70
355 'After the first toast': Ibid. 70
356 'Someone came out': Ivanovsky *Rockets and Space* 154
357 abyss: This and other details in SP and SW interview with Tamara Filatova September 29, 2019
358 'He froze on the threshold': Valentin Gagarin *My Brother Yuri* 174
359 'He's landed!': Anna Gagarina *A Word About My Son* 47

CHAPTER 34: TRIUMPH AND DEFEAT

361 'moved to tears' and subsequent quotations: GRSC April 13, 1961
362 'joy, wonder and admiration' and subsequent quotations: SP and SW interview with Volovich July 1, 2013
363 'You have made yourself immortal': Text of Khrushchev phone call from Pervushin *Yuri Gagarin* 379
364 'I, the undersigned, Sports Commissar': In museum display at the Museum of Cosmonautics, Moscow

365 'We never took our eyes off the sky': Pervushin *Yuri Gagarin* 411, citing Borisenko's audio interview from 1983
365 broke his arm: Jenks *Cosmonaut* 138
366 'It was an amazing flight': SP and SW interview with Volovich
366 'Kennedy's valet': Details of George Thomas and his wake-up call in Sidey *John F. Kennedy* 111–12
368 'The achievement by the USSR': *Life* April 21, 1961 27
368 Alan Shepard was at the Cape: Details in Thompson *Light This Candle* 282

CHAPTER 35 :THE BALL ON THE HILL

369 'a place of pilgrimage': K.A. Malyshev in Katz *Dawn of Gagarin's Spring* 9
370 'There were soldiers with machine-guns': SP and SW interview with Tatiana Makarycheva
370 'I couldn't hold myself back either': Margarita Butova in Katz *Dawn of Gagarin's Spring* 10
370 'Everybody was trying to break past' and additional Pallo quotations from *Notes by a member of the search and evacuation team, lead designer of OKB-1 A.V. Pallo, regarding meeting with Yu. A. Gagarin in the city of Engels* (1996) RGANTD. F. 107. Op. 5. D. 100. L. 1–5
370 'I remember tasting some': Doran and Bizony *Starman* 121
371 'We asked the soldiers': K.A. Malyshev in Katz *Dawn of Gagarin's Spring* 9
372 'One of them tried using a bicycle pump': SP and SW interview with Tatiana Makarycheva
372 'for several minutes': Gallai *With a Man on Board* 74
372 'You could fly it into space': Golovanov *Korolev* 710
372 'This is history now': Ivanovsky *Rockets and Space* 155
373 'There were people all along the runway': IL-14 pilot Viktor Malygin interviewed in Russian documentary *Yuri Gagarin: Three Days*
373 'I had worried and worried' and other Kamanin quotations: Kamanin diary April 12, 1961
373 '[Gagarin] was surrounded by generals' and other Titov quotations: Doran and Bizony *Starman* 125
375 'Of course we couldn't quite believe': Ibid. 110. It is possible that this phone call took place the following day on April 13 – if at all. Zoya's account appears to be the only evidence that Gagarin spoke to his wife and mother before meeting them at Vnukovo Airport in Moscow on April 14
375 'eyes were wet': Golovanov *Korolev* 712

375 four hundred members of the press: More exactly, 426. Details of press conference in John F. Kennedy Presidential Library, JFKPOF-054-011-p0001
376 'contraption' and 'toy store': Quoted by Theodore Sorensen in Logsdon *John F. Kennedy* 70
378 long-awaited publication: *Ad Hoc Mercury Panel Report*, 12 April 1961. Facsimile in John F. Kennedy Presidential Library, FKPOF-086a-002-p0001

CHAPTER 36: PARTY TIME

379 'They were lovely MiGs': Doran and Bizony *Starman* 128
380 'I am going to give him such a welcome!': Russian documentary *Yuri Gagarin: Three Days*
380 a decoy: Memoirs of Lieutenant General Tsedrik, who also describes the 'aggressiveness' of the crowds in Pervushin *Yuri Gagarin* 392–3
380 'The weather was wonderful': *Yuri Gagarin: Three Days*
381 'The Number One Event' and 'a Universal Good': Swiss and Vatican newspapers quoted in Logsdon *John F. Kennedy* 72; 'Greatest Feat of Science': *Evening Gazette* April 12, 1961
381 '"I Can See Everything"': *Vancouver Sun* April 12, 1961
381 'Why wasn't it a woman, eh?': *Daily Mirror* April 13, 1961
381 A twelve-thousand-foot mountain, glacier and Gagarinite: *New York Times* April 14, 1961
382 'Hymn' and 'a record day for the production of poetry': Jenks *Cosmonaut* 154
382 copied directly from ... *Life*: *Life* April 21, 1961
382 'America Is Stunned': *Pravda* April 13, 1961
382 '*HOW RUSS WON SPACE RACE*': *Chicago Daily News* April 12, 1961
382 '*KHRUSHCHEV BOASTS OF SPACE LEAD*': *Daily News* April 13, 1961
383 'Russia has won a great scientific victory': *Man into Space* NBC-TV April 12, 1961
383 'No contest in history': *The Race for Space* WPIX-TV April 13, 1961
383 'Tell me' and 'I want to see our country mobilized': Logsdon *John F. Kennedy* 73
383 '"*SO CLOSE YET SO FAR*"': *Huntsville Times* April 12, 1961
383 'heartiest congratulations': *New York Times* April 13, 1961
383 'They just beat the pants off us': Thompson *Light This Candle* 282
384 'a deep sense of personal disappointment': NASA press release April 12, 1961

384 'We had them': Thompson *Light This Candle* 283
384 'Can you imagine': SP and SW interview with Tamara Filatova
385 they had been rehearsing: Kamanin diary April 13, 1961
385 'Never – not even in the spacecraft': Riabchikov *Russians in Space* 42
385 'The rather short major': Sergei Khrushchev *Nikita Khrushchev* 434
386 'I am happy to report': Footage of speech and event in RGANTD
386 'Please don't cry, Mamma': Doran *Starman* 130
387 'I doubt if anybody in the world': Gagarin *Road to the Stars* 172
387 'was radiant': *Yuri Gagarin: Three Days*
387 secret decree: 'Note from D. Ustinov, A. Shelepin, K.Vershinin and others to Central Committee of USSR Communist Party April 11, 1961', in *Soviet Space* 338
387 'millions of yards of red bunting … photographs': *New York Times* April 15, 1961
387 'panic-stricken': Sergei Khrushchev *Nikita Khrushchev* 433
388 'who has done and seen': April 14, 1961 BBC archives
388 'There were tears of joy': BBC *Cold War* episode 8
389 'I am boundlessly happy' and also Khrushchev's speech: *New York Times* April 15, 1961
390 'It was astonishing to see': Doran and Bizony *Starman* 132
390 'We shouted, "*Yura! Yura!*"': SP and SW interview with Titova
391 his car was held up: This version of the story is from Korolev's daughter Natalya in *Father* 353. One of Korolev's assistants, Vladimir Shevalyov, has an alternative version in which the fan belt in Korolev's Chaika broke down. Harford *Korolev* 176
391 'We work underground': Koroleva *Father* 355
392 'I walked with the crowds': SP and SW interview with Koroleva
392 'from time to time Valya touched': Anna Gagarina *A Word About My Son* 50

EPILOGUE: ENDGAME

393 'If somebody can just tell me': Sidey *John F. Kennedy* 120. Sidey was in the White House cabinet room when the president made this statement
395 'Can we leapfrog': This and other quotations in this meeting from Sidey *John F. Kennedy* 118–21
396 ten times as much as the bomb: The wartime Manhattan Project's *estimated* cost was $2 billion or approximately $29 billion in 2021. Webb's higher-end estimate of $40 billion was substantially too high. The Apollo programme cost approximately $25.4 billion (Forbes *Apollo 11's 50th Anniversary: The Facts and Figures* July 20, 2019)

397 The bombing began within hours … on April 15: SW is indebted to Rasenberger *Brilliant Disaster*, a fascinating recent account of the Bay of Pigs operation
397 'the least covert military operation': Salinger *With Kennedy* 147
398 'provide the Cuban people': Sergei Khrushchev *Nikita Khrushchev* 435–6
398 114 Cuban fighters: Rasenberger *Brilliant Disaster* xiv
398 'the grimmest I can remember' and other quotations: Salinger *With Kennedy* 147
398 'in the most emotional, self-critical state': Logsdon *John F. Kennedy* 79
398 'sons of bitches' and 'How could I have been so stupid?': Caro *The Years of Lyndon Johnson* 183
398 'was not accustomed to failure': BBC *Cold War* episode 10
399 'scorn': Salinger *With Kennedy* 149
399 the loneliest of figures: Sidey *John F. Kennedy* 131
399 'a chance of beating the Soviets': 'Memorandum for Vice President' April 20, 1061. John F. Kennedy Presidential Library: JFKPOF-030-019-p0006
399 'With an all-out crash program': Logsdon *John F. Kennedy* 88. For Kennedy's moon decision SW is indebted to both Logsdon's excellent account and Douglas Brinkley's likewise excellent *American Moonshot: John F. Kennedy and the Great Space Race*
399 $20 billion: A figure in this ballpark was presented to Lyndon Johnson by April 22, 1961, by NASA's deputy administrator Hugh Dryden. The $20 billion was also a 'best guess' estimate in the words of NASA's public affairs chief Bill Lloyd shortly before Kennedy made his May 25 speech. Logsdon *John F. Kennedy* 110
400 'Dramatic achievements in space': Johnson memo to Kennedy April 28, 1961, Logsdon *John F. Kennedy* 90
400 'The Americans are licked': *Life* April 21, 1961 26–7
400 again his launch was delayed: According to CBS's live coverage the original launch time was 07.30 Eastern Time. Shepard left approximately two hours late
401 'I'm a wetback now': Thompson *Light This Candle* 294
401 'Everybody hit their knees': SW interview with Barbree
401 'I felt a shiver': Kranz *Failure Is Not an Option* 52
401 'Be sure to wave': Thompson *Light This Candle* xviii
401 'Please Daddy': SW interview with Laura Shepard Churchley
401 'But he didn't say mess': *NASA's Greatest Missions* episode 1
401 'ominous': Sidey *John F. Kennedy* 152

402 'We took our chances': Kraft *Flight* 141
402 'This was how free men': Sidey *John F. Kennedy* 152
402 'beautiful view': Burgess *Freedom 7* 252. Appendix 2 contains the entire radio communications transcript and precise timings
402 'Shit': Thompson *Light This Candle* 300
402 '*OUR SHEP DOES IT!*': *Newsday* May 5, 1961
402 '*SHEP DID IT!*': *Orlando Evening Star* May 5, 1961
402 '*SHEPHERD'S TRIP A-OKAY*': *Cincinnati Enquirer* May 5, 1961
402 '*BOY WHAT A RIDE!*': *Detroit Times* May 5, 1961
402 'inferiority': *Pravda* May 6, 1961
402 '*MILLIONS CHEER*': New York *Daily News* May 5, 1961
403 'a million pounds': Sidey *John F. Kennedy* 153
403 'The future of our entire manned': Neufeld *Von Braun* 363
403 'I believe that this nation': Special Message to Joint Session of Congress on Urgent National Needs, May 25, 1961. John F. Kennedy Presidential Library: JFKWHA-032
403 'before this decade is out': In an early draft of the speech the year 1967 was given as a target date. This was changed to the end of the decade at the request of NASA's chief James Webb, to give more breathing space. Neufeld *Von Braun* 363
404 'cold bloodedly': Logsdon *John F. Kennedy* 83
404 It was made because it had to be made: Boris Chertok, one of Korolev's most senior rocket engineers, made an interestingly provocative counter-factual point in his magisterial *Rockets and People* 79: 'I contend that if Gagarin's flight on 12 April 1961 had ended in failure, U.S. Astronaut Neil A. Armstrong would not have set foot on the Moon on 20 July 1969.' Logsdon *John F. Kennedy* 50 also notes that had Shepard flown first 'it is unlikely that the Soviet launch of Yuri Gagarin … would have had such a dramatic impact on U.S. space policy'. Perhaps if Ham's fuel had not run out half a second early, the moon landings might never have taken place
404 'a piece of cake': Johnson Space Center interview with Shepard February 20, 1998
404 'It's a long way': Thompson *Light This Candle* 420
404 'miles and miles and miles': Ibid. 426
404 'Mount Everest': Interview with Shepard, American Academy of Achievement February 1, 1991
405 'Daddy's pad': SW interview with Laura Shepard Churchley
405 'I realised how much I'd missed': Glenn *Glenn* 522
406 'World's First Astrochimp': Burgess and Dubbs *Animals in Space* 258

407 three complete Saturn Vs on display: At the Kennedy Space Center at Cape Canaveral, Florida, the Johnson Space Center in Houston, Texas, and the Davidson Center for Space Exploration in Huntsville, Alabama. These examples contain stages from different original Saturn Vs. In Huntsville there is also a replica Saturn V that is displayed vertically
407 A war crimes trial: In Essen, Germany between 1968 and 1970
407 'He has given valuable service': Neufeld *Von Braun* 471
408 worst offences: Ibid. 475
408 a crater on the moon: It lies at the western rim of the Oceanus Procellarum. See Mallon *Rocket Man* in the *New Yorker*, October 15, 2007
408 the first two-man spacecraft: The second Voskhod in March 1965 (when Gagarin's friend Aleksey Leonov also made the first spacewalk). The first Voskhod, carrying three cosmonauts, flew in October 1964. Voskhod means 'sunrise'. The first woman in space, Valentina Tereshkova, flew solo aboard Vostok 6 in June 1963. By comparison the first American female astronaut, Sally Ride, flew on the space shuttle *Challenger* in June 1983, almost exactly twenty years later
408 'aged ten years': Siddiqi *Challenge* 511
408 'I am in a constant state': Ibid. 512
409 The operation was performed by Boris Petrovsky: It would not be unusual for the Minister of Health to perform surgery on a figure of such major importance as Korolev. The details of the operation are from Korolev's daughter Natalya, a doctor herself: *Father* 402–4. On page 440 she describes the missing lucky coins
409 the *New York Times*: The short piece on page 82 is headlined 'Leading Space Scientist Dies' and states that Korolev 'was reputed to be the mysterious "chief designer" of the space program', but it does not verify the case
409 'an avalanche': Harford *Korolev* 281
409 'The country has lost': Siddiqi *Challenge* 516
410 'All I wanted': SP and SW interview with Natalya Koroleva
410 None of the Vanguard Six: Of the original class of twenty cosmonauts only Boris Volynov is still alive (as of December 2020)
410 'I no longer exist' and 'You were always better': From TV documentary *Nelyubov*
411 'To my little Tomochka': From Russian documentary *Gherman Titov: First after Yuri*
412 A secret decree: 'Decree of the Presidium of the Supreme Soviet of the USSR: On Gifts for Comrade Gagarin Yu. A. and his Family Members'. April 18, 1961. *Soviet Space* 354–5

412 'I was reminded': Doran and Bizony *Starman* 165
413 'alcoholic intoxication': Ibid. 159
414 'Father was so angry' and subsequent quotations: SP and SW interview with Elena Gagarina
414 the causes of that crash: See Siddiqi *Challenge* 628, also Doran and Bizony *Starman* 207–29 for a detailed account of the crash and subsequent investigations. Andrew Osborn's 'What Made Yuri Fall?' in *Air and Space Magazine* September 2010, offers an alternative explanation involving an open air vent in the cockpit. Aleksey Leonov's version stems from his own experience in the area on the day of the crash, when he heard what sounded to him like the supersonic bang of another aircraft just a couple of seconds before the sound of a plane crash (SW and SP interview with Aleksey Leonov). When he was finally allowed access to the files in 1986 Leonov discovered that his own evidence in the original 1968 investigation had been altered
415 'but without proof' and subsequent quotations: SW and SP interview with Tamara Filatova
416 His landing site: To this day no one knows the exact spots where either Gagarin or his Vostok sphere landed. Both places were originally marked with simple posts which soon disappeared. Some local witnesses believe that the commemorative statue and monument that supposedly mark the ploughed field into which Gagarin parachuted is in the wrong place, albeit close by. The Vostok landing site did not get its own monument until very much later, largely because of the secrets surrounding the manner of Gagarin's return to earth. According to SP's interview with local witness Viktor Solodkyi, the present tall, rather unimaginative, concrete pole was only erected on the presumed landing spot as late as 2017, thanks to the efforts of local children and history enthusiasts. Even that pole, however, may not have been placed in the correct location, but approximately 200m away. Nevertheless it commands a splendid and very Russian view of the River Volga
416 Baikonur: After Gagarin's flight the cosmodrome was officially identified to the world as Baikonur, the name of the tiny village 280 kilometres to the north, as had been agreed at the state commission on April 8. The town of Leninsk (changed from Leninsky in 1969) was itself renamed Baikonur in 1995. Gagarin's launch pad was used for the last time on September 25, 2019, when Soyuz MS-15 lifted off at sunset that day. On board were Oleg Skripochka, Hazza Al Mansouri and Jessica Meir, a doctor in marine biology and NASA astronaut. The author was privileged to witness this final launch from Gagarin's pad

416 a pet parrot called Lora: In an email to SW from Valentina's granddaughter Ekaterina Karavaeva (October 15, 2020) the parrot 'spent its last couple of months with me while grandmother was at the hospital and it outlived her for just a few days and at the very end it felt as if it simply lost the purpose of living any longer'

ACKNOWLEDGEMENTS

I BEGAN WRITING *BEYOND* in March 2020, one week before the UK went into its first national lockdown as a result of the Covid pandemic. But its genesis goes much further back to 2012, when I was commissioned by Working Title Films to develop and direct a documentary based on the raw, unseen, secret Soviet film footage surrounding Yuri Gagarin's flight. That commission resulted in three trips to Russia to interview witnesses to that spectacular moment in history, several of whom – like too many others – have since sadly passed away. Much of the film footage we discovered at the time was revelatory, but much of it too had deteriorated beyond repair in dusty vaults, or had simply been lost, and other than a thirty-minute 'taster' the full documentary was never made. But its spirit burned on, at least in my brain, to be given new life in its present form as a book. For that I must thank my daughter, Kitty, to whom this book is dedicated and who first suggested that I keep all my original notes and transcripts and encouraged me to write this story instead. I must also thank those who have been with me since that original film journey, especially Eric Fellner and Amelia Grainger at Working Title, both such inspired and inspiring producers, my fabulous film agent Jane Villiers, the fantastic filmmakers Christopher Riley and Duncan Copp, and my outstanding Russian film researcher, Alexander Kandaurov.

I owe too a special debt to Yuri Gagarin's daughter, Elena Gagarina, and to Elena's own daughter Ekaterina Karavaeva, both of whom have been bulwarks of support over what is now nearly ten years, while always respecting my editorial independence as a documentary director and as a writer. I thank them both for their many kindnesses and for their trust. In a similar vein, I must also thank Laura Shepard Churchley, Alan Shepard's daughter, who generously invited me into her home in Cocoa Beach and from whose roof I had a grandstand view of an unforgettable night launch of a Falcon Heavy rocket from the Kennedy Space Center just across the water. Many other witnesses to this history, both in Russia and the United States, have generously given me their time and allowed me to probe and record their memories. Their names are listed at the head of the bibliography section of this book and I thank every one. All of them gave me memories of my own which I shall forever cherish.

I am also enormously thankful to those many who, without any expectations for themselves, allowed me the benefit of their learning and experience and liberally opened their lists of contacts to me. If there are any I have in error omitted to name here I hope they will forgive me, but they include David Fairhead, Heather Walsh, Tony Reichhardt, Bert Vis, JoAnn Morgan, Gerry Griffin, Leo de Boer, Keith Haviland, Jennifer Ross-Nazal, Roger Launius, Dwayne Day and Rodney Krajca at NARA, Fort Worth. They all lit lamps which helped guide me on my way.

In the United States I am especially indebted to Manfred 'Dutch' von Ehrenfried and Terry Greenfield, both longstanding former NASA engineers who worked on the Mercury-Redstone missions and who suffered salvoes of my emails with equanimity and great good humour. I am also grateful to Jay Barbree for a wonderfully entertaining tour of Cocoa Beach in his very speedy convertible and to Christopher Jim Williams, chief of media relations at Cape Canaveral Air Force Station, who gave up two of his busy days to show me around the original launch pads and bunkers and, as an unexpected extra treat, allowed me thrillingly close-up access to watch the launch of an Atlas V carrying a Boeing Starliner into the Floridian dawn.

In Russia I must thank Inna Byikevich, Director of the People's Museum of Yuri Gagarin in Saratov, Elena Chernykh, my guide at the S.P. Korolev RKK Energia Museum, Elena Plakhova of the Engels Museum of Local History, Svetlana Medvedeva of the Uzmorie School Yuri Gagarin Museum, Nikolai Mironov, Director of the Museum of First Flight in Gagarin, and Tamara Filatova, Gagarin's niece and Director's Adviser at the Combined Memorial Gagarin Museum. I am also grateful to Nikolai Lovut, our hard-worked but always cheerful driver who took me to so many of these places and bore with my very muddy boots in his pristine car. In Baikonur, I am grateful to Antonina Bogdanova, Director of the Baikonur Cosmodrome Museum and to Galina Milkova of the Museum of Baikonur History, for her eye-opening two-day tour of this extraordinary town and cosmodrome. To their illustrious ranks must be added Vladislav Shevkunov, director of Vegitel Aero and Space Tourism at Star City, who enabled me to witness the last ever launch from Gagarin's pad in September 2019, when a Soyuz rocket carrying three astronauts thundered into the skies and made the earth tremble just as Gagarin's R-7 did almost sixty years before.

This book could not have been written without the expert advice of three of the finest historians in this field. I would especially like to extend my heartfelt thanks to Igor Lissov in Moscow, Colin Burgess in Sydney and Asif Siddiqi, Professor of History at Fordham University in New York, all of whom went many, many extra miles on my behalf, helped correct my litter of errors (albeit those that remain are of course my responsibility), suggested much-needed improvements and, in Colin's and Asif's cases, read and made very useful comments on the initial draft. Their contributions have been invaluable from first to last. I must also thank my admirable translators, Erik Alstad and Clair Walmsley, both of whom were willing to take time away from their own studies to render a thick pile of Russian documents into elegant English, and Anne Miles and Wendy

George, who each did such a superb job of transcribing my many, often shamefully badly recorded, interviews.

My excellent literary agents, Henry Dunow in the US and David Godwin in the UK, have both been rock-solid pillars of encouragement throughout this project and I thank them both for their guidance and enthusiasm, as I do too the terrific teams at my publishers HarperCollins on both sides of the Pond, among them Iain Hunt for his sensitive copyediting, and above all my brilliantly talented and amazingly patient editors, Terry Karten and Arabella Pike, who have been beside me every step of the way.

On a more personal note, I cannot omit to say a word here about my parents, both now in their nineties, who throughout the long months of lockdown and shielding have been a constant and amazing fount of love as I wrote, despite our not being able to see or hug each other in person. They well remember many of the events described in this book even if, as they have told me, they were somewhat distracted by my happening to be born just ten days after Gagarin's flight and thirteen days before Shepard's. Perhaps that's where the obsession began! In addition I am beyond moved by the love, care and support showered on me by Sally George, who has been absolutely there for me, cheering me on when I was stumbling and keeping me (almost) sane through a very challenging year. She was my first reader and has been an inspired critic. As with my earlier book *Shockwave*, she also came up with the title. But more than all that she has been an incredible and very special anchor. This book would simply not exist without her.

It would not, finally, exist without another person too. Svetlana Palmer first worked with me on the original documentary incarnation of this story, and she came back on this incarnation to work with me as my Russian researcher, my colleague and my friend. I owe her an incalculable debt of gratitude. Her tireless pursuit of often obscure Soviet sources, her graceful but always searching interviews with every Russian participant, her intelligence, professionalism, passion and unwavering commitment to the task in hand, however difficult or apparently impossible, have left me repeatedly stunned. More than that, she has been great fun along the way. Neither of us I think will ever forget our first night in a freezing Baikonur after a twenty-nine-hour journey from London with nothing but a bar of Cadbury's chocolate and a box of Pringles to keep us going. I thank her from the bottom of my heart.

ILLUSTRATIONS

Isolation chamber (Author's collection)
Mercury Control Center (NASA)
Soviet launch bunker (Science History Images/Alamy)
Mercury capsule cockpit (NASA)
Vostok capsule cockpit (123RF)
Vostok in orbit (Zoonar GmbH/Alamy)
Mercury capsule with engineers (Wartimepd/Alamy)
Gagarin medical (Sputnik/Science Photo Library)
Korolev's R7 rocket (SPUTNIK/Alamy)
Gagarin and Titov on bus ride to launch pad (Sputnik/Science Photo Library)
Gagarin's launch (Science History Images/Alamy)
Gagarin mobbed after landing (Anatoly Pekarsky)
Charred Vostok sphere after landing (SPUTNIK/Alamy)
Gagarin and Nikita Khrushchev celebrate at Moscow airport (Science History Images/Alamy)
Crowds go wild in Red Square (Sputnik/Science Photo Library)
Shepard in his Mercury capsule (NASA/VRS/Science Photo Library)
The Redstone launch (NASA)
President Kennedy and Jackie watch on live TV (World History Archive/Alamy)
Gagarin filmed with family (Elena Gagarina Photo Archive)
Gagarin with parents (Sputnik/Science Photo Library)
Heroic Gagarin statue Moscow (Felix Lipov/Alamy)

BIBLIOGRAPHY, FILMOGRAPHY AND ORIGINAL INTERVIEWS

Original Interviews

SW is Stephen Walker and SP is Svetlana Palmer
Telephone interviews marked with asterisk

Russia

Natalya Beregovaya (Popovich), SP Moscow December 16, 2019
Rostislav Bogdashevsky, SW Moscow April 27, 2012
Valentina Bykovskaya, SW & SP Moscow October 10, 2019
Tamara Filatova, SP & SW Gagarin September 29, 2019
Elena Gagarina, SP & SW Moscow September 28, 2019
Natalya Koroleva, SP & SW Moscow October 3, 2019
Adilya Kotovskaya, SP & SW Moscow July 2, 2013
Khionia and Vladimir Kraskin, SP & SW St Petersburg January 30, 2015
Alexei Leonov, SP & SW Cologne July 1, 2013
Igor Lissov, SP Moscow December 16, 2019
Tatiana Makarycheva, SP & SW Uzmorie village school October 1, 2019
Irina Ponomareva, SP Moscow December 17, 2019
Boris Smirnov, SW Moscow April 28, 2012
Viktor Solodkyi, SP & SW Uzmorie village school and Gagarin's landing site October 1, 2019
Tamara Titova, SP & SW Moscow October 3, 2019
Vitaly Volovich, SP & SW Moscow July 1, 2013

Boris Volynov, SP & SW Moscow July 1, 2013
Dmitry Zaikin, SW April 27, 2012

USA (all interviews SW)
Jay Barbree, Cape Canaveral December 17, 2019
Laura Shepard Churchley, November 18, 2019*
Manfred 'Dutch' von Ehrenfried, November 5, 2019*
Terry & Jean Greenfield, Titusville December 20, 2019
Terry Greenfield, William Chandler, Ike Rigell, Cape Canaveral December 18, 2019
Julie Shepard Jenkins, November 22, 2019*
Jackie Lovelace Johnson, November 1, 2019*
Sharon Lovelace, November 12, 2019*
Glynn Lunney, Houston December 14, 2019
JoAnn Morgan, Titusville December 19, 2019
Ike and Kathy Rigell, Titusville December 20, 2019

Primary Sources – Print

(Note: Published and unpublished memoirs of witnesses are included in Secondary Sources)

USSR/Russia
Cold War BBC/CNN Television Documentary Series 1995–1998
Original interviews archived at King's College, London:
Boris Chertok
Khionia Kraskina
Maria Stepanova
Oleg Troyanovsky
Sergei Khrushchev Parts 1 and 2
Gherman Titov

Clinical Physiological Description of the Cosmonaut Cadet of the First Division of the TsPK, Soviet Air Force First Lieutenant Gagarin, Yuri Alekseyevich. January 14, 1961. Baikonur Cosmodrome Museum archive.
Data from Psychoneurological Observations of Behaviour of Gagarin Yu. A. April 11, 1961. Baikonur Cosmodrome Museum archive.
Gagarin, Yuri. Letter to Valentina Gagarina and their two daughters April 10, 1961. *Komsomolskaya Pravda* March 9, 2017.

Instructions from the Communist Party Central Committee to the cosmonaut on use and control of the spacecraft 'Vostok-3A'. January 25, 1961. From copy in possession of Asif Siddiqi. Original in the Russian State Archive of Scientific-Technical Documentation (RGANTD), Moscow.

Kamanin, Nikolai P. *Skrytyi Kosmos (Hidden Cosmos).* Moscow: Infortext-IF, 1995.

Korolev, Sergei P. *Gorizont Sobytiy. Nezhnye pis'ma surovogo cheloveka 1947–1965* (*Timeline of Events. Tender Letters of a Fierce Man 1947–1965*). Moscow: OOO Boslen, 2019.

Korolev's List of Vostok Names. S.P. Korolev Rocket and Space Corporation Energia Museum display. Korolev city, Russia.

Pallo, Arvid V. *Notes by a Member of the Search and Evacuation Team, Lead OKB-1 Designer A.V. Pallo, on Meeting Yu.A. Gagarin in the city of Engels.* Moscow: RGANTD, f. 107, op. 5, d. 100, l. 1–5, 1996.

Pervyi Navsegda (First Forever). A Volume dedicated to the 80th Birthday of the First Cosmonaut of the Planet Earth Yu. A. Gagarin. Ed. Klimashevskaya, Olga L. Moscow: Institut Izucheniya Reform Predprinimatel'stva (Institute for the Study of Entrepreneurship Reform), 2014:

On Preparations for the Launch of the Manned Spaceship Vostok, September 10, 1960. 22.

On the results of the examination carried out with cosmonaut cadets at the Air Force Cosmonaut Training Centre. January 25, 1961. 129–32.

Decree of the Presidium of the Central Committee of the Communist Party of the Soviet Union. On the Launch of Korabl'Sputnik, April 3, 1961. 30.

Vostok Firing Officer's log. April 12, 1961. 95.

Report of Comrade Yu. A. Gagarin, April 13th 1961 at the meeting of the State commission after space flight. 87–94.

VHF Communications, Vostok. April 12, 1961. 42–60.

HF Communications Vostok and On-Board Tape Recording. April 12, 1961. 60–4.

Also included in this work are the following eyewitness accounts:

Chekunov, Boris S. 301–3.

Kirillov, Anatoly S. 526–36.

Ponomarev, Gennady P. 553.

Solodukhin, Vladimir N. 277–85.

Stadnyuk, Vladimir Ye. 371–2.

Yaropolov, Vladimir I. 308–17.

Core Data on product Rocket Carrier 'Vostok' and Korabl'-Sputnik 'Vostok'. April 11, 1961. *Pervyi Pilotirujemyi Polyot (First Manned Flight) Volumes 1–2.* Moscow: Rodina Media, 2011, Vol. 1, 431.

Report on the Landing of Korabl'-Sputnik Vostok, signed by I.G. Borisenko. April 12, 1961. Moscow Museum of Cosmonautics.

On the Results of the Examination carried out with Cosmonaut Cadets at the Cosmonaut Training Centre of the Air Force. January 25, 1961. Baikonur Cosmodrome Museum archive.

Secret Report to the State Commission (Gagarin, Yuri A.). May 9, 1961. Original in the Russian State Archive of Scientific-Technical Documentation (RGANTD), copy in possession of Asif Siddiqi.

Service Description of the cosmonaut cadet of the first division of the Centre for Preparation of Cosmonauts [TsPK] of the Air Force Senior Lieutenant Gagarin Yuri Alekseyevich V-879760. January 9, 1961. Baikonur Cosmodrome Museum archive.

Sovetsky Kosmos (Soviet Space). Special edition for the 50th anniversary of Yuri Gagarin's Flight. Vestnik Arkhiva Prezidenta Rossiyskoi Federatsii (Annals of the Presidential Archive of the Russian Federation). Moscow: OAO Mozhaisky Poligrafichesky Institut, 2011:

Strictly Secret Note from D. Ustinov, R. Malinovsky, K.Rudnev and others to the Communist Party of the Soviet Union (CPSU) Central Committee No. VP-3/1647. September 10, 1960. 285–6.

Top Secret Note from R. Malinovsky and M. Zakharov to F. Kozlov, L. Brezhnev and D. Ustinov No. 75204ob. February 15, 1961. 303–5.

Top Secret Cypher Telegram from M. Keldysh, S. Korolev, S. Zverev and others to the CPSU Central Committee. No. 799/sh. March 9, 1961. 311.

Top Secret Note from D. Ustinov, K. Rudnev, V. Kalmykov and others to the CPSU Central Committee No. VP-13/466. March 14, 1961. 311–12.

Top Secret Cypher Telegram from M. Keldysh, S. Korolev, N. Yurishev and A. Zakharov to the CPSU Central Committee No. 951/sh. March 25, 1961. 332.

Top Secret Note from F. Kozlov, L. Brezhnev, D. Ustinov to the CPSU Central Committee No. VP-1955ov. March 29, 1961. 332–3.

Top Secret Note from D. Ustinov, K. Rudnev, V. Kalmykov, and others to the CPSU Central Committee No. VP-13/534. March 30, 1961. 333–6.

Top Secret Note from D. Ustinov, A. Shelepin, K. Vershinin and others to the CPSU Central Committee. April 11, 1961. 338.

Decree of the Presidium of the Supreme Soviet of the USSR: On Gifts for Comrade Gagarin Yu. A. and his Family Members. April 18, 1961. 354–5.

Titov, Gherman A. Unpublished letters to Tamara Titova in her private archive.
Vis, Bert. Transcript of Interview with Pavel Popovich. Ottawa, September 30, 1996. Vis personal archive.

USA

CIA Archives: https://www.cia.gov/library
Sydney Finer, *The Kidnapping of Lunik* (1967). Sanitised for release September 1995.
Here are the Men and Formulas that Launched Sputnik, March 18, 1958. Released 1996.
Intelligence Warning of the 1957 Launch of Sputnik: https://www.cia.gov/library/readingroom/collection/intelligence-warning-1957-launch-sputnik.
Khrushchev's 'We Will Bury You'. February 7, 1962. Released January 22, 2001.
Moskva Scientific Research Institute NII Bolshevo 4 – Strategic Weapons Industrial Facilities USSR. January 1, 1971. Released July 29, 2003.
NIE (National Intelligence Estimate) 11-5-60 Soviet Capabilities in Guided Missiles and Space Vehicles. May 3, 1960.
NIE 11-4-60 Main Trends in Soviet Capabilities and Policies 1960–1965. December 1, 1960. Released May 17, 2012.
NIE 11-5-61 Soviet Capabilities in Guided Missiles and Space Vehicles. April 25, 1961.
Plaster, Henry G. 'Snooping on Space Pictures' *CIA Studies in Intelligence* Vol. 8 Fall 1964. Released 1995.
Soviet Version of the U-2 Incident. June 9, 1960. Released August 11, 2010.
Space Contribution of NIE 11-60. February 9, 1960. Released August 14, 2001.
The President's Intelligence Checklist. June 17, 1961. Released September 15, 2015.
Wheelon, Albert & Graybeal, Sidney. 'Intelligence for the Space Race' *CIA Studies in Intelligence* Vol. 5 Fall 1961. Released September 18, 1995.

Dryden, Hugh. *Discussion Notes by the Deputy Administrator of the National Aeronautics and Space Administration. April 22, 1961.* Available at Office of the Historian, US Department of State. https://history.state.gov/historicaldocuments/frus1961-63v25/d362.
Grimwood, J. *Mercury Pressure Suit Development. U.S. Government Memorandum to Lee N. McMillion, November 1, 1963.*

John F. Kennedy Presidential Library: www.jfklibrary.org:
Inaugural address, January 20, 1961. Digital identifier (DI): JFKPOF-034-002-p0001
Interview with WGN-TV, April 10, 1961. DI: JFKWHA-021-005
Low, George. Interview, May 1, 1964. DI: JFKOH-GML-01
Memorandum for Vice President, April 20, 1061. DI: JFKPOF-030-019-p0006
President's Science Advisory Committee (PSAC) *Ad Hoc Mercury Panel Report*, April 12, 1961. DI: FKPOF-086a-002-p0001
Press Conference, February 8, 1961. DI: JFKWHA-009
Press Conference, February 15, 1961. DI: JFKWHA-011
Press Conference, March 23, 1961. DI: JFKWHA-020
Press Conference, April 12, 1961. DI: JFKPOF-054-011-p0001
Special Message to Joint Session of Congress on Urgent National Needs, May 25, 1961. DI: JFKWHA-032

Khrushchev's Speech to the UN, September 23, 1960. Wilson Center Digital Archive: https://digitalarchive.wilsoncenter.org/document/155185.pdf?v=9f7ac7df82c2cf1162b9f845c67ef067.
Logsdon J., with Launius, R.D. Ed. *Exploring the Unknown: Selected Documents in the History of the US Civil Space Program, Vol 7: Human Spaceflight: Projects Mercury, Gemini and Apollo.* NASA History Series, 2008. Includes Report of the Ad Hoc Mercury Panel, April 12, 1961, 1–34.
NASA *The Mercury Redstone Project*, NASA, September 1961.
NASA *Proceedings of a Conference on Results of the First Manned Suborbital Spaceflight.* June 6, 1961. https://msquair.files.wordpress.com/2011/05/results-of-the-first-manned-sub-orbital-space-flight.pdf
NASA 'Report to Committee on Science and Astronautics US House of Representatives', *Aeronautical and Astronautical Events of 1961*, US Government Printing Office, 1962.
NASA *Results of the Project Mercury Ballistic and Orbital Chimpanzee Flights.* Eds. Henry, James P. & Mosely, John D., NASA 1963.
NASA Lyndon B. Johnson Space Center, Houston (JSC) History Portal for Project Mercury: https://historycollection.jsc.nasa.gov/JSCHistoryPortal/history/mercury.htm
NASA JSC Oral History Project: https://historycollection.jsc.nasa.gov/JSCHistoryPortal/history/oral_histories/oral_histories.htm
Charles A. Berry, interviewed by Carol Butler. Houston, Texas April 29, 1999.
Manfred H. 'Dutch' von Ehrenfried, interviewed by Rebecca Wright. Houston, Texas, March 29, 2009.

Maxime A. Faget, interviewed by Jim Slade. Houston, Texas. June 18–19, 1997.
Maxime A. Faget, interviewed by Carol Butler. Houston, Texas. August 19, 1998.
Dr Robert Gilruth, interviewed by Linda Ezell, Howard Wolko, Martin Collins. National Air and Space Museum, Washington DC. June 30, 1986.
John H. Glenn Jnr., interviewed by Sheree Scarborough. Houston, Texas. August 25, 1997.
Caldwell C. Johnson, interviewed by Michelle Kelly. League City, Texas. April 1, 1998.
Christopher Kraft, interviewed by Rebecca Wright. Houston, Texas. May 23, 2008 and August 5, 2014.
Christopher Kraft, interviewed by Jennifer Ross-Nazzal. Houston, Texas. April 14, 2009.
Eugene F. Kranz, interviewed by Roy Neal. Houston, Texas. March 19, 1998.
Glynn S. Lunney, interviewed by Roy Neal. Houston, Texas. March 9, 1998.
Glynn S. Lunney, interviewed by Carol Butler. Houston, Texas. January 28, 1999 and February 8, 1999.
Dee O'Hara, interviewed by Rebecca Wright. Mountain View, California. April 23, 2002.
Alan B. Shepard, Jr., interviewed by Roy Neal. Pebble Beach, Florida. February 20, 1998.
Dr Robert B. Voas, interviewed by Summer Chick Bergen, Vienna, Virginia. May 19, 2002.
Guenter F. Wendt, interviewed by Catherine Harwood. Titusville, Florida. February 25, 1999.

National Archives and Records Administration (NARA). Guide to locations of NASA records by Field Center: https://history.nasa.gov/nara1.html
National Archives and Records Administration (NARA), Fort Worth, Texas, containing records from the Lyndon B. Johnson Space Center. Mercury Series. Relevant Boxes include:
Boxes 1–18: Chronological Files 1951–June 1961
Boxes 24–31: Mercury Flight Related Documents, including MR-1, MR-1A, MR-2, MR-BD & MR-3
Box 47: Project Mercury Status Reports (PMSR):
PMSR Number 8, period ending October 31, 1960.
PMSR Number 9, period ending January 31, 1961.
PMSR Number 10, period ending April 30, 1961.

PMSR Number 11, period ending July 31, 1961.
Box 74: Familiarization Manuals including *Astronaut's Handbook Project Mercury*
Box 88: Astronaut Selection and Training

Mercury Press Conferences (transcripts):
January 31, 1961. Cape Canaveral. *Following the Launch of Mercury-Redstone 2* (Entry 60)
February 21, 1961. Cape Canaveral. *Mercury-Atlas 2* (Box 28)
February 22, 1961. Cape Canaveral. *Project Mercury/Astronauts* (Box 17)
April 25, 1961. Cape Canaveral. *Mercury-Atlas III*

Mercury Press Releases:
January 28, 1961. Release 61-14-3. *Animal Flight Program* (Box 27)
February 28, 1961. *Summary Statement: Program Results and Objectives by Hugh Dryden before the Senate Committee on Aeronautical and Space Sciences* (Box 17)
March 22, 1961. Release 61–57. *Mercury Redstone Booster Development Test*
April 23, 1961. Release 61–85. *Mercury-Atlas 3 Spacecraft Flight Test*

NASA Information Plans and Memoranda:
Memorandum on Mercury-Redstone Booster Problems. March 20, 1961
Information Plan: Redstone Development Test MR-BD. March 21, 1961
Memorandum on Mercury-Redstone Booster Development Flight (MR-BD). March 26, 1961
Memorandum on Mercury-Redstone Booster Development Test. March 27, 1961
Public Information Operating Plan: Project Mercury MR-3. April 21, 1961

NASA Working Papers:
Project Mercury Working Paper no. 174: Astronaut Preparation and Activities Manual for Mercury-Redstone Mission No. 3 (MR-3). February 6, 1961 (Box 18)
Project Mercury Working Paper no. 192: Postlaunch Report for Mercury-Redstone No.3 (MR-3). June 16, 1961

NASA Press Releases at NASA News: https://www.nasa.gov/centers/kennedy/pdf/744715main_1961.pdf
March 27, 1961. (No title.) Subject regarding news coverage for upcoming first manned suborbital Project Mercury flight.

April 12, 1961. *Statement by Mr. Robert R. Gilruth, NASA Space Task Group.*
April 12, 1961. *Comment by Mercury Astronaut Virgil I. Grissom at 10:30 a.m.*
April 12, 1961. *Comment by Mercury Astronaut Alan B. Shepard, Jr.*
April 12, 1961. *Comment by Mercury Astronaut John H. Glenn.*

Proceedings of House Committee on Science and Astronautics. Full text (all 1961): March 13, 14, 22, 23, April 10, 11, 14, 17. https://babel.hathitrust.org/cgi/pt?id=uc1.31822015358211&view=1up&seq=201
Ohio State University Libraries: Transcript of *Mercury Seven Press Conference*, April 9, 1958. https://library.osu.edu/site/friendship7/selection-and-training/

Newspapers & Magazines
British Library (BL)
British Newspaper Archive (BNA)
New York Public Library (NYPL)

Aviation Week 1959–61 (https://archive.aviationweek.com)
Daily Herald April 1961 (BNA)
Daily Mirror 1961 (BNA)
Daily Worker April 1961 (BL)
Izvestiya January–May 1961 (BL)
Life Magazine 1959–61 (NYPL)
New York *Daily News* 1960–1 (NYPL)
New York Times January–May 1961 (NYPL & *Times Machine*)
Ogonyok January–May 1961 (BL)
Pravda January–May 1961 (BL)
Time Magazine January–May 1961 (NYPL)
Washington Post 1960–1 (NYPL)

Secondary Sources
Abramov, Isaac and Skoog, Ingemar. *Russian Spacesuits.* New York: Springer, 2003.
Andrews, James and Siddiqi, Asif (eds). *Into the Cosmos: Space Exploration and Soviet Culture.* Pittsburgh: University of Pittsburgh Press, 2011.
Applebaum, Anne. *Gulag: A History.* New York: Random House, 2003.
Baker, David. *The History of Manned Spaceflight.* New York: Crown, 1982.
Barbree, Jay. *Live from Cape Canaveral: Covering the Space Race from Sputnik to Today.* London: Collins/Smithsonian Books, 2007.

Belyakova, L.A. *Oral history account.* Recorded and edited by Sokolov, G.I., Star City, February 2016.

Blagov, D. *Korolev's Breakthrough into the Universe: Memories of a Space Industry Veteran.* November 21, 2018. https://www.i-podmoskovie.ru/history/korolyevskiy-proryv-vo-vselennuyu-vospominaniya-veterana-kosmicheskoy-otrasli/

Borisenko, Ivan G. & Romanov, Aleksander P. *Where All Roads into Space Begin.* Moscow: Progress Publishers, 1982.

Brantz, Dorothee (ed.). *Beastly Natures: Animals, Humans, and the Study of History.* University of Virginia, 2010.

von Braun, Wernher. *History of Rocketry and Space Travel.* Nelson, 1967.

Brinkley, David. *American Moonshot: John F. Kennedy and the Great Space Race.* New York: HarperCollins, 2019.

Brugioni, Dino. *Eyes in the Sky: Eisenhower, the CIA and Cold War Aerial Espionage.* Annapolis: Naval Institute Press, 2010.

Burgess, Colin. *Selecting the Mercury Seven: The Search for America's First Astronauts.* New York: Springer Praxis Books, 2011.

Burgess, Colin. *Freedom 7: The Historic Flight of Alan B Shepard, Jr.* New York: Springer, 2014.

Burgess, Colin and Dubbs, Chris. *Animals in Space: From Research Rockets to the Space Shuttle.* Chichester: Praxis, 2007.

Burgess, Colin and Hall, Rex. *The First Soviet Cosmonaut Team: Their Lives, Legacy and Historical Impact.* Chichester: Praxis, 2009.

Burnett, Thom. *Who Really Won the Space Race?* London: Collins & Brown, 2005.

Burrows, William. *This New Ocean: The Story of the First Space Age.* Modern Library, 1999.

Cabled radio in the USSR. http://kommunalka.colgate.edu/cfm/essays.cfm?ClipID=340&TourID=910

Cadbury, Deborah. *Space Race: The Battle to Rule the Heavens.* London: Harper Perennial, 2006.

Caro, Robert. *The Years of Lyndon Johnson. Vol.4: The Passage of Power.* Bodley Head, 2014.

Carpenter, Scott and Stoever, Chris. *For Spacious Skies: The Uncommon Journey of a Mercury Astronaut.* New York: Harcourt, 2002.

Cassutt, Michael. 'Star City at 50'. *Air & Space Magazine*, March 2011.

Catchpole, John. *Project Mercury: NASA's First Manned Space Programme.* Chichester: Praxis, 2001.

Chalkin, Andrew. 'How the Spaceship got its Shape'. *Air & Space Magazine*, November 2009.

Chambers, Joseph. *A Century at Langley. The Storied Legacy and Soaring Future of NASA Langley Research Center.* Government Printing Office, 2017.

Chentsov, Nikolai. 'World Famous but Secret in Every Way', *Nauka i Zhizn'*. Issue 2, February 1991.

Chertok, Boris. *Rockets and People, Vol. III: Hot Days of the Cold War.* Washington: NASA History Series, 2009.

CIA Historical Collections: *Penetrating the Iron Curtain: Resolving the Missile Gap with Technology.* United States Printing Office 2014.

Cole, Michael D. *Vostok 1: First Human in Space.* Springfield, N.J.: Enslow Publishers, 1995.

Collins, Michael. *Carrying the Fire.* W.H. Allen, 1975.

Computers in Spaceflight: The NASA Experience. https://history.nasa.gov/computers/Ch8-2.html

Cooper, Gordon with Henderson, Bruce. *Leap of Faith: An Astronaut's Journey into the Unknown.* HarperCollins, 2000.

Day, Dwayne. 'In Space No One Can Hear You Sigh'. *Space Review*, September 22, 2014.

Day, Dwayne. 'Pay No Attention to the Man with the Notebook: Hugh Sidey and the Apollo Decision'. *Space Review*, December 5, 2005.

Day, Dwayne. 'Those Magnificent Spooks and their Spying Machines'. *Spaceflight*, Vol. 39, March 1997, pp. 98–100.

Dikhtyar, Adolf. *Do togo kak prozvouchalo 'Poyekhali' (Before We Heard the Word 'Let's Go!').* Moscow: Politizdat, 1987.

Doran, Jamie & Bizony, Piers. *Starman: The Truth Behind the Legend of Yuri Gagarin.* London: Bloomsbury, 1998.

Dunar, Andrew & Waring, Stephen. *Power to Explore: A History of the Marshal Space Flight Center 1960–1990.* The NASA History Series, 1999. https://archive.org/stream/nasa_techdoc_20000031366/20000031366_djvu.txt

Early History of Star City.
Star City website http://zato-zvezdny.ru/?page_id=55

von Ehrenfried, Manfred. *The Birth of NASA: The Work of the Space Task Group, America's First True Space Pioneers.* Springer, 2016.

Eicher, David J. & May, Brian. *Mission Moon 3D: Reliving the Great Space Race.* London Stereoscopic Company, 2018.

Elliott, Richard G. 'On a Comet Always'. *New Mexico Quarterly*, Vol. 36, issue 4, 1966.

French, Francis & Burgess, Colin. *Into That Silent Sea: Trailblazers of the Space Era 1961–1965.* University of Nebraska, 2007.

Gaddis, John. *The Cold War: A New History.* London: Penguin, 2007.

Gagarinday.ru Website of the Social and State Committee for the Commemoration of Yu. A. Gagarin's 85th Anniversary. http://gagarinday.ru/articles/17-polet-yu-a-gagarina-v-kosmos-vzgljad-na-realnye-sobytija-teh-let-iz-xxi-veka.html

Gagarin, Yuri A. *Road to the Stars.* Honolulu: University Press of the Pacific, 2002.

Gagarin, Valentin A. *Moi brat Yuri (My Brother Yuri).* Moscow: Moskovsky Rabochy, 1979.

Gagarina, Anna T. *Slovo o syne (A Word about my Son).* Moscow: Progress Publishers, 1983.

Gallai, Mark L. *S chelovekom na bortu (With a Man Onboard).* Moscow: Sovetsky Pisatel', 1985.

Garber, Megan. 'Astro Mad Men: NASA's 1960s Campaign to Win America's Heart'. *The Atlantic*, July 31, 2013.

Garber, Megan. 'Kennedy, before Choosing the Moon: "I'm not that interested in Space"'. *The Atlantic*, September 12, 2012.

Gibbons, R.F. & Clark, S.P. 'The Evolution of the Vostok and Voshkod Programs'. *Journal of the British Interplanetary Society (JBIS)*, Issue 38, 1985, 3–10.

Glenn, John with Taylor, Nick. *John Glenn: A Memoir*. New York: Bantam Books, 1999.

Golovanov, Yaroslav K. *Korolev: Fakty I Mify (Korolev: Facts and Myths).* Moscow: Nauka, 1994.

Golyakhovsky, Vladimir Yu. *Oral History Memoir.* New York: William E. Wiener Oral History Library of the American Jewish Committee at New York Public Library, 1979.

Golyakhovsky, Vladimir Yu. *Russian Doctor. A Surgeon's Life in Contemporary Russia and Why He Chose to Leave.* New York: St Martin's Press, 1984.

Grahn, Sven. *An Analysis of the Flight of Vostok.* http://www.svengrahn.pp.se/histind/Vostok1/Vostok1X.htm

Grahn, Sven. *Radio Systems of Soviet/Russian Manned Spacecraft.* http://www.svengrahn.pp.se/radioind/mirradio/mirradio.htm

Grahn, Sven. *Sputnik 6 and the Failure of December 22.* http://www.svengrahn.pp.se/histind/sputnik6/sputnik6.html

Grahn, Sven. *TV from Vostok.* http://www.svengrahn.pp.se/trackind/TVostok/TVostok.htm#Shemya

Grahn, Sven. *Vostok Retrofire Attitude.* http://www.svengrahn.pp.se/histind/vostokretro/vostokretro.html

Granath, Bob. *NASA Kennedy Space Center Historic Hangar S.* NASA, March 2016. https://www.nasa.gov/feature/historic-hangar-s-was-americas-cradle-of-human-space-exploration

Gunter's Space Page. https://space.skyrocket.de/doc_sdat/vostok-3k.htm

Hall, Rex & Shayler, David. *The Rocket Men: The First Soviet Manned Spaceflights.* Springer, 2001.

Hardesty, V. and Reisman G. *Epic Rivalry: The Inside Story of the Soviet and American Space Race.* National Geographic, 2007.

Harford, James. *Korolev: How One Man Masterminded the Soviet Drive to Beat America to the Moon.* New York: John Wiley & Sons, 1997.

Harpole, Tom. 'Saint Yuri'. *Air & Space Magazine*, January 1999.

Hendrickx, B. 'The Kamanin Diaries 1960–1963'. *Journal of the British Interplanetary Society*, Vol. 50, No. 1, January 1997.

Ivanovsky, Oleg G. *Rakety i kosmos v SSSR: Zapiski sekretnogo-konstruktora (Rockets and Space in the USSR: Notes of a Secret Designer).* Moscow: Molodaya gvardya, 2005.

Jenks, Andrew L. *The Cosmonaut Who Couldn't Stop Smiling.* Ithaca: NIU Press, 2012.

Kamanin, Lev N. 'People Aren't Born Heroes, They Become Heroes'. *Unpublished article*, 2000.

Katz, Vladislav. *Zarya Gagarinskoi Vesny (The Dawn of Gagarin's Spring).* Jerusalem: AT International, 2019.

Khrushchev, Nikita S. *Khrushchev Remembers.* Boston: Little, Brown and Co., 1970.

Khrushchev, Sergei N. (ed.) *Memoirs of Nikita Khrushchev: Statesman 1953–1964*, Vol. 3. Pennsylvania State University Press, 2013.

Khrushchev, Sergei N. *Nikita Khrushchev and the Creation of a Superpower.* Pennsylvania State University Press, 2000.

King, Andrew. *Spinning Out Heroes: The Johnsville Centrifuge.* Pennsylvania Center for the Book. https://www.pabook.libraries.psu.edu/literary-cultural-heritage-map-pa/feature-articles/spinning-out-heroes-johnsville-centrifuge.

Koppel, Lily. *The Astronaut Wives Club.* Kindle Edition, Headline, 2016.

Koroleva, Natalya. *Father. Volumes 1 & 2.* Moscow: Restart, 2008.

Kouprianov, Valery N. *Kosmicheskaya odisseya Yuria Gagarina* (*The Space Odyssey of Yuri Gagarin*). St Petersburg: Polytechnica, 2011.

Kraft, Chris. *Flight: My Life in Mission Control.* New York: Dutton, 2001.

Kranz, Gene. *Failure Is Not an Option: Mission Control from Mercury to Apollo and Beyond.* New York: Berkley Books, 2000.

Lamb, Lawrence. *Inside the Space Race: A Space Surgeon's Diary.* Synergy Books, 2006.

Lebedev, L., Lyk'yanov, B., Romanov, A. *Sons of the Blue Planet.* Moscow: Political Literature Press, 1971.

Leonyonok, Ye. S. *Oral History account.* Recorded and edited by Sokolov, G.I., February 2015.

Link, Mae Mills. *Space Medicine in Project Mercury.* NASA, SP 4003, 1965.

Lissov, Igor. *Instructions for the setting up of PVU (Timer Programming Device) 'Granit-5B'.* Lissov personal archive.

Logsdon, John. *John F. Kennedy and the Race to the Moon.* New York: Palgrave Macmillan, 2010.

Logsdon, John (ed.). *The Penguin Book of Outer Space Exploration.* New York: Penguin, 2018.

Loktev, Aleksandr. 'Four meetings with Pallo'. *Herald*, No. 18 (225), August 31, 1999. http://www.vestnik.com/issues/1999/0831/win/loktev.htm

Mallon, Thomas. 'Rocket Man'. *New Yorker*, October 22, 2007.

Maybury, Jim. 'A Tribute to Edward C. Dittmer'. New Mexico Museum of Space History, Curation Paper 8, Summer 2012.

McClesky C. & Christensen D. 'Dr Kurt H. Debus: Launching a Vision'. 52nd International Astronautical Congress, Toulouse October 1–5, 2001. https://www.nasa.gov/centers/kennedy/pdf/112024main_debus.pdf

McQuaid, Kim. 'Sputnik Reconsidered: Image and Reality in the Early Space Age'. *Canadian Review of American Studies* 37, 2007.

Meeter, George. *The Holloman Story: Eyewitness Accounts of Space Age Research.* New Mexico: University of New Mexico Press, 1967.

Menand, Louis. 'Ask Not, Tell Not: Anatomy of an Inaugural'. *New Yorker*, November 1, 2004.

The Mercury Seven. *We Seven: by the Astronauts Themselves.* New York: Simon & Schuster 1962.

Mieczkowski, Yanek. *Eisenhower's Sputnik Moment: The Race for Space and World Prestige.* Ithaca, New York: Cornell University Press, 2013.

Hal G. Miller. *The Early Days of Simulation and Operations*, oral history memoir. Great Falls Virginia, June 30, 2013.

Mitroshenkov, V. (ed.) *Pioneers of Space.* Moscow: Progress Publishers, 1989.

Mullane, Mike. *Riding Rockets.* New York: Scribner, 2006.

Murphy, Alan. 'The Losing Hand: Tradition and Superstition in Space Flight'. *Space Review*, May 28, 2008.

Nelson, Craig. *Rocket Men: The Epic Story of the First Men on the Moon.* London: John Murray, 2010.

Neufeld, Michael. *Von Braun, Dreamer of Space, Engineer of War.* New York: Vintage, 2008.

Nilson, Thomas. 'The Cold War Airbase Will be Developed into High-end Residential Neighbourhood'. *Barents Observer*, June 10, 2018. https://thebarentsobserver.com/en/node/4037

Oberg, J. *Uncovering Soviet Disasters.* New York: Random House, 1988.

On MEI (Moscow Electronics Institute) Contribution to the First Manned Space Flight on the 50th Anniversary of Yu. A. Gagarin's Space Flight. Moscow: Glasnost, 2011.

Ordway III, Frederick & Sharpe, Mitchell. *The Rocket Team.* New York: Thomas Crowell, 1979.

Osborne, Ray. *Images of America: Cape Canaveral.* Charleston S.C.: Arcadia, 2008.

Parfitt, Tom. 'How Yuri Gagarin's Flight Was Almost Grounded'. *Guardian*, April 6, 2011.

Pervushin, Anton I. *Yuri Gagarin: Odin polyot i vsia zhiz'n* (*Yuri Gagarin: One Flight and a Whole Life*). Moscow: OOO Pal'mira & AO T8 Izdatel'skie Tekhnologii, 2017.

Poroshkov, Vladimir V. (ed.) *Neizvestny Baikonur (Unknown Baikonur: Collection of Accounts by Baikonur Veterans).* Moscow: Globus, 2001.

Powers, Francis Gary & Gentry, Curt. *Operation Overflight: A Memoir of the U-2 Incident.* London: Hodder & Stoughton, 1970.

Rafikov, Makhmud. 'Gagarin Felt Very Unwell on Return from Orbit. I Think Lemonade was to Blame'. *Natsia* (*Nation: Journal of The Russian Common Sense Team*), April 11, 2017.

Rasenberger, Jim. *The Brilliant Disaster: JFK, Castro, and America's Doomed Invasion of Cuba's Bay of Pigs.* Simon & Schuster, 2011.

Rebrov, Mikhail F. 'A Star Traversing Cape Horn'. *Krasnaya Zvezda*, April 12, 1994.

Reichhardt, Tony. 'Light this Candle'. *Air & Space Magazine*, April 2012.

Rhea, John (ed.). *Roads to Space: An Oral History of the Soviet Space Program.* New York: Aviation Group Week, 1995.

Riabchikov, Yevgeny & Kamanin, Nikolai P. (eds) *Russians in Space.* New York: Doubleday, 1971.

Rigell, Ike. *Ike: The Memory of Isom 'Ike' Rigell.* Kindle edition, Koehler Books, 2017.

Romanov, Aleksandr P. *Korolev.* Moscow: Molodaya Gvardiya, 1996.

Rose, Andrea (ed.). *Gagarin in Britain.* London: British Council, 2011.

Ryan, Amy & Keeley, Gary. 'Sputnik and US Intelligence: The Warning Record'. *Studies in Intelligence*, Vol. 61, No 3, September 2017.

Salinger, Pierre. *With Kennedy.* London: Jonathan Cape and Weidenfeld & Nicolson, 1967.

Schlesinger, Arthur. *A Thousand Days: John F. Kennedy in the White House.* Boston: Mariner, 2002.

Scott, David & Leonov, Alexei with Toomey, Christine. *Two Sides of the Moon.* London: Simon & Schuster, 2004.

Shepard, Alan & Slayton, Deke with Barbree, Jay. *Moon Shot.* New York: Open Road, 2011.

Siddiqi, Asif A. *Challenge to Apollo: The Soviet Union and the Space Race 1945–1974.* Washington, D.C.: National Aeronautics and Space Administration, 2000.

Siddiqi, Asif A. 'The Man behind the Curtain'. *Air & Space Magazine*, November 2007.

Siddiqi, Asif A. 'Mourning Star: The Nedelin Disaster'. *Quest*, Vol. 3, No. 4, Winter 1994.

Siddiqi, Asif A. 'Russia's Long Love Affair with Space'. *Air & Space Magazine*, August 2007.

Siddiqi, Asif. A. 'Soviet Design Bureaux'. *Spaceflight*, Issue 39, August 1997. 277–80.

Siddiqi, Asif. A. 'Declassified Documents Offer a New Perspective on Yuri Gagarin's Flight'. *Space Review*, October 12, 2015.

Sidey, Hugh. *John F. Kennedy, President.* New York: Crest, 1963.

Slayton, Donald K. with Cassutt, Michael. *Deke!* New York: Tom Doherty, 1994.

Solodukhin, Anatoly N. *Brief Memoirs of a Test Engineer.* (Unpublished).

Spacecraft Vostok and Soyuz 7K-OK Control and Instrument Panels. https://vostoksupersite.weebly.com

Spielmann, Karl. *Analyzing Soviet Strategic Arms Decisions.* London: Routledge, 2019.

Stone, Robert & Andres, Alan. *Chasing the Moon.* London: William Collins, 2019.

Suvorov, Vladimir & Sabelnikov, Alexander. *The First Manned Spaceflight: Russia's Quest for Space.* New York: Nova Science Publishers, 1997.

Swenson, Loyd & Grimwood, James & Charles Alexander. *This New Ocean: A History of Project Mercury.* NASA History Series, St Petersburg, Florida, 1989.

Thielmann, Greg. 'The Missile Gap Myth and its Progeny'. *Arms Control Today*, Vol. 1, No. 4, May 2011.

Thompson, Neal. *Light this Candle: The Life & Times of Alan Shepard.* New York: Three Rivers Press, 2004.

Titov, Gherman, as told to Borzenko, S. *700,000 kilometres through Space.* Moscow: Foreign Languages Publishing House, 1961.

Titova, Tamara N. *Zhena Kosmonavta (A Cosmonaut's Wife).* Barnaul: Altai State Memorial Museum of G.S. Titov, 2014.

Tsiolkovsky, Konstantin E. 'The Investigation of Universal Space by Means of Reactive Devices'. *Vestnik Vozhdukhoplavania (Herald of Aeronautics),* Issue 3, 1912, p. 16.

Turkina, Olesya. *Soviet Space Dogs.* London: Fuel, 2014.

Villain, Jacques. *A Brief History of Baikonur.* Presented at 45th Congress of the International Astronautical Federation. Jerusalem, October 1994.

Vostok panels and information on tracking. https://vostoksupersite.weebly.com.

Ward, Bob & Glenn John. *Dr Space: The Life of Wernher von Braun.* Naval Institute Press, 2013.

Wade, Mark. *R-7 Technical Details.* http://www.astronautix.com/r/r-7.html

Wade, Mark. *Vostok Technical Characteristics.* http://www.astronautix.com/r/r-7.html.

'We Are Simply Surviving Here Today. Why they Feel Homesick for Russia in Baikonur'. *Argumenty i Fakty,* April 25, 2018.

Wendt, Gunter. *The Unbroken Chain.* Canada: Apogee, 2001.

Wolfe, Tom. *The Right Stuff.* San Francisco: Black Swan Press, 1989.

Wright, Rebecca. 'A Home for Heroes'. *Houston History Magazine.* Center for Public History at the University of Houston, March, 2013.

Yazdovsky, Vladimir I. *Na tropakh vselennoi (Along the Pathways of the Universe).* Moscow: Slovo, 1996.

Yazdovsky, Vladimir I. 'They Were the First'. *Aerospace Journal,* No. 2, March–April 1966.

Zak, Anatoly. *Preparing Sputnik 2 for Flight.* http://www.russianspaceweb.com/sputnik2_preflight.html

Zak, Anatoly. *Russian Ground Control.* http://www.russianspaceweb.com/kik.html

Zak, Anatoly. *Site 2 in Baikonur.* http://www.russianspaceweb.com/baikonur_r7_2.html

Zak, Anatoly. *Vostok in Preparation.* http://www.astronautix.com/v/vostok.html

Film & Television

Alan Shepard interview. American Academy of Achievement. February 1, 1991.

Chasing the Moon. PBS/BBC, 2019.

Cosmic Journey. USSR, 1936. https://www.youtube.com/watch?v=hDhJKzuOb2w
Cosmonauts: How Russia Won the Space Race. BBC4, 2014.
Days That Shook the World: Reach for the Stars. BBC2, 2004.
Freedom 7: Full Mission. NASA, 1961.
From the Earth to the Moon. Episode 1. HBO, 2006.
Gagarin's Field. Saratov TV Studio, 1974.
Gherman Titov. Pervyi posle Yuria (Gherman Titov. First after Yuri). Russian Channel One, 2010.
History of Mercury Mission Control. NASA, 2013.
Horizon: The Race to Ruin. BBC2, 1981.
Horizon: Red Star in Orbit. BBC2, 1990.
Man in Space. Tomorrowland series, *Season 1, ep.20.* Disney TV, 1955.
Marina Popovich Interview. Film with English transcript. Flying Heritage Collection, Washington. http://video.flyingheritage.com/v/116998872/general-lieutenant-marina-lavrentyevna-popovich.htm
Mercury 13. David Sington & Heather Walsh, Netflix, 2018.
Mission Control: The Unsung Heroes of Apollo. David Fairhead, USA, 2017.
NASA's Greatest Missions, Episode 1, *Ordinary Supermen*; Episode 3, *Friends and Rivals.* Discovery Channel, 2009.
Nelyubov. On mog byt' pervym (Nelyubov. He Could Have Been the First). Russian Space Studio, 2007.
One Small Step: The Story of the Space Chimps. History Channel UK and CBC Television, 2003.
Pioneers of the Vertical Frontier: Aeromedical Research Laboratory, 1967. https://www.youtube.com/watch?v=Pq_C26JG6FQ
Razlouchennye nyebom (Separated by the Sky). Studio Vstrecha, Channel One Russia, 2010.
Red Stuff: The True Story of the Russian Race for Space, Leo de Boer, 1999.
Reputations Special: Yuri Gagarin – Starman. BBC2, 1998.
RGANTD, Moscow: Extensive film archive of the Soviet space programme 1957–1961, including material shot by Vladimir Suvorov and his colleagues.
Son & Moon: Diary of an Astronaut (Manuel Huerga). 2009.
Soviet 1950s Science Fiction Rocket Launch. https://www.youtube.com/watch?v=LFKLMgYrCaw
Space Race: Superpowers, Secrets and Soaring Ambition. BBC2, 2005.
The Astronauts, United States Project Mercury. NASA, 1960.
Yuri Gagarin Remembered. Ekran Documentary Studios, USSR, 1969.
Yuri Gagarin: Three Days and a Whole Life. History Channel Russia, 2011.

INDEX